TRANSPORTATION AND TRAFFIC THEORY

PREVIOUS SYMPOSIA AND PROCEEDINGS

I   General Motors Research Laboratory, Warren, Michigan, 1959: THEORY OF
TRAFFIC FLOW, Robert Herman, Editor, Elsevier Publishing Company,
Amsterdam, The Netherlands, 1961

II   Road Research Laboratory, London, England, 1963: PROCEEDINGS OF THE
SECOND INTERNATIONAL SYMPOSIUM ON THE THEORY OF ROAD TRAFFIC FLOW,
LONDON 1963, Joyce Almond, Editor. Organisation for Economic Co-
Operation and Development (OECD), Paris, 1965

III   Transportation Science Section ORSA, New York, 1965: VEHICULAR TRAFFIC
SCIENCE, L.C. Edie, R. Herman and R. Rothery, Editors. American
Elsevier Publishing Company, Inc., New York, 1967

IV   Technical University, Karlsruhe, Germany, 1968: BEITRAEGE ZUR THEORIE
DES VERKEHRSFLUSSES, Wilhelm Leutzbach and Paul Baron, Editors.
Strassenbau und Strassenverkehrstechnik Heft 86, 1969

V   University of California, Berkeley, 1971: TRAFFIC FLOW AND
TRANSPORTATION, Gordon F. Newell, Editor. American Elsevier Publishing
Company, Inc., New York, 1972

VI   University of New South Wales, Sydney, Australia, 1974: TRANSPORTATION
AND TRAFFIC THEORY, D.J. Buckley, Editor. A.H. & A.W. Reed Pty Ltd.,
Artarmon, New South Wales, 1974

VII   The Institute of Systems Science Research, Kyoto, Japan, 1977:
PROCEEDINGS OF THE SEVENTH INTERNATIONAL SYMPOSIUM ON TRANSPORTATION AND
TRAFFIC THEORY, Tsuna Sasaki and Takeo Yamaoka, Editors. The Institute
of Systems Science Research, Kyoto, 1977

VIII University of Toronto, Canada, 1981: PROCEEDINGS OF THE EIGHTH
INTERNATIONAL SYMPOSIUM ON TRANSPORTATION AND TRAFFIC THEORY,
V.F. Hurdle, E. Hauer and G.N. Steuart, Editors. University of Toronto
Press, Canada, 1983

Proceedings of the
NINTH International Symposium on
TRANSPORTATION AND
TRAFFIC THEORY

Delft, The Netherlands
11–13 July 1984

Edited by
J. VOLMULLER and R. HAMERSLAG

CRC Press
Taylor & Francis Group
Boca Raton London New York

CRC Press is an imprint of the
Taylor & Francis Group, an **informa** business

First published 1984 by VNU Science Press

Published 2019 by CRC Press
Taylor & Francis Group
6000 Broken Sound Parkway NW, Suite 300
Boca Raton, FL 33487-2742

ISBN-13: 978-0-367-45175-2 (pbk)
ISBN-13: 978-90-6764-008-4 (hbk)

Visit the Taylor & Francis Web site at
http://www.taylorandfrancis.com

and the CRC Press Web site at
http://www.crcpress.com

CIP DATA
Proceedings of the Ninth International Symposium on Transportation and Traffic Theory: Delft, The Netherlands, 11–13 July 1984/ed. by J. Volmuller and R. Hamerslag; Utrecht: VNU Science Press.–Ill.
With index
ISBN 90-6764-008-5 geb.
SISO 377 UDC 656
Key words: transportation/traffic theory

CONTENTS

SPONSORS

NINTH INTERNATIONAL SYMPOSIUM ON TRANSPORTATION AND TRAFFIC THEORY

Under the auspices of

DELFT UNIVERSITY OF TECHNOLOGY

Sponsors

DELFT UNIVERSITY OF TECHNOLOGY
ROYAL TOURING CLUB ANWB - on the occasion of the hundredth anniversary
in 1983
NETHERLANDS RAILWAYS COMPANY
ROYAL INSTITUTE OF ENGINEERS

PREFACE

Twenty-five years ago, in 1959, the first International Symposium on Traffic
Flow Theory was organized in Detroit, Michigan, U.S.A.
An impressive series of symposia has been held since, each characterized by
the high standard of the papers read and discussed and by the informal
atmosphere which prevailed both during and after the formal sessions.
The impact which the symposia had on the development of theoretical thinking
in transportation and traffic flow has been obvious. The proceedings of each
symposium have been among the best sellers in their field. Papers submitted
to even the earliest symposia are still often referred to in research work and
in practical transportation and traffic planning and engineering.
The need for more knowledge and insight in the theoretical background of
transportation and traffic remains unabated. The socio-economic situation in
all parts of the world does not allow for the "easy" solution. Financial
means fall short and socio-political sensitivity for the environmental
consequences of any proposed solution increases by the day. Finding the
acceptable solution requires an ever increasing amount of skill from planners
and designing engineers, fed by a better theoretical insight and understanding
of the phenomena involved. The original aim which inspired the initiators to
organize the first symposium, has lost nothing of its urgency.
Prospects are good that the Ninth Symposium can be held in the tradition of
the previous ones. Many excellent papers have been submitted and it has been
hard work for the referees and Selection Committee to sift out the 27 papers
which the symposium program allowed for. They deserve our sincere
appreciation.
I wish also to express thanks to the International Committee and the National
Committee for their stimulating contribution to the organization of this Ninth
Symposium, and to the sponsors who, through their financial contributions,
made this symposium possible.
Our special appreciation goes to Mrs. Sylvia J. Wamsteker-Andriessen. The
symposium's success is due to a large extent to the unremitting enthusiasm,
creativity and punctuality with which she conducted the symposium's
administration.

J. Volmuller
April 1984

COMMITTEES

INTERNATIONAL COMMITTEE

R.E. Allsop, Transport Studies Group, United Kingdom
P.H. Fargier, Traffic Research Institute, France
N.H. Gartner, University of Lowell, U.S.A.
E. Hauer, University of Toronto, Canada
R. Herman, University of Texas at Austin, U.S.A.
W. Leutzbach, University of Karlsruhe, West-Germany
G.F. Newell, University of California, U.S.A.
H.G. Retzko, Technical University Darmstadt, West-Germany
J.C. Tanner, Transport & Road Research Laboratory, United Kingdom
R.J. Vaughan, University of Newcastle, Australia
T. Sasaki, Kyoto University, Japan
J. Volmuller, Delft University of Technology, The Netherlands
S. Yagar, University of Waterloo, Canada

NATIONAL COMMITTEE

J. Volmuller, Delft University of Technology, chairman
B. Beukers, Rijkswaterstaat
R. Hamerslag, Delft University of Technology
J. Kuiper, N.V. Nederlandse Spoorwegen
J.H. Papendrecht, Delft University of Technology, secretary
G.R. de Regt, Royal Dutch Touring Club ANWB
V. Vidakovic, Delft University of Technology
T. de Wit, Royal Institution of Engineers in The Netherlands

Secretariat: Mrs. S.J. Wamsteker-Andriessen

SELECTION COMMITTEE

R. Hamerslag, Delft University of Technology, chairman
S. Bexelius, Rijkswaterstaat
H. Botma, Delft University of Technology, secretary
J.A. Bourdrez, Netherlands Economic Institute
P.H.L. Bovy, Delft University of Technology
J. van Est, Research Center for Physical Planning TNO
G.R.M. Jansen, Delft University of Technology
J.J. Klijnhout, Rijkswaterstaat
M.F.A.M. van Maarseveen, Traffic and Transportation Group TNO
J.P. Roos, DHV Consulting Engineers
C.J. Ruijgrok, Traffic & Transportation Group TNO
J.G. Smit, Bureau Goudappel Coffeng B.V.
V. Vidakovic, Delft University of Technology
H.J. van Zuylen, Verkeersakademie Tilburg

Ninth International Symposium on
Transportation and Traffic Theory
© 1984 VNU Science Press, pp. 1–20

AN APPROXIMATIVE ANALYSIS OF THE HYDRODYNAMIC THEORY ON TRAFFIC FLOW AND A FORMULATION OF A TRAFFIC SIMULATION MODEL

TSUNA SASAKI,[1] MASAHARU FUKUYAMA[2] and
YOSHIHARU NAMIKAWA[1]
[1] *Department of Transportation Engineering, Kyoto University, Kyoto 606, Japan*
[2] *Social Systems Department, Mitsubishi Research Institute, Inc., Tokyo 100, Japan*

ABSTRACT

An approximative analysis is introduced to the Hydrodynamic Theory.
For an analysis of traffic phenomena in the time-distance space, waves
which have continuous property over the space are approximated by some
discrete dislocations (referred to as quasi shock waves), the behavior
of which is similar to that of shock waves.
The graphic construction, successive computation of the queue tail
trajectory and a procedure are presented. The results from the approximative analysis are compared with those from direct application
of the Hydrodynamic Theory.
Principal conclusions of the analysis are: the traffic phenomena derived from the approximative analysis comes closer to that from the
direct application of the Hydrodynamic Theory as the number of quasi
shock waves is increased. In practice, however, sufficient accuracy
can be obtained if two quasi shock waves are assigned to represent the
waves for each under- and over-saturated region.
On the basis of these analytical results, a simulation model was developed for traffic flow on a signalyzed network. The model was applied to a corridor with a network of 118 links and 44 signalized intersections. The model can well describe the actual traffic phenomena,
especially those during periods of heavy traffic congestion.

1.   INTRODUCTION

The Hydrodynamic Theory, which was developed by Lighthill and Whitham
(1955) and Richards (1956), has been recognized as one of the most powerful tools in analyzing traffic phenomena. Its area of direct application, however, has been limited to somewhat simple cases such as an
isolated intersection (Rorbech, 1968; Stephanopoulos, Michalopoulos and
Stephanopoulos, 1979) and freeway traffic (Inoue, 1973; Okutani and
Inoue, 1973; Stock, Blankernhorn and May, 1973). In this study an approximation is introduced to the Hydrodynamic Theory. The objective is
to discuss the nature of the approximation in detail and also to develop
a simulation model for traffic flow on a signalized road network.

One aspect of the traffic behaviour which causes difficulties in traffic
control is the evolution of a queue which starts to grow at a signal
intersection and backs up to the upstream links penetrating another
signal intersection on the way. In this study we attempt to develop
a model to describe such phenomena. Michalopoulos, Stephanopoulos and
Pisharody (1980) discussed the traffic behaviour in links between signal
intersections on the basis of the Hydrodynamic Theory. They, however,
excluded queues which reached the upstream end of the link and backed
up into other links.

2.   APPROXIMATIVE MODEL

2.1   Approximation of Wave Behavior through Quasi Shock Waves

Figures 1a, b and c show distributions of the traffic density, k, along
a road section (x axis) in which traffic is moving left to right. Figure 1a is at time, $t=t_0$, which includes a point, $x=x_0$, where the density
is discontinuous, with the density in the upstream region, k_u, being
less than in the downstream region, k_d. In such case a shock wave (SW)

forms at the point x_0 which propagates with a speed w_{ud} given by

$$w_{ud} = \frac{q_u - q_d}{k_u - k_d} \quad (1)$$

where q_u and q_d are the flows in the up- and downstream regions respectively. The speed of the SW corresponds to a slope of a chord which connects two points, (k_u, q_u) and (k_d, q_d), on the flow-density curve, which is shown in Fig.1a'.

A case in which $k_u > k_d$ is depicted with a broken line in Fig. 1b. Such a distribution is observed when the lane is blocked partially by a disabled car, etc. The

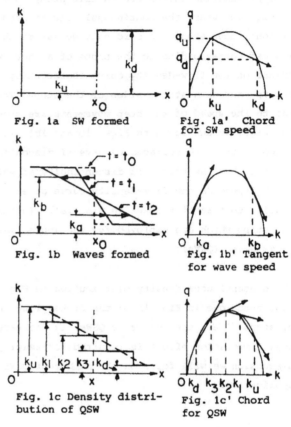

Fig. 1a SW formed

Fig. 1a' Chord for SW speed

Fig. 1b Waves formed

Fig. 1b' Tangent for wave speed

Fig. 1c Density distribution of QSW

Fig. 1c' Chord for QSW

Fig. 1 Density distribution and flow-density curve for introduction of QSW

upstream traffic could be in a queue which is formed by a lane closure or a red signal in case $k_u = k_J$ and $k_d = 0$, where k_J is the jam density. After removal of the blockage (suppose that it occurs at $t = t_0$), the density distribution is deformed continuously as time passes. Configurations of density distribution at $t = t_1$ and $t = t_2$ ($t_0 < t_1 < t_2$) are depicted with solid lines in Fig. 1b. The Hydrodynamic Theory employs a concept of waves to analyze the deformation of the density distribution. The wave is interpreted as a propagation of a position with a constant density. For example, the propagations of positions for the density k_a and k_b are shown in the figure. The speed of any wave for density k and the corresponding flow q is equal to a tangent at the point (k,q) on the flow-density curve. Those for densities k_a and k_b are shown in Fig.1b'.

An approximation introduced in this study is conceptually demonstrated
in Fig. lc, where the continuously distributed density is replaced by
discontinuous steps. Accordingly we assume that each step moves with
a speed corresponding to the slope of a chord which connects such
points on the flow-density curve representing the state of traffic on
the up- and down stream sides of the step. The chords corresponding to the speed of the step movements are shown in Fig. lc'. Comparing Figs. lc and lc' with Figs. lb and lb', we can see that the continuous waves are replaced by several discontinuous steps. Also we
can see that the nature in discontinuity as well as representation
of the speed on the flow-density curve of each step in Fig. lc' is
similar to that of the SWs in Fig. la'. In this sense we refer to
the steps which are employed for the approximation of waves as quasi
shock waves (QSWs).

The discontinuous density distribution in Fig. lc comes closer to the
continuous one in Fig. lb or the broken line in Fig. lc as the number
of steps (i.e., the number of QSWs) is increased. In practice, however,
less computation effort is required for fewer QSWs. In the following
discussion we will focus on the number of QSWs required for practical
application.

2.2 Behavior of QSWs at an Isolated Intersection

If the capacity of an approach at a signal intersection is sufficient
to handle the flow on the approach, a queue which forms during one
red phase dissipates during the next green phase. If the flow exceeds
the capacity of the approach, the queue grows upstream and accumulates
as the cycles proceed. First we analyze a case with no queue accumulation and later one with queue accumulation.

Figs. 2a and b show a graphic construction of the density distribution
in the time-distance space. In Fig. 2a, the abcissa x and ordinate t
designate distance and time respectively. It is supposed that an intersection is located at x=0 and a green phase starts at t=0. It is
also supposed that the traffic moves from negative x to positive x.
Figure 2b shows a flow-density curve on which some relevant informa-

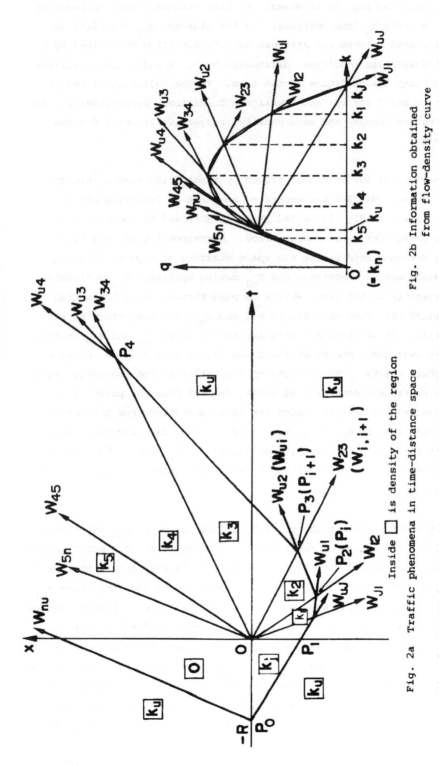

Fig. 2b Information obtained from flow-density curve

Inside ☐ is density of the region

Fig. 2a Traffic phenomena in time-distance space

Fig. 2 Graphic construction of traffic phenomena in time-distance space

tion to construct Fig. 2a is drawn. We also suppose a uniform incoming
flow of density k_u from upstream. In the figures, W_{XY} signifies SWs/
QSWs generated between its upstream density condition designated by X,
and the downstream condiiton, designated by Y. X and Y are subscripts
which explain the condition of the term. In the following, X and/or
Y can be u and d for up- and downstream conditions, respectively; J and
n for the jam density and zero density, respectively; and i for the
density k_i.

The behaviour of SW/QSW can be traced by identifying events through
which some new SWs/QSWs are generated and then by following their
trajectories. In Fig. 2a, a red phase is supposed to begin at t=-R,
i.e., R designates a red phase period. Corresponding to this event,
two SWs are generated in the t-x space starting at a point (-R, 0),
W_{nu} towards moving downstream and W_{uJ} moving upstream; W_{nu} designates
the trajectory of the last vehicle to pass through the intersection
just before the onset of a red phase, and W_{uJ}, the trajectory of the
queue tail. These lines are parallel to the lines W_{nu} and W_{uJ} in Fig.
2b. The next event occurs at the point (0, 0) in the space, where a
green phase starts. When the QSW approximation is not employed, waves
for the continuous density k of $0 \leq k \leq k_J$ fan out from the point (0, 0).
Here, instead, we set QSWs which are lines emanating from the point
(0, 0) specified by W_{J1}, W_{12}, W_{23},...etc. in Fig. 2a. The QSWs partition the space into regions with densities k_J, k_1,..etc. The slopes of
the trajectories are parallel to those of the chords for W_{J1}, W_{12},...etc.
in Fig. 2b.

The next event occurs at the point P_1 in Fig. 2a where W_{uJ} intersects
W_{J1}. Since the region of density k_J is dissipated at this point, a new
state is generated in which a flow density k_1 is adjacent to that with
density k_u and a new SW, W_{u1} appears. The slope of the trajectory W_{u1}
is parallel to the line W_{u1} in Fig. 2b. The same discussion is developed for points P_2, P_3,..and so forth. The sequence of solid lines in
Fig. 2a obtained in this way gives the trajectory of the queue tail. In
the above discussion we may call W_{u1}, W_{u2},...etc. QSWs, since they differ
from those derived from the direct application of the Hydrodynamic Theory. In this paper, however, we will call them SWs because the density
distribution associated with W_{u1} is similar to that shown in Fig. 1a.

The co-ordinates of the points $P_i : (t_i, x_i)$, i=0,1,2,..,n in Fig. 2a(where n is the number of QSWs used for the approximation; in Fig. 2a n=6) can be obtained by successive computation. The following equations are derived looking at two successive points $P_i : (t_i, x_i)$ and $P_{i+1} : (t_{i+1}, x_{i+1})$.

$$w_{i,i+1} = x_{i+1} / t_{i+1} \tag{2}$$

$$w_{ui} = (x_{i+1} - x_i) / (t_{i+1} - t_i) \tag{3}$$

where w_{XY} denotes the speed of W_{XY}. The relative locations of P_i and P_{i+1}, as well as the SWs $W_{i,i+1}, W_{ui}$, are noted in parentheses in Fig. 2a. Solving Eqs. (2) and (3), we obtain a solution in the following matrix form.

$$\begin{pmatrix} x_{i+1} \\ t_{i+1} \end{pmatrix} = \frac{1}{w_{i,i+1} - w_{ui}} \begin{pmatrix} w_{i,i+1} & -w_{ui} w_{i,i+1} \\ 1 & -w_{ui} \end{pmatrix} \begin{pmatrix} x_i \\ t_i \end{pmatrix} \tag{4}$$

Applying Eq. (4) iteratively one can compute the sequence of $P_i : (t_i, x_i)$ (i=1,2,..,n) starting at the point $P_0 : (-R, 0)$.

The total travel time T and distance D and also average vehicle speed $\bar{v}$ in the region $x_1 \leq x \leq x_2$, $t_1 \leq t \leq t_2$ can be computed from

$$T = \int_{x_1}^{x_2} \int_{t_1}^{t_2} k(t,x) \, dt dx \tag{5}$$

$$D = \int_{x_1}^{x_2} \int_{t_1}^{t_2} q(t,x) \, dt dx \tag{6}$$

$$\bar{v} = D / T \tag{7}$$

where k(t,x) and q(t,x) are the distributions of density and flow in the t-x space. When the approximation is employed, the space is partitioned by SWs/QSWs. In each partitioned area, density (and the corresponding flow) is constant. Consequently, the above integrals are equivalent to summing the products of the area and the corresponding density (or flow) for each of the polygonal partitions. An increase in the number of QSWs reduces the area of each polygon and improves the computational accuracy of the integral (in the sense that the computed value comes closer to that from the direct application of the Hydrodynamic Theory).

A computational example is presented below. Here we assume a quadratic
flow-density relationship which is written as

$$q = v_f k \ (\ 1 - k/k_J) \qquad\qquad\qquad (8)$$

where v_f is the free flow speed. From Eqs. (1) and (8), SW/QSW speed
w_{XY} is written as

$$w_{XY} = v_f \ (1 - k_X/ \ k_J - k_Y \ / \ k_J \ ) \qquad\qquad\qquad (9)$$

Applying Eq. (9) in Eq. (4), one can trace a trajectory of the queue
tail for a quadratic flow density relationship.

Figures 4a, b and c show one of the computational results. They also
demonstrate a difference in the configuration of density distribution
in accordance with the number of QSWs. The value of parameters used
are: R=30 sec; k_J=0.15 veh/m; v_f=12 m/sec; k_u= 0.030 veh/m.

The number of QSWs employed to draw Figs. 3a, b and c are two, four and
eight, respectively, assigning half of them for the propagation downstream and half for that upstream. The queue tail trajectory obtained
from the direct application of the Hydrodynamic Theory, e.g., one computed from formulas developed by Richards (1956), is also shown in the
figures with broken lines. It can be easily seen from these figures
that the angles at which the linear segments meet are rounded off and
the queue tail obtained approximative analysis comes closer to the broken lines as the number of QSWs is increased.

In reviewing the process of drawing Fig. 2a, we can see that SWs/QSWs
which are generated at the onset of a red phase and a green phase col-
lide with each other and form new SWs, etc. This process of tracing the
trajectories of QSWs is decomposed into a sequence of two principal iterative steps: first, identify the event which occurs in the next earliest moment and , second, set the SWs/QSWs in accordance with the event.
For the example shown in Fig. 2, the events are the onset of red and
green phases and collision of the SWs/QSWs. A procedure to simulate the
trajectories of the SWs/QSWs as well as to compute the total travel time
and distance is given below.

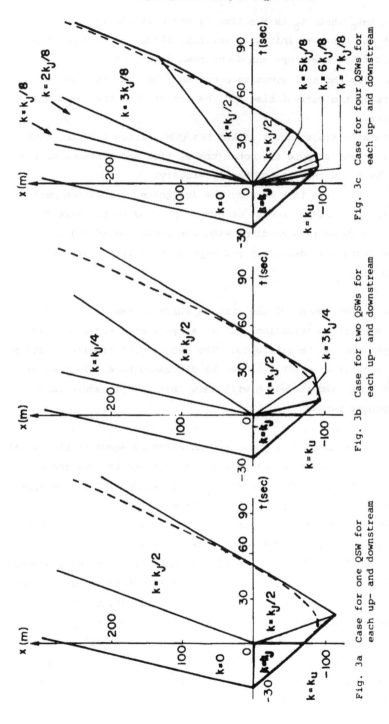

Fig. 3a Case for one QSW for each up- and downstream

Fig. 3b Case for two QSWs for each up- and downstream

Fig. 3c Case for four QSWs for each up- and downstream

Fig. 3 Improvement of approximation with increase in number of QSWs

   i) Set $t=t_0$ where t_0 is the time to start simulation.

  ii) Set the SWs/QSWs initially existing along the road section
      specifying their up- and downstream densities.

 iii) Identify the next event to occur and the time interval δt
      between the current time and the moment when the event will
      occur.

  iv) Determine trajectories of all SWs/QSWs within the time inter-
      val δt and compute the total travel time and distance during
      δt by the methods described previously.

   v) Set $t=t+\delta t$ and if t exceeds the designated simulation period
      terminate the procedure, otherwise proceed to the next step.

  vi) Set SWs/QSWs in accordance with the event specifying their up-
      and downstream densities and repeat steps iii), iv), v),
      and vi).

This is just a restatement of the process which we have discussed
through the graphic construction, but in a systematic way from which
a computer program can be developed. The programming requires certain
techniques, details of which will not be discussed here. Instead we
will present some examples which will show that the procedure functions in a proper manner.

A computational example for a queue accumulation is shown in Fig. 4 (a),
(b), (c) and (d). The values of parameters employed to draw the figures are: v_f=12 m/sec; k_J=0.15 veh/m; k_u=0.045 veh/m ; T_c=60 sec; G=R=
30 sec. Part (a) is obtained from the direct application of the Hydrodynamic Theory. The figure differs from broken lines in Fig. 3 in the
sense that the queue tail never passes through the intersection but
grows upstream. Parts (b), (c) and (d) are obtained from the procedure.
In applying the procedure the length of the road section is set to be
sufficiently large. In drawing parts (b), (c) and (d), the number of
QSWs employed are one, two and four, respectively, for the upstream region of the intersection. The trajectory of the queue tail in part (a)
is also shown in the figure with the broken lines. It is clearly seen
from the figure that the results from the approximative analysis come
closer to broken lines as the number of QSWs increased.

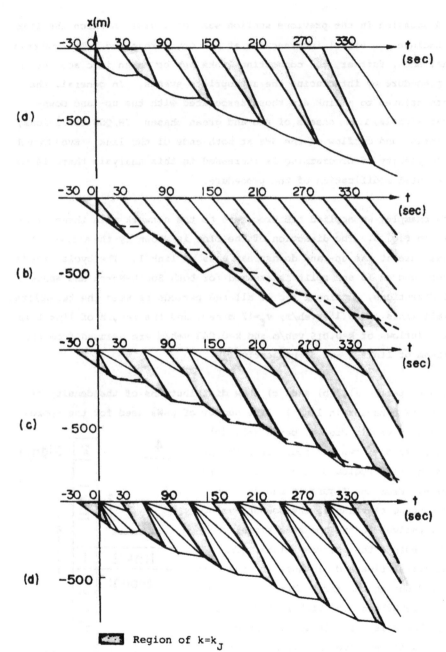

Region of $k=k_J$

Fig. 4 Accumulation of queue: Effect of increase in number
of QSWs: part (a) is drawn without approximation,
parts (b),(c) and (d) are for one, two and four QSWs
respectively

2.3 Analysis of Links

The discussion in the previous section was for a case in which the link
may include one signal intersection at the downstream end. The upstream
signal and , further, the connecting links can be taken into account in
the procedure by introducing the appropriate events. In general, the
events related to a link are those associated with the up- and downstream signals, i.e., onsets of red and green phases, SW/QSW collisions,
and inflow and outflow of the SWs at both ends of the link. Even though
the complexity of programming is increased in this analysis there is no
fundamental modification of the procedure.

As an example, we applied the procedure to the network with three links
shown in Fig. 5. The direction of the flow is shown by the arrows. The
signals are at the up- and downstream ends of link 1. The cycle time is
40 sec, and an 18 sec split is assumed for both South-North and West-
East directions, putting 2 sec of all red periods between the two splits.
For all links, k_J=0.150 veh/m; v_f=12 m/sec, and the length of link 1 is
200m. Inflows of k=0.015 veh/m and k=0.045 veh/m are assumed from the
upstream of links 2 and 3, respectively.

Figure 6, parts (a), (b) and (c) show distributions of the density as
well as SW behavior in link 1. The number of QSWs used for the approximation of waves in Fig. 6, parts (a), (b)
and (C) are one, two and four for each up-
and downstream region of the signal intersection. From the figure it can be seen
that a queue starting at the downstream
signal evolves to the upstream signal
and then enters the upstream links. The
total travel time and distance in link 1 is
computed during the first three cycles for
several lengths of the link from 200m to
500m. There is very little difference in
the total travel time and distance between
the cases in parts (b) and (c). The maximum difference is 1.2% in the total travel
distance for the link length of 500m. As-

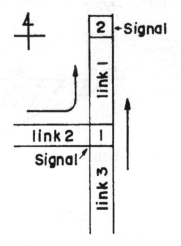

Fig. 5 Example of con-
nected links

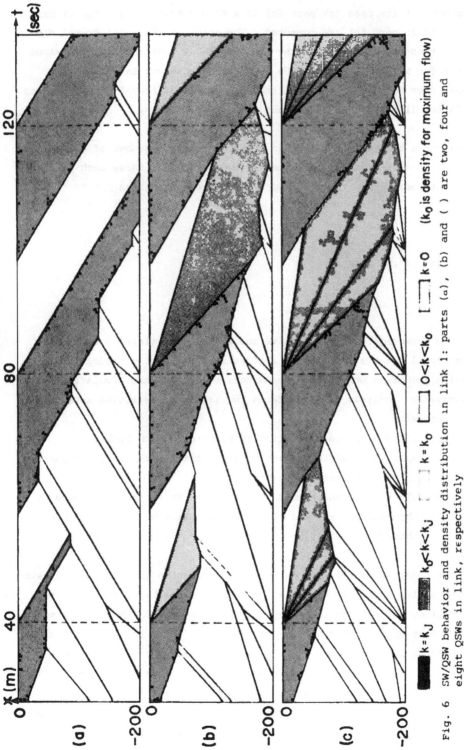

Fig. 6 SW/QSW behavior and density distribution in link 1: parts (a), (b) and (c) are two, four and eight QSWs in link, respectively

■ $k = k_J$ ▨ $k_0 < k < k_J$ ▧ $k = k_0$ □ $0 < k < k_0$ ▭ $k = 0$ (k_0 is density for maximum flow)

suming that the case for part (c) is a true configuration, the errors in
the case in part (a) are computed. The error in the total travel distance tends to be larger as the link length is increased. The maximum
error for a link length of 500m is 8.6%. The errors in the average
speed have the same pattern as those in the total travel time: 2.5% and
9.7% for link lengths of 200m and 500m, respectively.

From these observations, one can conclude that the number of QSWs required for practical application are two or at most three each to represent whole waves in over- and under-saturated regions.

3.   SIMULATION MODEL

Taking advantage of approximative analysis, we developed a simulation model for traffic flow in a signalized network. In addition
to the phenomena which have been discussed so far, the simulation model
involves an analysis of the effects required for describing actual traffic phenomena:

   i) attraction and generation of traffic on each link
   ii) split of traffic at each intersection (i.e., consideration for
       turning and through traffic)
   iii) blockage of the through traffic lane by a backed up queue in
        the turning lane and of the turning lane by a queue in the
        through traffic lane

In the model we divided the duration of simulation into a number of successive time intervals (referred to as time slice hereafter). At the
same time, the network is decomposed into directed links which connect
successive signal intersections. Further, in order to analyze the split
of traffic into approaches, a link is decomposed into five (or less if
fewer are adequate) sections as shown in Fig. 7. We call a region in
the time-distance space associated with the section and time slice a
"simulation unit". The simulation proceeds as we look at the continuity
of the in- and outflow between adjacent simulation units within a time

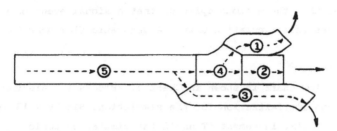

Fig. 7 Decomposition of link into sections

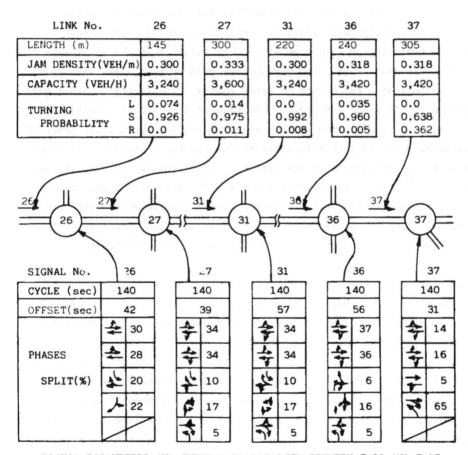

LINK No.		26	27	31	36	37
LENGTH (m)		145	300	220	240	305
JAM DENSITY(VEH/m)		0.300	0.333	0.300	0.318	0.318
CAPACITY (VEH/H)		3,240	3,600	3,240	3,420	3,420
TURNING PROBABILITY	L	0.074	0.014	0.0	0.035	0.0
	S	0.926	0.975	0.992	0.960	0.638
	R	0.0	0.011	0.008	0.005	0.362

SIGNAL No.	26	27	31	36	37
CYCLE (sec)	140	140	140	140	140
OFFSET(sec)	42	39	57	56	31
PHASES SPLIT(%)	30	34	34	37	14
	28	34	34	36	16
	20	10	10	6	5
	22	17	17	16	65
		5	5	5	

SIGNAL PARAMETERS AND TURNING PROBABILITY BETWEEN 7:00 AND 7:15

Fig. 8 Input data for several links and signals

slice and between successive time slices. The length of the time slice
must be smaller than the minimum split so that a signal event occurs
only once or less in a simulation unit. A quadratic flow-density curve
is employed.

The simulation flow chart is shown in Chart 1. Blocks 1-3 are for data
input, output and initialization for the simulation. Blocks 4-13 are
repeated with the time increment ∆T until the simulation period T_{max} is
over. Blocks 14-19 are called if needed. Attracted and generated trips
on links during simulation period must be given as input data. In
block 4, they are treated in terms of density which is subtracted from
and added to (in the case of attraction and generation respectively)
the density distributed along the section 5 (see Fig. 7) of each link
at the beginning of each time slice. In blocks 6 and 7, all the signal
events are dealt with. Two QSWs are assigned to each over- and under-
saturated region for onsets of green phases. In block 8 together with
blocks 17, 18 and 19, SW split, blockage of flow and also release of
the blockage are treated as the events by setting appropriate SWs/QSWs.
Vertical queue is partially employed for the backed-up queue in block
19. For an estimation of the travel time between two points, pseudo-
cars are run along the specified route. The position of each car is
computed in each time-slice, and the travel time is accumulated until
it arrives at the destination. This computation is performed in block
10. The model is programmed in FORTRAN with 3785 statements.

The simulation model was applied to a corridor 6.43km long, part of
National Road No. 192, in Tokushima City. This is a main street which
connects the suburbs and the city center and experiences heavy traffic
congestion during every peak period.

A field survey, in June 15, 1982, provided the travel time on the
6.43km corridor and the traffic flow near the principal intersections.
The simulation was performed under the same circumstance as that of
the survey. The network had 118 links and 44 signal intersections.
The simulation is for 2 hours from 7:00 to 9:00 a.m. towards the city
center. Input data for some links and intersections are shown in Fig.

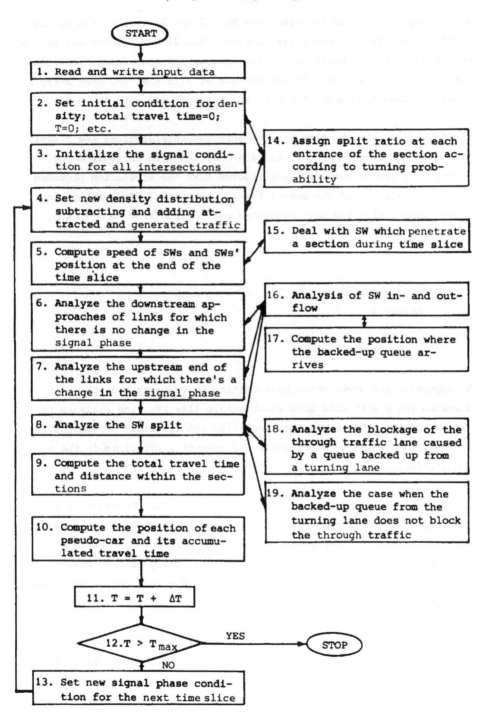

Chart 1 Simulation flow chart

8.   Through traffic at intersections No. 27 and 31 was blocked by the
right turning traffic every peak period. No. 31 was an especial bottle
neck because of its small capacity. Information concerning the signal
was for 7:00 - 7:15 a.m. It was changed at 7:15 a.m. and then every 5
minutes. The CPU time for the simulation was 60 sec with FACOM M382.

In Figs. 9 and 10, the flow for the selected intersections and travel
time obtained from the simulation are compared with those from the
survey. It can be seen from the figures that the simulation provided
a good description of the actual traffic behaviour. Especially the
travel time during the most congested period from 7:30 to 8:30 a.m.

4.   CONCLUSION

By approximating waves which have continuous property over the time-
distance space with some QSWs which behave like SWs, the analysis of
the Hydrodynamic Theory can be simplified and its field of application
widened. A procedure demonstrates a fundamental principle in the approximation.

Through the analysis, it was found that the traffic phenomena derived
from the approximative analysis comes closer to that from the direct
application of the Hydrodynamic Theory as the number of QSWs is increased. In practice, however, sufficient accuracy can be obtained if
two QSWs are assigned to represent the waves for each under- and over-
saturated region.

The practicability of the approximative analysis is confirmed through
applying the theory to a formulation of a simulation model for traffic
flow on a signalized network. The model has been applied to a corridor. The simulation result shows that the model can well describe
traffic phenomena, especially for heavily congested flow.

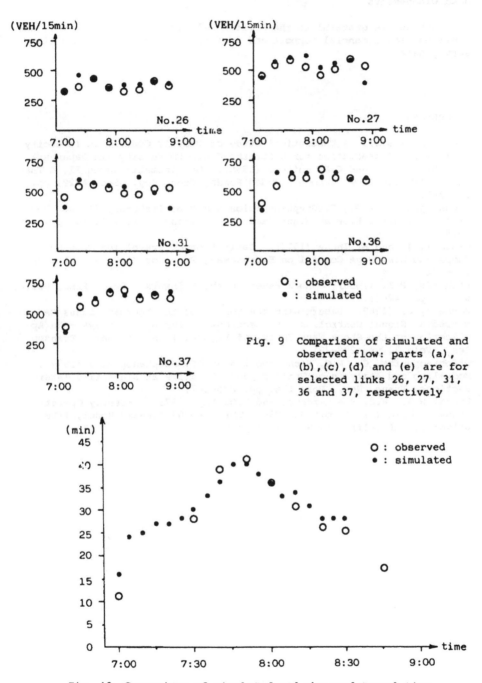

Fig. 9 Comparison of simulated and observed flow: parts (a), (b),(c),(d) and (e) are for selected links 26, 27, 31, 36 and 37, respectively

Fig. 10 Comparison of simulated and observed travel time

Acknowledgements

The authors ?.re grateful to the Sumitomo Electric Industries, Ltd., for their partial financial support and assistance in collecting the field survey data.

REFERENCES

Inoue, N. (1973). A Theoretical Study of Traffic Control on Intercity Expressway, Dissertation for Doctrate, Kyoto University (in Japanese).
Lighthill, M.J. and G.B. Whitham (1955). On Kinematic Waves II, A Theory of Traffic Flow on Long Crowded Roads, Proc. Roy. Soc. A229, pp. 317-345.
Michalopoulos, P.G., G. Stephanopoulos and V.B. Pisharody (1980). Modeling of Traffic Flow at Signalized Links, Trans. Sci. Vol. 14, pp. 9-14.
Okutani, I. and N. Inoue (1973). Estimation of Travelling Time between Ramps and Discharge Control on Expressway, Proc. of JSCE, No. 211, pp. 100-102.
Richards, P.I. (1956). Shock Waves on the Highways. Oper. Res., Vol.4, No. 1, pp. 42-51.
Rorbech, J. (1968). Determining the Length of the Approach Lanes Required at Signal Controlled Intersections on Through Highways - An Application of the Shock Wave Theory of Lighthill and Whitham. Transpn Res., Vol. 2, pp. 283-291.
Stephanopoulos, G., P.G. Michalopoulos and G. Stephanopoulos (1979). Modelling and Analysis of Traffic Queue Dynamics at Signalized Intersections, Trans. Res., Vol. 13A, pp. 295-307.
Stock, W.A., R.C. Blankernhorn and A.D. May (1973). Freeway Operation Study - Phase III, Report No. 73-1, The FREQ 03 Freeway Model, ITTE, University of California Special Report.

Ninth International Symposium on
Transportation and Traffic Theory
© 1984 VNU Science Press, pp. 21–42

MACROSCOPIC FREEWAY MODEL FOR DENSE TRAFFIC – STOP–START WAVES AND INCIDENT DETECTION

REINHART D. KÜHNE
AEG-Telefunken Forschungsinstitut, Ulm, F.R.G.

ABSTRACT

A continuum model for freeway traffic flow is described which includes relaxation of the equilibrium speed of the static speed-density relation ("Fundamental Diagram") and anticipation of traffic conditions downstream. Stability analysis shows that in light traffic the equilibrium solution given by the Fundamental Diagram is stable, while in dense traffic, jams with stop–start waves occur. The formation of stop–start waves is in full analogy to the creation of roll waves in inclined open channels with a suitable water height. The methods to derive such roll wave solutions from the basic hydrodynamic equations of shallow water theory are used to describe the stop–start waves in the freeway model.
As in fluid dynamics, where the change from laminar flow to turbulent flow is announced by critical fluctuations, the change from steady traffic to traffic with jams and stop–start waves is indicated by large fluctuations. From theory and measurement of the broadening of the speed distribution due to the critical fluctuations an algorithm of traffic classification and incident detection is derived. On the basis of a simplified version of this algorithm using only local measurements of the speed distribution an automatic traffic jam warning system is developed.

1.   INTRODUCTION

1.1 Macroscopic Traffic Variables

The most impressive way to describe the traffic flow on highways certainly is a diagram which shows distance-time curves of a set of vehicles pursued over a long street segment. Such diagrams have been
produced by Treiterer (1975) in the United States by aerial photogrammetry techniques (Fig. 1)

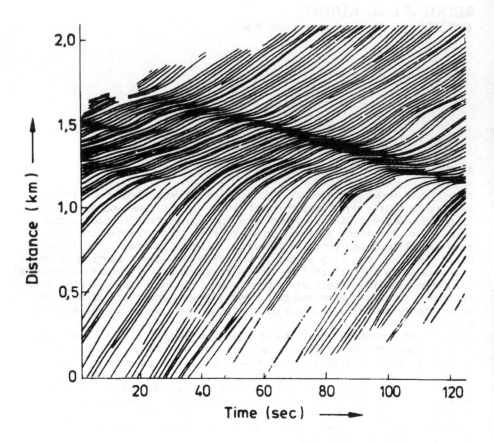

Fig. 1.   Distance-time diagrams of lane 1 of a highway at Columbus, Ohio
          (Treiterer, 1975)

The distance-time diagrams show that the traffic flow develops in a
manner which allows a macroscopic description in terms of hydrodynamics.

The macroscopic traffic variables are:

traffic volume or flow rate	q	(vehicles per hour)
traffic density	ρ	(vehicles per lane per km)
mean speed	v	(speeds averaged over a suitable set of vehicles)

Within a macroscopic description the relation

$$q = \rho \cdot v \qquad \text{(flow rate = density} \cdot \text{velocity)}$$

always holds.

In simple macroscopic freeway models another equilibrium

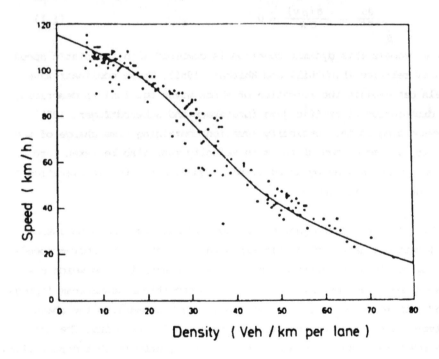

Fig. 2. Equilibrium speed-density relation. The curve is fitted to
measured traffic data (Duncan, 1979)

relationship e.g. between mean speed and density

$$V = V(\rho) \qquad (\text{"Fundamental Diagram"})$$

is given. As a rule this "Fundamental Diagram" is an empirical fit to
measured traffic data; it summarizes the microscopic headway behaviour
at different densities (Fig. 2).

1.2 Dynamic Freeway Model

To record spatial and temporal variations the static relations between
flow-rate, density and mean speed presented so far must be enlarged to
a dynamic description. Assuming that on the street segment considered
there is no on-ramp and off-ramp traffic the equation of continuity is
(x = space coordinate along the highway, t = time coordinate)

$$\frac{\partial \rho}{\partial t} + \frac{\partial (\rho v)}{\partial x} = 0 \qquad (1.1)$$

In some models this dynamic equation is combined with the static speed-
density relation (Lighthill and Whitham, 1955). These nonlinear wave
models can explain the formation of shock waves but fail in describing
the dissolution of traffic jams (Leutzbach and Schwerdtfeger, 1981).
Corresponding to the continuity equation containing time change of the
density, a time change of the mean velocity must also be taken into
account. This is done by an additional equation for the acceleration of
substantial volume element.

In this "higher order" dynamic continuum formulation the substantial
acceleration consists of a relaxation to the static equilibrium speed-
density relation (relaxation time γ^{-1}) and an anticipation which expresses the effect of drivers reacting to conditions downstream ("pressure" coefficient c_o^2). The relaxation time γ^{-1} summarizes the characteristic times for acceleration and brake and for reacting. The anticipation term, like in compressible gas flow, effects that disturbtions
as congested traffic move along the highway as waves. The pressure
coefficient c_o^2 herein has the meaning of the square of the congestion
velocity. This interpretation is discussed in detail by Baker (1983).

$$\frac{dv}{dt} = \frac{\partial v}{\partial t} + v \frac{\partial v}{\partial x} = \gamma (V(\rho) - v) \frac{c_o^2}{\rho} \frac{\partial \rho}{\partial x} \qquad (1.2)$$

The substantial acceleration therin is decomposed in the usual manner
into a local acceleration and a convection term.

Equations (1.1) and (1.2) present a slight modification of the macroscopic freeway model proposed by Payne (1979). They form the basis of
the following investigations on the formation of traffic jams and on
the influence of fluctuations in the traffic flow.

The system of equations (1.1) and (1.2) can be used as a boundary value
problem to calculate from the evaluation of density ρ and mean speed v
in time at one site the time development of these traffic variables at
another site. Cremer (1979) has done this by a discretization of the
starting equations given above and he has compared the shape with measurements. There is an excellent agreement between calculation and
measurement. In the present paper the system of equations is explored
directly by methods of theoretical physics. In this way one gets a general indication of how a given starting condition develops in space and
time.

2. STABILITY ANALYSIS FOR LIGHT AND DENSE TRAFFIC

2.1 Linear Stability

Equations (1.1) and (1.2) have an equilibrium solution

$$\rho = \rho_0 \qquad v = V(\rho_0) \qquad (2.1)$$

which is constant in space and time. To determine the stability of this
solution the trial solution

$$\rho = \rho_0 + \tilde{\rho}\, e^{ikx' + \omega t'} \qquad v = V(\rho_0) + \tilde{v}\, e^{ikx' + \omega t'} \quad (2.2)$$

is substituted in the model equations and only terms up to first order
in $\tilde{\rho}$ and $\tilde{v}$ are considered (the space coordinate and the time coordinate
are normalized with γ/c_0 and γ^{-1} respectively, i.e. $x' = xc_0/\gamma$ and
$t' = \gamma t$). This linear stability analysis leads to a time constant ω which
depends on the wave number k and a traffic parameter

$$a = -1 - \frac{\rho_0}{c_0} \frac{dV(\rho_0)}{d\rho} \tag{2.3}$$

This traffic parameter is a dimensionless parameter which characterizes
the traffic conditions like the Reynolds number characterizes a fluid
flow. If the real part of ω is greater than zero then density and mean
speed will increase exponentially - at least in the linear approximation - which means instability of the equilibrium solution. The real
part of ω is presented in Fig. 3 for different values of the traffic
parameter a as a function of the wave number k.

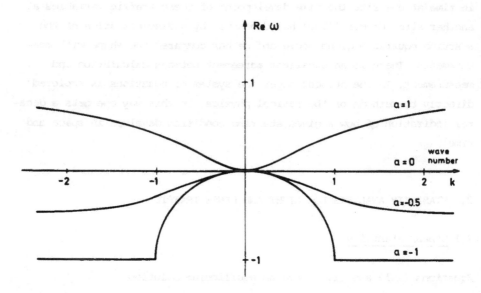

Fig. 3. Real part of the time constant ω in the linear stability analysis of the equilibrium solution as a function of the wave number
k for different values of the traffic parameter a. Apparently
instabilitity for a > 0 is independent of the wave number.

It is easy to see from this figure that for a = 0 the equilibrium solution becomes unstable whatever the wave number. In the unstable region
(a > 0) the graph shows that high wave numbers are more unstable than
low wave numbers. This means that jumps of density over small distances
will be amplified more than those over long distances. This is also a
hint that in dense traffic shock waves are created. Associated with
these shock and roll waves is a high accident rate.

The traffic parameter a can be expressed in terms of the traffic density ρ with a model fit for the speed-density relation (Cremer, 1979)

$$V(\rho) = v_f \left(1 - (\rho/\rho_m)^{n_1}\right)^{n_2} \tag{2.4}$$

(v_f = 120 km/h; ρ_m = 200 veh/km/lane; n_1 = 1.4; n_2 = 4 for German highways); this is shown in Fig. 4.

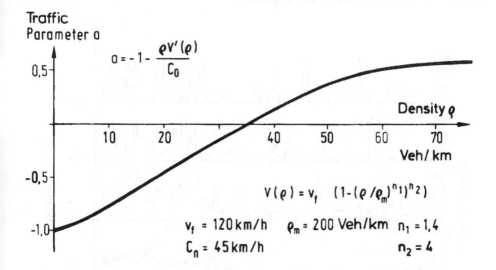

Fig. 4. Conversion of the traffic density ρ into the traffic parameter a

The point of change from stable to unstable traffic with jams and stop-start waves lies in the above case at 35 veh/km/lane. The speed-density relation is affected by speed limits, by the proportion of trucks and by the curvature of the road; therefore under suitable conditions the threshold for unstable traffic can be shifted into a more dense traffic regime.

2.2 Stop-Start Waves

Measurements of the mean speed upstream of a toll gate (Fig. 5) and the corresponding simulation of traffic behaviour by microscopic car following models (Wiedemann, 1974; Leutzbach and Haas, 1981) show that the state beyond the threshold is characterized by periodic stop-start traffic waves.

The change from the equilibrium solution with a temporal and spatial
constant distribution to a solution with periodic traffic breakdown and
following restoration is analogous to the formation of roll waves in an
open inclined channel when water height has exceeded a fixed amount.
Such phase transitions are well known in physics and in spite of the
great variety of phenomena such as evaporating of water, magnetization
of a ferro magnet, and formation of convection rolls in a heated viscous
fluid, they all have the same behaviour.

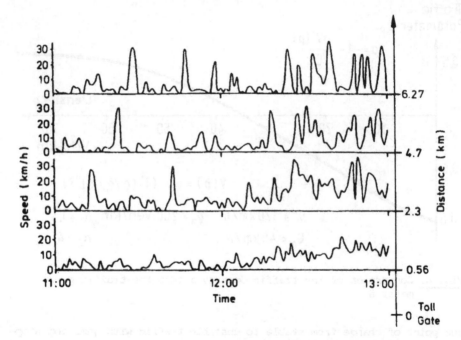

Fig. 5. Time development of mean speed upstream of a toll gate at different distances from the toll gate. Clearly recognizable are the
periodic stop-start waves, exceptionally distinct in the stationary region far away from the toll gate (after Koshi, 1982).

2.3 Wave Solutions

To look for solutions of the model equations (1.1) and (1.2) in the dense
traffic regime it is convinient to use dimensionless space and time coor-
dinates (compare section 2.1) and to make a nonlinear space coordinate
transformation

$$x' = x'(x,t) \qquad (2.5)$$
$$t' = t \qquad (2.6)$$

with

$$\frac{\partial x'}{\partial t} = -\rho v \qquad \frac{\partial x'}{\partial x} = \rho \qquad (2.7)$$

and

$$\partial_t = \partial_{t'} - \rho v \, \partial_{x'} \qquad \partial_x = \rho \partial_{x'} \qquad (2.8)$$

in order to simplify the equation of continuity. Going over to the inverse density $\chi = 1/\rho$ the model equations read (for simplicity the primes ' are suppressed in the following)

$$\chi_t - v_x = 0 \qquad (2.9)$$

$$v_t + \left(\frac{1}{\chi}\right)_x = V(1/\chi) - v \qquad (2.10)$$

where the subscripts mean partial derivatives with respect to time t and space x respectively.

To look for periodic solutions of the model equations it is necessary to make the following disposition

$$\chi(x,t) \overset{!}{=} \chi(y) \qquad (2.11)$$

$$v(x,t) \overset{!}{=} v(y) \qquad (2.12)$$

with

$$y = x + Qt \qquad (2.13)$$

i.e. we look for solutions where a profile is shifted along the highway. By transformation to the collective coordinate y the model equations, primary partial differential equations, become ordinary differential equations

$$(Q\chi - v)_y = 0 \qquad (2.14)$$

$$Qv_y + \left(\frac{1}{\chi}\right)_y = V(1/\chi) - v \qquad (2.15)$$

The continuity equation (2.14) can be directly integrated

$$v = Q\chi + v_g \qquad (2.16)$$

where v_g is an integration constant. Returning for a moment to the density ρ instead of the inverse density χ the integral (2.16) of the continuity equation can be written as

$$\rho(v - v_g) = Q \qquad (2.17)$$

This means density and speed in a frame running with group velocity v_g must always serve as supplements: Wave solutions with a profile moving along the highway are only possible if the density at one site increases at the same proportion as the mean speed with respect to the group velocity v_g decreases and vice versa. The constant Q has the meaning of a net flow.

Insertion of the integral of the continuity equation leads directly to a destination equation for the distribution of the inverse density along the highway in the case of a running profile:

$$(Q^2 - \frac{1}{\chi^2}) \, \chi_y = V \, (1/\chi) - Q \, \chi - v_g \qquad (2.18)$$

Since also in the continuum description a singularity of the density must be excluded, the vanishing of coefficient of the density slope χ_y must be connected with a vanishing of the right hand side to prevent absurd results $\chi_y = \infty$. The coefficient is zero for

$$\chi = 1/Q \quad \text{(inflection point)} \qquad (2.19)$$

which leads to the relation

$$v_g = V(Q) - 1 \qquad (2.20)$$

between group velocity v_g and net flow Q (keep in mind that all velocities are normalized to c_o). With a group velocity determined in this way the profile equation (2.18) has been integrated. The results are shown in Fig. 6 for different values of the net flow Q. For the speed-density relation a linear relation has been used which is valid in the vincinity of the inflection point. The collective coordinate y serves as a space coordinate. The inverse density shape is then represented in a frame running with group velocity v_g. In this running frame the inverse density increases with increasing distance. In a fixed system the density itself decreases with increasing time.

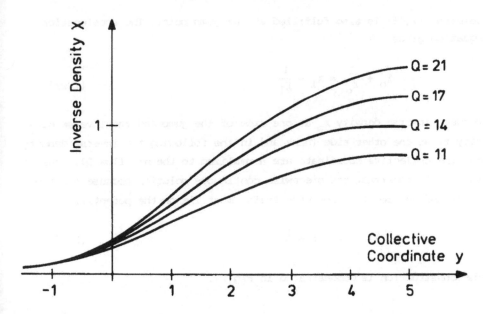

Fig. 6. Profile shape of the normalized inverse density as a function of the collective coordinate y

2.4 Construction of Piecewise Continuous Roll Waves

So far no periodic solutions are available by the wave disposition. This situation is identical with that in shallow water theory where the profile equations for a steady flow do not show periodic solutions although periodic roll waves are observed in every inclined open channel with suitable water height. To resolve this discrepancy pieces of continuous solutions as in Fig. 6 must be put together by jumps (Dressler, 1949). To look for such piecewise continuous periodic solutions is by no means misleading. Firstly, the starting point for the equation of continuity is a conservation law for the vehicle number, that is an integral law which allows finite jumps in the density. Secondly, the linear stability analysis of the equilibrium solution shows that high wave numbers get more unstable than lower ones. This leads to a steeper and steeper shape and to the formation of shock fronts.

Integration over the discontinuity with an infinitesimal integration range leads to a jump condition. The equation of continuity with the solution (2.16) is also fulfilled at the jump point. The acceleration equation gives

$$X_0 + \frac{1}{X_0} = X_1 + \frac{1}{X_1} \qquad (2.21)$$

between inverse density X_0 at one side of the jump and the inverse density X_1 at the other side (here and in the following the inverse density and the collective coordinate are normalized to the net flow Q). Thus it is easy to construct the piecewise continuous solution because starting point and end point of one wave train must lie on the potential

$$U(X) \quad = \quad X + \frac{1}{X} \qquad (2.22)$$

The construction is demonstrated in Fig. 7.

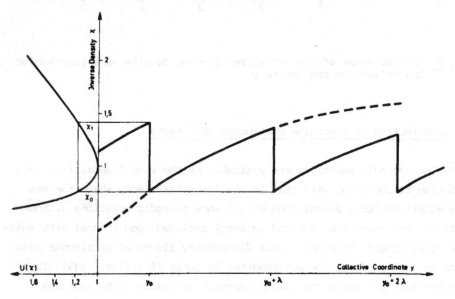

Fig. 7. Construction of a piecewise continuous roll wave solution for dense traffic by putting together continuous pieces using the jump condition that start and end must lie on $U(X)$. All variables are in normalized representation.

2.5 Inclusion of Dissipation and Representation of the Roll Waves Solutions in a Phase Portrait

Actually, traffic jumps with sharp shock fronts are smeared out because of statistical undulations. If is therefore plausible to enlarge the acceleration equation by a small dissipation term $\nu\, v_{xx}$; as in hydrodynamics the acceleration equation contains a dissipation term $\nu \nabla^2 v$ with dynamical viscosity ν (Novik, 1971; Bykhovskii, 1966). Extensions with second order derivatives are also discussed by Ashton (1966).

The profile equation for the self similar wave solution with this additional viscosity reads

$$(1 - \frac{1}{x^2})\, x_y = V\, (Q/x) - V(Q) + 1 - x + \nu x_{yy} \tag{2.23}$$

Restriction to the vicinity of the inflection point leads to the approximate expression

$$2\, (x - 1)\, x_y = a\, (x - 1) + \nu x_{yy} \tag{2.24}$$

Again the whole behaviour is described by the traffic parameter a

$$a = -1 - \frac{Q/c_o}{c_o}\, \frac{dV(Q/c_o)}{d\rho} \tag{2.25}$$

The only difference in the case of light traffic is that the parameter a is now related to the net density Q/c_o (in usual dimensions) instead of the equilibrium density in the former case.

Starting with the inverse density at the inflection point

$$X = x - 1 \tag{2.26}$$

and using the abbreviation

$$Y = X_y \tag{2.27}$$

for the slope, the profile equation becomes a evolution equation

$$X_y = Y \tag{2.28}$$

$$Y_y = \frac{1}{\nu}\, (-a\, X + XY) \tag{2.29}$$

which can be presented in a phase portrait (compare Fig. 8).

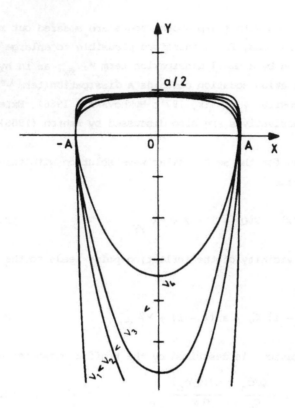

Fig. 8. Phase trajectories of the roll wave solution in the presence
of a small dissipation. For $\nu \to 0$ one regards break-off at
$X = \pm A$. In this case the oscillations take on a saw-tooth form.

2.6 Stability Analysis of the Roll Wave Solution

As in the stability analysis of the static equilibrium solution the roll
wave stability is studied by inserting the trial solution

$$X(y,t) = X_{st}(y) + \tilde{X}_k(y)\, e^{\omega t} \tag{2.30}$$

$$v(y,t) = v_{st}(y) + \tilde{v}_k(y)\, e^{\omega t} \tag{2.31}$$

in the full time-dependent model equations (2.9) and (2.10). Again the
deviations $\tilde{X}_k$ and $\tilde{v}_k$ are labeled by an index k with the meaning of a
quasi wave number. Namely, because of the periodicity of the stationary

roll wave solution x_{st} and also v_{st}, the deviations $\hat{x}_k$ and $\bar{v}_k$ must be periodic until a factor of absolute value 1 (Floquet theorem, see Flügge 1974):

$$x_k \ (y + \lambda) = e^{ik} \ \hat{x}_k(y) \qquad \hat{v}_k \ (y + \lambda) = e^{ik} \ \hat{v}_k(y) \qquad (2.32)$$

Using the relations for the stationary roll wave solution and regarding only terms up to first order in $\hat{x}_k$ and $\bar{v}_k$, the time constant ω is only a function of the traffic parameter taken at the net density ϱ/c_0 and the quasi wave number k

$$\omega = \omega \ (a,k) \qquad (2.33)$$

The real part of the time constant ω is plotted in Fig. 9.

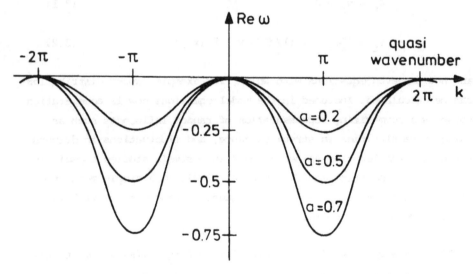

Fig. 9. Real part of the time constant ω in a linear stability analysis of the roll wave solution as a function of the quasi wave number k for different values of the traffic parameter a. There is stability for all positive values of a.

The real part of the time constant ω is always negative for positive values of the traffic parameter (regime of instability of the static equilibrium solution). Therefore the piecewise continuous periodic stop-start wave solution is stable. In dense traffic the stable state is characterized by a more or less periodic traffic breakdown with subsequent restoration.

3. FLUCTUATIONS IN TRAFFIC FLOW

3.1 Model Equations with Acceleration Noise

In hydrodynamics near the turnover point, where the stable solution
changes from the static equilibrium solution to the roll wave solution,
critical fluctuations occur: fluctuations which are usually rapidly dissolved do not remain microscopic but become macroscopic due to the interaction of the system particles, which in the region of the turnover
point, becomes long range. It is obvious to also look in traffic flow
for such critical fluctuations which predict the change of state. The
acceleration equation therefore will be enlarged by a fluctuating force.
The model equations (in normalized and transformed notation)

$$x_t - v_x = 0 \tag{3.1}$$

$$v_t + (\tfrac{1}{x})_x = V(1/X) - v + \Gamma(x,t) \tag{3.2}$$

are now Langevon equations with which, for example, speed distributions
can be calculated. Included in the model equations now is acceleration
noise as a summarizing representation of random influences such as
bumps, irregularities in street guidance, and fluctuations in drivers'
attention. Acceleration noise has been intensively studied especially
by analysing the trip recorder (Winzer, 1980). In a simple model the
fluctuating force is a δ-correlated gaussian random force with vanishing mean value (white noise)

$$< \Gamma(x,t) > = 0 \qquad < \Gamma(x',t') \; \Gamma(x,t) > \; = 2 \, \sigma_o^2 \quad \delta(x - x') \, \delta(t - t')$$

σ_o^2 is the standard variation of the acceleration noise of free traffic flow.

Using the model equations linearized around the static equilibrium solution the speed distribution in the stationary case gets a gaussian normal
distribution with a width depending on the required equilibrium density.
In Fig. 10 the standard variation with respect to the standard variation
of free traffic flow is shown.

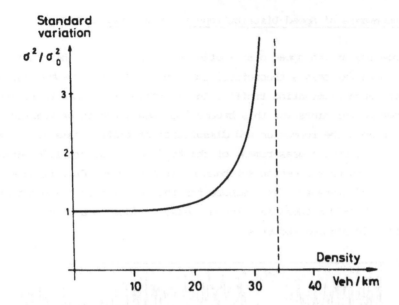

Fig. 10. Standard variation of the speed distribution as a function of
the equilibrium density. Near the instability point the distribution broadens more and more. This broadening announces the
formation of instabilities with jams and stop-start waves.

Even in this simple linearized model which is not valid beyond the critical density, the calculations show that the formation of jams and
stop-start waves in dense traffic is announced by a broadening of the
speed distribution.

The above inclusion of acceleration noise leads to a statistical description of traffic, which has as a basis the macroscopic description in terms
of the macroscopic traffic variables and the equilibrium speed-density
relation. The description is especially qualified for dense traffic. It
is quite a different approach from that of Prigogine (1961) and Edie et
al. (1980) in which they start from speed distribution as a consequence
of drivers' different wishes for appropriate speeds. In their approach
the distribution changes in space and time due to hydrodynamic convection
and overtaking, as in the Boltzmann equation of the kinetic gas theory.
Although in principle there are no restrictions, the Boltzmann approach
is more suitable for light traffic with nearly unlimited possibilities
for overtaking.

3.2 Measurements of Speed Distributions in Dense Traffic

The broadening of the speed distribution when approaching the critical
density is so far only a theoretical consequence of the freeway traffic
model for dense fluctuating traffic. The question is whether there are
also experimental hints for this broadening. Measurements of traffic
data which show the formation and dissolution of traffic jams are relatively scarce. Here a measurement of the traffic at Eastern 1976 between
10.30 and 13.50 on the German Autobahn A5 is presented. Fig. 11 shows
the mean speed averaged over 1 minute for the vehicles in the overtaking
lane. At mark 4600 s there is a speed breakdown (due to a traffic jam)
which after 10 minutes recovers.

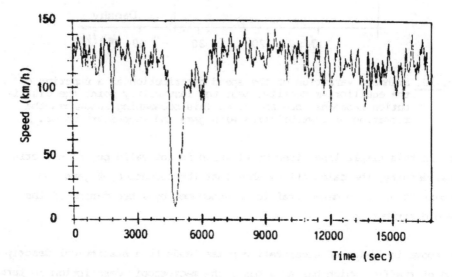

Fig. 11. Curve of mean speed on the Autobahn A5 Bruchsal - Karlsruhe
at km 617 from 15 April 1976, 10.30 to 13.50 (Leutzbach 1981)

In Fig. 12, the corresponding speed distributions are recorded. Vehicle
speeds are divided into groups of 5 km/h width. The number of vehicles
in each speed class is determined during a 5 minute measurement period.
The distribution 5 minutes before the beginning of the traffic jam is
indeed nearly gaussian. 3 minutes later, but still 2 minutes before the
beginning of the jam there is a clear broadening of the distribution:
there are more slow- but also more fast-moving vehicles and the traffic
flow becomes more erratic. Five minutes later just at the beginning of
the jam the speed distribution is even broader. Eight minutes later

there is an extremely broad distribution. This distribution also includes nonstationarities. Ten minutes later the distribution represents
barely congested traffic. This distribution is relatively small; critical fluctuations have dissolved.

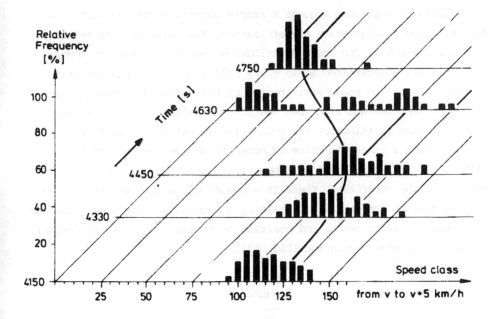

Fig. 12. Speed distribution during the formation of a traffic jam.

4. APPLICATIONS OF THE TRAFFIC THEORY

4.1 Incident Prediction by Speed Distribution Classification

The proposed dynamic traffic flow model can be used to detect incidents:
The time development of the mean speed at one site upstreams is measured;
the corresponding mean speed development at a site downstreams is calculated by the model equations and compared with the actual measurements at
the downstream site. If there are strong deviations between measurement
and calculation an incident has occurred on the street segment under consideration.

In the case of only one local measurement, e.g. for a transportable
jam warning system, the correlation method described above is not applicable. Until now in this case a simple threshold comparison for the
mean speed is used, but in many cases this system is too slow.

The results of section 3 suggest a simple algorithm for incident pre-
detection and traffic state classification. The basis of this method
is a measurement of the speed distribution over an interval of 2 minutes, this being determined again every half minute. In this way it is
guaranteed that the speed distribution is taken from a representative
number of vehicles and also that the actual traffic state is reproduced.
From the speed distribution the mean value is calculated at first and
compared with a threshold value in order to decide whether, on the one
hand, there is a jam caused by stop-start waves or by an accident, or
whether on the other hand, there is a danger of jams or the traffic is
flowing freely. The decision is made by analysing the speed distribution
with respect to the number and position of maxima and threshold comparison of the standard variation (Fig. 13).

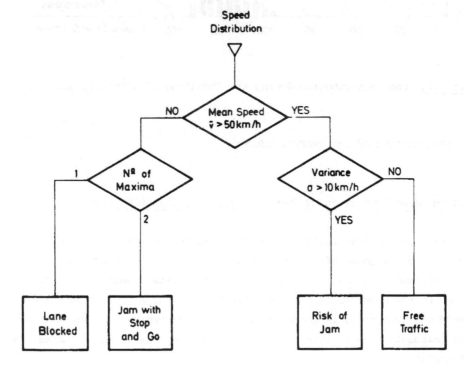

Fig. 13. Flow diagram for traffic state classification.

4.2 Transportable Jam Warning System

On the basis of the traffic classification algorithm described above
AEG-TELEFUNKEN has developed a mobile jam warning system. The speed
distribution is measured with mm-wave radar and by means of variable
traffic signs the drivers warned and suitable speed limitations announced. The principle of this system is shown in Fig. 14.

Fig. 14. Principle of a transportable jam warning system
(after Kühne, 1982)

5. REFERENCES

Ashton, W.D. (1966). The Theory of Road Traffic Flow (John Wiley & Sons,
London).
Baker, R. (1983). On the Kinematics and Quantum Dynamics of Traffic Flow,
Transpn. Res. B, 17B, pp. 55 - 66.
Bykhovskii, E.B. (1966). On Self-Similar Solutions of a System of Equa-
tions for Water Flow in a Sloping Channel, Journal Appl. Math. Mech. 30,
pp. 370 - 380.
Cremer, M. (1979). Der Verkehrsfluß auf Schnellstraßen (Springer Verlag,
Berlin), pp. 85 - 114.
Dressler, R.F. (1949). Mathematical Solution of the Problem of Roll-Waves
in Inclined Open Channels, Commun. Pure Appl. Math. 2, pp. 149 - 194.
Duncan, N.C. (1979). A Further Look at Speed-Flow-Concentration Character-
istics, Traff. Eng. Contr. 21, pp. 482 - 483.

Edie, L.C., Hermann, R., Lam, T.N. (1980). Observed Multilane Speed Distribution and the Kinetic Theory of Vehicular Traffic, Trans. Science 14, pp. 55 - 76.
Flügge, S. (1974). Practical Quantum Mechanics (Springer Verlag, Berlin) p. 62.
Koshi, M., Iwasaki, M., Ohkura, I. (1981). Some Findings and an Overview on Vehicular Flow Characteristics. In: Proc. of the 8'th Int. Symp. on Transportation and Traffic Theory, Toronto, pp. 295 - 306.
Kühne, R. (1982). Neue Ansätze zur Beschreibung des Verkehrsablaufs auf Schnellstraßen. In: Verkehrstheorie (Haus der Technik, Essen), pp. 117 - 129.
Leutzbach, W. (1981). Ermittlung von zeitabhängigen Geschwindigkeitsverteilungen aus aktuellen Verkehrsdaten, Research order for the Research Institute of AEG-TELEFUNKEN in Ulm, Germany.
Leutzbach, W., Haas, M. (1981). Simulation des Verkehrsablaufs auf zweistreifigen Tunnelstrecken, Institut für Verkehrswesen, Universität Karlsruhe, Germany.
Leutzbach, W., Schwerdtfeger, Th. (1981). Description of the Dissolution of Traffic Jams Using Continuity Theory, Institut für Verkehrswesen, Universität Karlsruhe, Germany.
Lighthill, M.J., Whitham, G.B. (1955). A Theory of Traffic Flow on Long Crowded Roads, Proc. Roy. Soc. A229, pp. 317 - 345.
Novik, O.B. (1971). Model Description of Roll-Waves, Journal Appl. Math. Mech. 35, pp. 938 - 951.
Payne, H.J. (1979). A Critical Review of a Macroscopic Freeway Model. In: Research Directions in Computer Control of Urban Traffic Systems (Am. Soc. Cov. Eng., New York), pp. 251 - 265.
Prigogine, J. (1961). A Boltzmann-like Approach to the Statistical Theory of Traffic Flow. In: Hermann, R. (ed), Theory of Traffic Flow (Elsevier, New York), pp. 158 - 164.
Treiterer, I., (1975). Investigation of Traffic Dynamics by Aerial Photogrammetry Techniques, Techn. Rep. PB 246 094, Columbus, Ohio, U.S.A.
Wiedemann, R. (1974). Simulation des Straßenverkehrsflusses, Schriftenreihe des Instituts für Verkehrswesen, Universität Karlsruhe, Germany.
Winzer, Th. (1980). Messung von Beschleunigungsverteilungen, Forschung Straßenbau und Straßenverkehrstechnik 319.

Ninth International Symposium on
Transportation and Traffic Theory
© 1984 VNU Science Press, pp. 43–63

HEADWAY DISTRIBUTION MODEL BASED ON THE DISTINCTION BETWEEN LEADERS AND FOLLOWERS

TAKESHI CHISHAKI[1] and YOUICHI TAMURA[2]

[1] *Department of Civil Engineering, Faculty of Engineering, Kyushu University, Hakozaki, Fukuoka 812, Japan* and [2] *Department of Civil Engineering, Faculty of Engineering, Yamaguchi University, Tokiwadai, Ube, Yamaguchi 755, Japan*

ABSTRACT

One approach to the formulation of elementary headway distribution model based on the distinction between leaders and followers is investigated, and a model of actual headway distribution in a given time period is derived by assembling the elementary model. The variance of relative speed distribution of successive vehicles is used for estimating the proportion of leaders and followers at a given headway. The elementary model of headway distribution is formulated based on this proportion of leaders and followers, and the relationships between model parameters and 1 minute traffic volume are examined by using observed data of traffic flow in two-lane roads. Then, the elementary model is applied to propose a practical distribution model of headway in a given time period, which may reflect the pattern of fluctuation of 1 minute traffic volume and the presence of congested flow. Our practical model can be calculated only by giving the distribution of 1 minute traffic volume and can readily be foreseen.

1.   INTRODUCTION

The formulation of headway distribution is the most basic and important
traffic matter to solve some traffic problems, e.g., problems of gap
acceptance, traffic flow control at intersections and various traffic
simulations, in which the solution may be depend on the reappearence or
estimation of headway distribution. Therefore, it is expected that model
of headway distribution is found with exactitude.

Several models of headway distribution have been proposed in previous
investigations.[1)~6)] Typical models in those studies were composite
models based on the assumptions that vehicles could be classified into
leaders and followers, and that headway distributions for leaders and
followers were different each other. However, the definition of leaders
and followers were ambiguous in those models, and the models were not
apparently distinguished between leaders and followers. Furthermore, the
models have neglected fluctuation of traffic flow, which seems to be
reflect on the headway distribution considerably. Thus, the applicability
of those models are still limited.

The purpose of this paper is to find a reasonable method of classifying
vehicular headways into those of leaders and followers, to develop a
model of elementary headway distribution for 1 minute traffic flow which
can evaluate the headway distribution of leaders and followers individually, and to derive a practical model of headway distribution.
The practical model reflects the fluctuation of traffic flow by assembling the elementary models according to the distribution of 1 minute
traffic volume. The time of 1 minute is empirically selected as the unit
time interval, since this interval is short enough to obtain detailed
pattern of fluctuation of traffic flow.

To develop our models of headway distribution in a complete form, data
of traffic flow observed in two-lanes roads are used. Two-lanes roads
are popular as arterial highways in our country and have many problems
on control of traffic flow or improvement of traffic facilities.

Basic matters of observation are summarized in Table 1.

2. COMPOSITIONS OF LEADER AND FOLLOWER IN TRAFFIC FLOW

2.1 Relative speed distribution

In this paper, vehicles that they are running with their own free speed
are defined as leaders, and ones that they are constrained by the foregoing vehicles and running with speed below their free speed are defined
as followers. In other words, vehicles are judged as leaders or followers according to whether their running speeds are slower than their own
free speed or not. When they are running close to the foregoing vehicles,
they would often be assumed as followers. However, those vehicles may
be leaders, if their free speed are equal or below the speed of the foregoing vehicles.

Therefore, it is considered that a set of vehicles running with the same
headway includes both of leaders and followers, and that the occupying
rate of leaders or followers should be estimated in order to develop a
precise headway distribution model. Since it is difficult to evaluate
the free speed of each vehicle, a method to estimate the rate of leaders
and followers in vehicles with same headway will be developed by the use
of relative speeds of successive vehicles. That is, we take note of the
relation between relative speed dis-
tribution and headway in traffic flow.

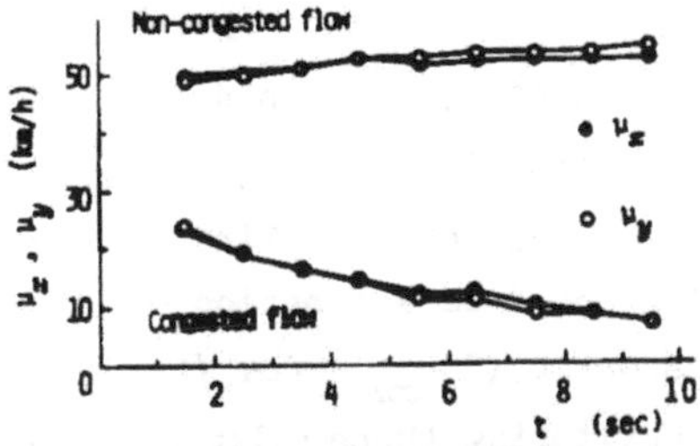

Let the speeds of two successive vehicles be denoted by X and Y, respectively. If X and Y are normal
variates with the means and the
standard deviations, μ_x , σ_x and
μ_y , σ_y , respectively, relative
speed Z (= X - Y) is also
normal variate with the mean,
μ_z , and the standard deviation,
σ_z , as given by following eq-
ations:

Fig.1 Relationships between
mean speed and headway
(● , ○ : data)

Table 1 Summary of observations

Routes(National roads)	R.190	R.202	R.2
Number of observation points	3	1	2
Number of observed vehicles	9078	1845	2626
Duration of observation (hours)	13	2	4
Road conditions	level,straight,two-lane and no-passing		
Devices	16 and 8 mm cine camera		

$$\mu_z = \mu_x - \mu_y \qquad (1)$$

$$\sigma_z^2 = \sigma_x^2 - 2Cov_{xy} + \sigma_y^2 \qquad (2)$$

where Cov_{xy} is the covariance of X and Y.
The relationships of Eqs.(1) and (2) may also be adopted for the non-normal variates approximately. The relationships between the parameters in Eqs.(1),(2) and headways are obtained from the observed data as shown in Figs.1 ~ 4. In this analysis, vehicles running with speed below the critical velocity of 35 km/h, which is the upper limit of speed for congested flow obtained by our previous observations[7] in Japanese highways with two-lane, are judged as vehicles in congested flow.

Consequently, for non-congested flow, $\mu_x = \mu_y = \mu_n$, $\sigma_x = \sigma_y = \sigma_n$ and covariance of X and Y is a function of headway, t ($Cov_{xy} = Cov_n(t)$). Substituting these relations into Eq.(1) and (2), μ_z is equal to 0 and the variance of relative speed in non-congested flow is expressed as follows:

$$\sigma_{zn}^2(t) = 2\{ \sigma_n^2 - Cov_n(t) \} \qquad (3)$$

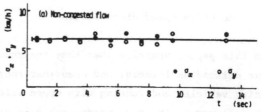

Fig.2 Relationships between standard deviation of speed distribution and headway (● , ○ : data)

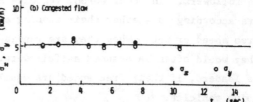

Fig.3 Relationships between Cov_{xy} and headway (● , ○ : data)

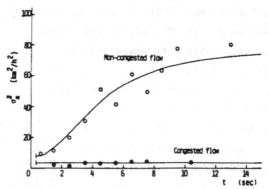

Fig.4 Relationships between variance of relative speed distribution and headway (● , ○ : data)

$Cov_n(t)$ is a decreasing function of t as shown in Fig.3, and takes the maximum value, $Cov_{zn.max}$, at $t = t_0$ (t_0 is the minimum headway), and zero at $t \to \infty$. Therefore, $\sigma_{zn}^2(t)$ takes the minimum value $\sigma_{zn.min}^2$ at $t = t_0$, and the maximum value, $\sigma_{zn.max}^2$, for $t \to \infty$. Using Eq.(3), the following equations for the facts as mentioned above are derived

$$\sigma_{zn.max}^2 = 2\sigma_n^2 \qquad (4) \qquad\qquad \sigma_{zn.min}^2 = 2\{ \sigma_n^2 - Cov_{n.max} \} \qquad (5)$$

The regression equation of $Cov_n(t)$ for the data as shown in Fig.4 is obtained as

$$Cov_n(t) = \frac{B}{A (t - t_0)^2 + 1} \qquad (6)$$

in which, $t_0 = 0.35$ second, $A = 0.055$ second^{-2} and $B = 36.00$ (km/h)^2. Substituting Eq.(6) into Eq.(3), $\sigma_{zn}^2(t)$ can be obtained.

For congested flow, it is found that $\mu_x = \mu_y = \mu_c$, $\sigma_x = \sigma_y = \sigma_c$ and that Cov_{xy} is constant ($Cov_{xy} = Cov_c$). Accordingly, the mean of the relative speed is equal to zero and the variance, σ_{zc}^2 , is constant. Then, Eq.(3) yields as

$$\sigma_{zc}^2 = 2 \{ \sigma_c^2 - Cov_c \} \qquad (7).$$

Fig.1 shows that μ_n and μ_c are respectively the increasind and decreasing function of t. The following equations were also derived by the regression analysis of data:

$$\mu_n = 48.9 + 2.5 \ln(t - t_0) \qquad\qquad \text{(km/h)} \qquad (8)$$
$$\mu_c = 25.6 - 8.1 \ln(t - t_0) \qquad\qquad \text{(km/h)} \qquad (9)$$

The theoretical values σ_n and σ_c are shown by solid lines in Fig.2, the regression curve of Eq.(6), one as in Fig.3, and the curve of $\sigma_{zn}^2(t)$, one as in Fig.4. The regression curves of Eqs.(7),(8) and (9) are also demonstrated in Fig.4 and Fig.5, respectively. It can be found that these regression curves fit for the

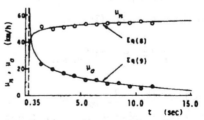

Fig.5 Regression curves of mean speed ($\bullet$, $\circ$: data)

data very well.

2.2 Rate of leaders (followers) in traffic flow

The rate of leaders is defined as the rate of leaders in all vehicles
running with the same headway.

As mentioned in Section 2.1, the correlation between speeds of two successive vehicles in non-congested flow becomes larger as headway becomes
smaller. Therefore, the relative speed distribution is apparently
affected by the rate of leaders in traffic flow, and leader's rate can
be formulated by using the variance of relative speed distribution. In
the other hand, the coefficient of correlation between speeds of two
successive vehicles in congested flow is about 0.932 constantly. So, all
vehicles in congested flow may be assumed to be followers.

The probability density function, $\phi(z)$, of relative speed for successive
vehicles is assumed as follows:

$$\phi(z) = \alpha\phi_l(z) + \beta\phi_f(z) \tag{10}$$

where, $\phi_l(z)$ and $\phi_f(z)$ are the p.d.fs. of relative speed in the case
that all vehicles are leaders and the case that all vehicles are followers, respectively. The parameters, α and β , are the rate of $\phi_l(z)$
and $\phi_f(z)$ occupying in $\phi(z)$, where $\alpha + \beta = 1.0$.

As discussed in Section 2.1, the mean of those p.d.fs. are equal to zero,
and the variances are $\sigma_{zn}^2(t)$, $\sigma_{zn.max}^2$ and $\sigma_{zn.min}^2$, respectively.
Then, the variance, $\sigma_{zn}^2(t)$, of relative speed in non-congested flow is
derived from Eq.(10) as

$$\sigma_{zn}^2(t) = \alpha\sigma_{zn.max}^2 + \beta\sigma_{zn.min}^2 \tag{11}$$

In Eq.(11), $\sigma_{zn.max}^2$ and $\sigma_{zn.min}^2$ are constants. Therefore, α and
β can be considered to be functions of t, ($\alpha = \alpha(t)$, $\beta = \beta(t)$).
Substituting $\beta = 1 - \alpha$ into Eq.(11) and solving for α , we get

$$\alpha(t) = \frac{\sigma^2_{zn}(t) - \sigma^2_{zn.min}}{\sigma^2_{zn.max} - \sigma^2_{zn.min}} \qquad (12)$$

Substituting Eqs.(3),(4),(5) and (6) into Eq.(12), we obtain Eq.(13). In addition, using the relation of $\alpha(t) = 1 - \beta(t)$, Eq.(14) is obtained as the expression of $\beta(t)$.

$$\alpha(t) = \frac{A(t - t_0)^2}{A(t - t_0)^2 + 1} \quad (13) \qquad \beta(t) = \frac{1}{A(t - t_0)^2 + 1} \quad (14)$$

From the process of derivation, it can be understood that $\alpha(t)$ and $\beta(t)$ are the rates of leader and follower in all vehicles running with headway t.

3. ELEMENTARY MODEL OF HEADWAY DISTRIBUTION FOR 1 MINUTE TRAFFIC FLOW

Using $\alpha(t)$ and $\beta(t)$ for non-congested flow, the occupying rates of leader and follower in all vehicles in 1 minute traffic flow, P_l and P_f, can be expressed as follows:

$$P_l = \int_0^\infty \alpha(t)h_n(t)dt \qquad (15) \qquad P_f = \int_0^\infty \beta(t)h_n(t)dt \qquad (16)$$

where, $h_n(t)$ is the p.d.f. of headway in non-congested flow, and $P_l + P_f = 1.0$. Assuming the equations

$$\alpha(t)h_n(t) = P_l h_l(t) \qquad (17), \quad \text{and} \quad \beta(t)h_n(t) = P_f h_f(t) \qquad (18),$$

a composite model of headway distribution can be obtained as follows:

$$h_n(t) = P_l h_l(t) + P_f h_f(t) \qquad (19)$$

where, $h_l(t)$ and $h_f(t)$ are the p.d.fs. of headway of leaders and followers in non-congested

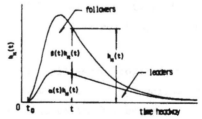

Fig.6 Leaders-followers composition of headway distribution

flow. When Eq.(19) is assumed as a headway distribution model, P_l and P_f are expressed by

$$P_l = \frac{T - T_f}{T_l - T_f} \quad (20) \qquad P_f = \frac{T_l - T}{T_l - T_f} \quad (21)$$

where, T, T_l and T_f are the means of $h_n(t)$, $h_l(t)$ and $h_f(t)$, respectively.

In congested flow, all vehicles are assumed to be followers and headway distribution, $h_c(t)$, is formulated by a simple distribution model.

The headway distribution model of Eq.(19) may be derived by determing the headway distribution of leaders and followers individually. As it is considered that 1 minute traffic volume is suitable to evaluate the variation of traffic flow, the original data are separate into every 1 minute volume and the headway distribution data are rearranged according to the level in volume of traffic flow. Using Eqs.(13) and (14), the rearranged data can be classified into those of leaders and followers. Subsequently, the relations between these statistical parameters and 1 minute traffic volume are found as follows:

$$T_l = 66.314 \; q^{-0.7460} \quad (22) \qquad V_l = 2133.4 \; q^{-1.1558} \quad (23)$$

$$T_f = 3.0887 \; q^{-0.1336} \quad (24) \qquad V_f = 5.3727 \; q^{-0.5614} \quad (25)$$

$$T_c = 60 \; q^{-1.0} \quad (26) \qquad V_c = 1928.8 \; q^{-2.4746} \quad (27)$$

in which q is the 1 minute traffic volume; T_l, T_f and T_c are the means of $h_l(t)$, $h_f(t)$ and $h_c(t)$; V_l, V_f and V_c are the variance of $h_l(t)$, $h_f(t)$ and $h_c(t)$.

These regression curves fit very well for the data, as shown in Fig.7. Substituting Eqs.(22) and (24) into Eqs.(20) and (21), P_l and P_f can be expressed as a function of 1 minute volume q (See Fig.8).

Several theoretical distribution were examined for each headway distribution of leaders and followers in non-congested and congested flows, and the best fitness in all traffic flow levels is obtained when the log-normal distribution is assumed for the each p.d.f. of the headway distribution.[8) 9)]   That is,

$$h_l(t) = \frac{1}{\sqrt{2\pi}\ \zeta_l(t - t_0)} \exp\left[-\frac{1}{2}\left(\frac{\ln(t - t_0) - \xi_l}{\zeta_l}\right)^2\right] \tag{28}$$

Using the mean and the variance of headway distribution, the parameters in Eq.(28) are expressed as follows:

$$\xi_l = \ln(T_l - t_0) - \frac{1}{2}\ln\left[\frac{V_l}{(T_l - t_0)^2} + 1\right] \tag{29}$$

$$\zeta_l = \sqrt{\ln\left[\frac{V_l}{(T_l - t_0)^2} + 1\right]} \tag{30}$$

The model mentioned above is the headway distribution of leaders in non-congested flow. The headway distribution of followers in non-congested flow is expressed by replacing $h_l(t)$, ξ_l, ζ_l, T_l and V_l with $h_f(t)$, ξ_f, ζ_f, T_f and V_f in Eqs.(28),(29) and (30), respectively.   The headway distribution for congested flow is also expressed by replacing $h_l(t)$, ξ_l, ζ_l, T_l and V_l with $h_c(t)$, ξ_c, ζ_c, T_c and V_c in Eqs.(28),(29) and (30).   Substituting Eqs.(22),(23),(24),(25),(26) and (27) into Eqs.(29) and (30), the parameters in headway distribution models are expressed as a function of 1 minute traffic volume, and

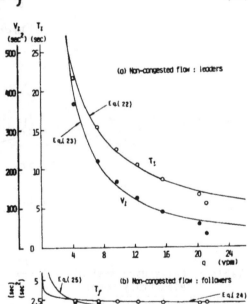
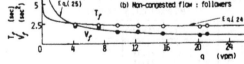
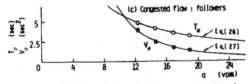

Fig.7 Relationships between statistics of headway distribution and 1 minute traffic volume (• , ○ : data)

those curves are shown as solid lines in Fig.9.

The model formulated above has been compared with each headway distribution data rearranged by every 1 minute traffic volume and the proposed models for both of non-congested and congested flow in wide range of traffic flow level were accepted at 5 % significant level of goodness-of-fit test.   (See Table 2)

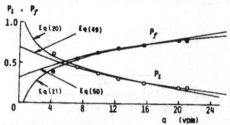

Fig.8 Relationships between P_l, P_f
and 1 minute traffic volume
(● , ○ : data)

Table 2   Chi-square test of elementary model

No.	q	d.f.	x^2 values	Good-fit
1	4.2	10	11.39	Yes
2	7.3	12	19.32	Yes
3	9.8	13	21.51	Yes
4	12.5	14	17.67	Yes
5	15.6	12	13.71	Yes
6	20.3	11	10.34	Yes
7	21.1	9	14.24	Yes
8	12.4	7	8.96	Yes
9	14.7	7	6.92	Yes
10	17.0	8	4.89	Yes
11	19.1	8	3.34	Yes

No.1–7 : Non-congested flow
No.8–11: Congested flow
q : 1-minute traffic volume
d.f.: degree of freedom
Significance level = 5 %

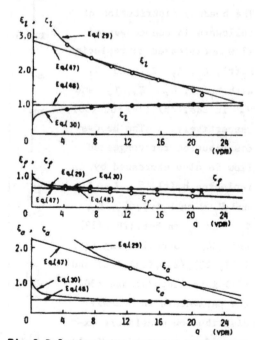

Fig.9 Relationships between parameters
of elementary model and 1 minute
traffic volume (● , ○ : data)

4. PRACTICAL MODEL OF HEADWAY DISTRIBUTION OF TRAFFIC FLOW IN A GIVEN PERIOD

4.1 Concept of model

In the case that the fluctuation of traffic flow in a given time period is small, the elementary model of headway distribution derived in the previous chapter may be used as it is. However, the elementary model may not always provide satisfactory results in the case that the fluctuation of traffic flow can be considered.   Such a case requires development of a new model which reflects the fluctuation of traffic flow. Since the elementary model is based on the 1 minute traffic flow, the practical headway distribution in arbitrary duration is approximately obtained by overlapping them according to composition of 1 minute traffic volume. A concept of this practical model is shown in Fig.10.

The p.d.f. of weighted 1 minute traffic volume (for short, weighted volume distribution), $\Psi(q)$ is defined by

$$\Psi(q) = \frac{q\psi(q)}{\int_0^\infty q\,\psi(q)\,dq} \tag{31}$$

where $\psi(q)$ is the p.d.f. of 1 minute traffic volume distribution (for short, volume distribution). That is, the weighted volume distribution is defined by the rate of vehicles belonging to each level of 1 minute traffic volume in all vehicles.

A new headway distribution model for a traffic flow in a given interval, $H(t)$, can be expressed by following equation:

$$H(t) = \int_0^\infty \Psi(q)h(t)\,dq \tag{32}$$

where $h(t)$ is the p.d.f. of elementary headway distribution. Replacing $H(t)$, $\Psi(q)$ and $h(t)$ in Eq.(32) with $H_n(t)$, $\Psi_n(q)$ and

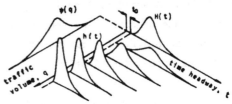

Fig.10 Concept of practical model

$h_n(t)$ for non-congested flow, and with $H_c(t)$, $\Psi_c(q)$ and $h_c(t)$ for congested flow, respectively, the following equations are obtained:

$$H_n(t) = \int_0^\infty \Psi_n(q) h_n(t) dq \qquad (33)$$

$$H_c(t) = \int_0^\infty \Psi_c(q) h_c(t) dq \qquad (34)$$

where $H_n(t)$ and $H_c(t)$ are headway distributions of traffic flows which composed only of non-congested and congested flows; $\Psi_n(q)$ and $\Psi_c(q)$ are the p.d.fs. of the weighted volume distribution for non-congested and congested flows. Generally, traffic flow in an arbitrary duration consists of non-congested and congested flows. Hence, the p.d. f. of headway distribution for actual traffic flow, $H_m(t)$, is expressed as follows:

$$H_m(t) = R_n H_n(t) + R_c H_c(t) \qquad (35)$$

where R_n and R_c are the respective rates of non-congested and congested vehicles in all vehicles. The model of Eq.(35) represents the headway distribution of actual traffic flow in a given duration, which is called a practical model hereafter.

4.2 Weighted volume distribution and rate of non-congested (congested) vehicles

In order to obtain the weighted traffic volume distribution, $\Psi_n(q)$ and $\Psi_c(q)$, and the rate of non-congested (congested) vehicles R_n (R_c), a cumulative distribution of 1 minute traffic volume for the data No.8 shown in Table 4 is plotted on a normal probability paper, as shown in Fig.11(a). A straight line in Fig.11(a) is drawn by using the mean and the variance of data. The resulting graph of data points shows linearity, and the p.d.f. of 1 minute traffic volume can be assumed by the following normal distribution:

$$\psi(q) = \frac{1}{\sqrt{2\pi}\,\nu} \exp\left\{-\frac{1}{2}\left(\frac{q-\kappa}{\nu}\right)^2\right\} \qquad (36)$$

where κ and ν^2 are the mean and the variance of $\psi(q)$. Substituting Eq.(36) into Eq.(31), the following equation is derived for the p.d.f. of weighted volume:

$$\Psi(q) = \frac{q}{\sqrt{2\pi}\ \nu\kappa} \exp\left\{ - \frac{1}{2}\left(\frac{q-\kappa}{\nu}\right)^2 \right\}$$

(37)

The mean, κ_w, and the variance, ν_w^2, of Eq.(37) are expressed by using the mean and the variance of volume distribution as follows:

$$\kappa_w = \frac{\nu^2}{\kappa} + \kappa$$

(38)

$$\nu_w^2 = \nu^2 \left(1 - \frac{\nu^2}{\kappa^2}\right)$$

(39)

Cumulative weighted distribution of traffic volume for the same data as shown in Fig.11(a) is plotted in Fig.11(b). Fig.11(b) shows that weighted volume distribution can also be assumed as the normal distribution.

A comparison of κ_w and κ_{wd}, and a comparison of ν_w and ν_{wd} are shown as in Fig.12, in which, κ_w and ν_w are determined from Eqs.(38) and (39), and κ_{wd} and ν_{wd} are the mean and the standard deviation determined directly from

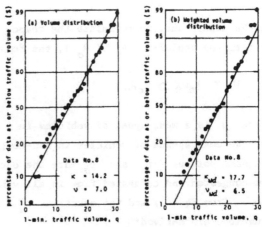

Fig.11 Cumulative distribution of 1 minute traffic volume and weighted volume

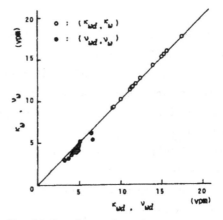

Fig.12 Results of estimation of mean and standard deviation of 1 minute weighted volume distribution

weighted volume distribution data. From these results, it can be found
that the means and the standard deviation of theoretical weighted volume
distribution are well agreeable with actual data. From the view point of
analytical advantage and direct connection of traffic volume distribution, the following normal distribution for the p.d.f. of weighted volume
may be assumed

$$\Psi(q) = \frac{1}{\sqrt{2\pi}\ \nu_w}\ \exp\left[-\ \frac{1}{2}\ \left(\frac{q\ -\ \kappa_w}{\nu_w}\right)^2\right] \tag{40}$$

On the other hand, to formulate the respective rate of non-congested
(congested) vehicles, R_n (R_c), the following relation is assumed:

$$\mu = r_n\mu_n + r_c\mu_c \tag{41}$$

where μ is a mean speed of vehicles in 1 minute volume q, μ_n and μ_c
are the mean speeds of vehicles in non-congested and congested flow of
the same volume. r_n and r_c are the occupying rate of vehicles in
non-congested and congested flow in all vehicles of 1 minute traffic
flow. From Eq.(41) and the equation $r_n + r_c = 1.0$, the following
equations are derived;

$$r_n = \frac{\mu\ -\ \mu_c}{\mu_n\ -\ \mu_c} \tag{42} \qquad\qquad r_c = \frac{\mu_n\ -\ \mu}{\mu_n\ -\ \mu_c} \tag{43}$$

Dividing the observed duration into sequential 1 minute periods, the
number of arrival of vehicles in the i-th interval, q_i , is obtained.
Using this q_i , the mean headway of traffic flow in the i-th interval is
evaluated instantly and substituting it into t in Eqs.(8) and (9), then
the mean speed of non-congested and congested traffic flow in the i-th
interval, μ_{ni} and μ_{ci} are obtained. Substituting these values into
Eqs.(42) and (43), the respective rate of non-congested and congested
vehicles in the i-th interval, r_{ni} and r_{ci} , can be obtained, since
μ is obtained from data. Consequently, the numbers of non-congested
and congested vehicles in q_i are given by $r_{ni}q_i$ and $r_{ci}q_i$, respectively.
Then, R_n and R_c are formulated as

$$R_n = \left.\sum_{i=1}^{M} r_{ni}q_i \middle/ \sum_{i=1}^{M} q_i\right. \tag{44}$$

$$R_c = \sum_{i=1}^{M} r_{ci} q_i \bigg/ \sum_{i=1}^{M} q_i \tag{45}$$

where M is the number of time intervals, $\sum_{i=1}^{M} q_i$ is the total number of arrival for the present duration, and $R_n + R_c = 1.0$.

4.3 Practical model of headway distribution for traffic flow in a given interval

4.3.1 Model I

Using the elementary model, the weighted volume distribution and the rate of non-congested (congested) vehicles, a model of headway distribution for traffic flow in a given interval, $H_m(t)$, is approximately expressed as follows:

$$H_m(t) = R_n \int_0^\infty \Psi_n(q)\{P_l h_l(t) + P_f h_f(t)\}dq + R_c \int_0^\infty \Psi_c(q) h_c(t)dq \tag{46}$$

Applying Eqs.(20),(21),(28) and (31) to Eq.(46), the practical model of headway distribution is obtained. This model is most precise one among three models proposed in this paper. However, we must use a numerical method of integration because the analytical integration of Eq.(46) is difficult. This fact leads to use directly the data for $\Psi_n(q)$ and $\Psi_c(q)$. This model is named as Model I.

4.3.2 Model II

Model I can be simplified, assuming that the weighted volume distribution is a normal distribution as expressed in Eq.(40), that parameters in the elementary model, ξ_l, ξ_f and ξ_c, are linear with respect to traffic volume, and that ζ_l, ζ_f and ζ_c, are approximately constants. That is, the following equations for the relationships between these parameters and 1 minute traffic volume may be assumed:

$$\xi_l = a_l q + b_l \tag{47} \qquad \zeta_l = \text{constant} \tag{48}$$

where a_l and b_l are constants. Parameters, ξ_f and ξ_c are obtained by replacing a_l and b_l with a_f, b_f and a_c, b_c in Eq.(47), respectively. Parameters, ζ_l, ζ_f and ζ_c have different values for each other. The relationships between the rates, P_l and P_f, and 1 minute volume is also assumed to be

$$P_l = A_p \exp(-B_p q) \quad (49) \qquad\qquad P_f = 1 - A_p \exp(-B_p q) \quad (50)$$

where A_p and B_p are constants. The constants in Eqs.(47),(48),(49) and (50) are shown in Table 3. Figs.8 and 9 show the comparison of between the data and these assumed regression curves. These approximation are agreeable in practical range of traffic volume.

Substituting Eqs.(40),(47),(48),(49) and (50) into Eqs.(33) and (34), $H_n(t)$ and $H_c(t)$ are given by

$$
\begin{aligned}
&H_n(t) \\
&= \frac{A_p \exp\{-\frac{1}{2}B_p(2\kappa_{wn}-B_p \nu_{wn}^2)\}}{\sqrt{2\pi}\sqrt{\zeta_l^2 + a_l^2\nu_{wn}^2}\,(t-t_0)} \exp\left(-\frac{1}{2}\left(\frac{\ln(t-t_0)-\{b_l + a_l(\kappa_{wn}-B_p\nu_{wn}^2)\}}{\sqrt{\zeta_l^2 + a_l^2\nu_{wn}^2}}\right)^2\right) \\
&+ \frac{1}{\sqrt{2\pi}\sqrt{\zeta_f^2 + a_f^2\nu_{wn}^2}\,(t-t_0)} \exp\left(-\frac{1}{2}\left(\frac{\ln(t-t_0)-(b_f + a_f\kappa_{wn})}{\sqrt{\zeta_f^2 + a_f^2\nu_{wn}^2}}\right)^2\right) \\
&- \frac{A_p \exp\{-\frac{1}{2}B_p(2\kappa_{wn}-B_p\nu_{wn}^2)\}}{\sqrt{2\pi}\sqrt{\zeta_f^2 + a_f^2\nu_{wn}^2}\,(t-t_0)} \exp\left(-\frac{1}{2}\left(\frac{\ln(t-t_0)-\{b_f + a_f(\kappa_{wn}-B_p\nu_{wn}^2)\}}{\sqrt{\zeta_f^2 + a_f^2\nu_{wn}^2}}\right)^2\right)
\end{aligned}
$$

$$(51)$$

$$
H_c(t) = \frac{1}{\sqrt{2\pi}\sqrt{\zeta_c^2 + a_c^2\nu_{wc}^2}\,(t-t_0)} \exp\left(-\frac{1}{2}\left(\frac{\ln(t-t_0)-(b_c + a_c\kappa_{wc})}{\sqrt{\zeta_c^2 + a_c^2\nu_{wc}^2}}\right)^2\right) \quad (52)
$$

A practical model obtained from Eqs.(35),(51) and (52) is named as Model II.

4.3.3 Model III

The mean and the variance of weighted volume distribution in Model II can be evaluated by Eqs.(38) and (39). However, using the mean and the variance estimated by the data, Eqs.(51) and (52) are simplified in computation. That is, the model of headway distribution can be obtained by replacing κ_{wn}, ν_{wn}^2, κ_{wc} and ν_{wc}^2 with κ_{wnd}, ν_{wnd}^2, κ_{wcd} and ν_{wcd}^2, which are the means and the variances obtained directly by the data of weighted volume distribution. This model is called Model III.

4.4 Testing and results

Three practical models for headway distribution are applied to 13 sets of data which are given in Table 4. Typical cases of comparison of cumulative distribution between data and each model are shown in Fig.13. Figure 13(a) is a case of low level of traffic volume which contains a little congested flow. Figure 13(b) is a case of which level of traffic flow is high and has no congested flow. Figure 13(c) is a case where the level in volume of traffic flow exceeds its capacity and the rate of congested flow is high. Since there are some difficulties to judge the superiority of one model to the others only by looking at Fig.13, the maximum differences between data and models are plotted in Fig.14. The x axis of the graph is the maximum difference for Model I ($D_{I\ max}$), and y axis is the one for Model II ($D_{II\ max}$) and Model III ($D_{III\ max}$). Number attached to plots is the data number given in Table 4. If plotted points are in the upper part of the straight line, of which slope is unity, Model I is better than the other models. As the results,

Table 3   Values of constants in Eqs.(47) to (50)

Non – congested flow						Congested flow		
leaders			followers			followers		
ξ_l	a_l	-0.07496	ξ_f	a_f	-0.01033	ξ_c	a_c	-0.06947
	b_l	2.8770		b_f	0.5827		b_c	2.2664
ζ_l	0.8917		ζ_f	0.5619		ζ_c	0.4012	
P_l , P_f	A_p		0.6850					
	B_p		0.06050					

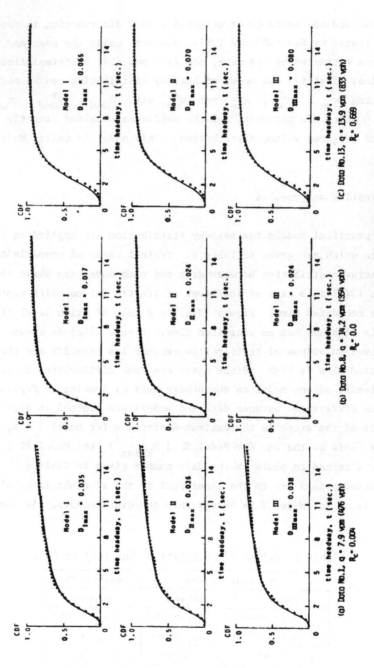

Fig.13 Comparisons of data and practical models

Fig.14 shows that Model I is the best and that Model II is better than Model III.

Results of Kolmogorov-Smirnov goodness-of-fit test are given as in Table 4. At 1 % significant level, 10 cases of 13 cases are accepted in Model I, and 9 cases are accepted in Model II and Model III.
The differences among the models and data are not so large, even for the rejected cases.

The special features of the case, in which all models are rejected, are as follows. In the case of No.10, the point of observation was located near the urban district and traffic flow was significantly influenced by signal control of intersection within 500 m upstream. In the case of No.12, the occupying rate of large vehicle in traffic flow (about 23 %) is considerably high as compared to the ones (about 5 %) in other traffic flows. In the case of No.13, the congested flow occupied the major portion of traffic flow and some headways for relatively high speed vehicles could not be recorded due to long

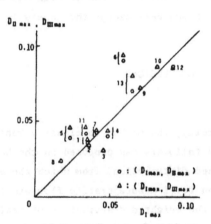

Fig.14 Comparisons of maximum difference of practical models

Table 4 Data and Kolmogorov-Smirvov test of practical model

No.	N(n)	T (min)	Q (vph)	R_σ (%)	$D_{I max}$	Good fit	$D_{II max}$	Good fit	$D_{III max}$	Good fit
1	416(405)	53	476	0.4	0.035	Yes	0.036	Yes	0.038	Yes
2	1353(1350)	98	819	0.1	0.036	Yes	0.034	Yes	0.034	Yes
3	1004(989)	105	549	6.3	0.036	Yes	0.032	Yes	0.032	Yes
4	615(603)	66	565	1.0	0.040	Yes	0.040	Yes	0.044	Yes
5	1114(1102)	121	542	0.0	0.023	Yes	0.039	Yes	0.046	Yes
6	459(450)	62	435	0.2	0.059	Yes	0.090	No	0.094	No
7	1010(1001)	73	822	0.6	0.041	Yes	0.042	Yes	0.044	Yes
8	1201(1195)	84	854	0.0	0.017	Yes	0.024	Yes	0.024	Yes
9	517(501)	48	624	2.6	0.071	Yes	0.072	Yes	0.072	Yes
10	1336(1334)	105	737	11.9	0.082	No	0.086	No	0.086	No
11	306(302)	40	453	0.0	0.034	Yes	0.042	Yes	0.046	Yes
12	718(712)	71	589	1.5	0.092	No	0.086	No	0.085	No
13	1845(1537)	133	833	66.9	0.065	No	0.070	No	0.080	No

N : Number of observed vehicles, n : Number of analyzed vehicles, T : Time interval of observation, Q : Traffic volume, R_σ : Rate of congested vehicles

intervals of photographing (1 frame / 2 seconds). Therefore, it seems
that these factors give some influences on the results.

4.5 Characteristics of headway distribution

The mean, coefficient of variation, skewness and kurtosis of headway
distribution can be calculated for each level in volume of traffic flow
by using the proposed models.   Fig.15 shows the calculated results by
using the Model II when ν is assumed $\kappa/3$. As traffic volume increase,
the skewness and kurtosis are remarkably increase, however, the coefficient of variation can be regarded almost constant. The model evaluate
the means well except the case which traffic volume is extremely low.

5.   CONCLUSIONS

Firstly, the method to classify vehicular headways into those of leaders
and followers was proposed on the basis of analysis of the relative
speed distribution, from which the elementary model of headway distribution for 1 minute traffic flow was formulated.   Secondly, the concepts
of the weighted distribution of traffic volume and the rate of non-
congested (congested) vehicles were discussed. Finally, the practical
model of headway distribution for traffic flow in a given period, which
could reflect the fluctuation of 1 minute traffic flow, was derived by

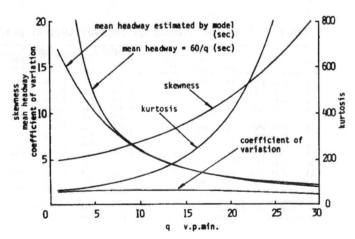

Fig.15
Characteristics
of headway
distribution

the use of the elementary model of headway distribution, the weighted
traffic volume distribution and the rate of non-congested (congested)
vehicles. The practical model of headway distribution can be now estimated by giving only the distribution of traffic volume.

Three practical models of headway distribution were proposed and applied
to the observed data of no-passing roads with two-lanes. It was found
that these models sufficiently fit to the data and that these are practically useful. From the view point of easiness of analytical treatment
and the direct connection with the distribution of 1 minute traffic
volume, the Model II will be recommended as a practical model for headway distribution.

Further investigation is necessary for the cases of traffic flows with
a lot of large vehicles and near the signal control.   The headway
distributions in those cases will be different from the ones in this
paper.

Acknowledgement

The authors would like to acknowledge the continuing encouragement and
kind assistance of Dr. T. Aida, Professor of Yamaguchi University.

REFERENCES

1) Greenberg.I.(1966). The log-normal distribution of headways,Austral.
   Road Res.,Vol.2,pp.14-18
2) Tolle,J.E.(1971). The lognormal headway distribution model,Traffic
   Eng.and Control,Vol.13,pp.22-24
3) Buckley,D.J.(1962). Road traffic headway distributions,Proc.Austral.
   Road Res.Bd.,Vol.1,pp.153-187
4) Buckley,D.J.(1968). A semi-Poisson model of traffic flow,Transpn.
   Sci.,Vol.3,pp.107-133
5) Tolle,J.E.(1976). Vehicular headway distribution:testing and results,
   Tanspn.Res.Rec.,No.456,pp.56-64
6) Wasielewski,P.(1979). Car-following headways on freeways interpreted
   by the semi-Poisson headway distribution model,Transpn.Sci.,Vol.13,
   pp.36-55
7) Tamura,Y.and Chishaki,T.(1892). Study on the phenomena in non-
   congested and congested traffic flow,Proc.34th Annual Meeting of
   Chugoku-Shikoku Branch of JSCE,pp.248-249,(In Japanese)
8) Tamura,Y.and Chishaki,T.(1982). Headway distribution models based on
   the classification of free and constrained flowing vehicles,Tech.Rep.
   of the Kyushu Univ.,Vol.55,No.2,pp.89-95,(In Japanese)
9) Tamura,Y.and Chishaki,T.(1983). Time headway distribution model
   based on the composition of free and constrained flowing vehicles,
   Proc.of JSCE,No.336,pp.159-168,(In Japanese)

Ninth International Symposium on
Transportation and Traffic Theory
© 1984 VNU Science Press, pp. 65–87

DYNEMO: A MODEL FOR THE SIMULATION OF TRAFFIC FLOW IN MOTORWAY NETWORKS

THOMAS SCHWERDTFEGER
Institut für Verkehrswesen, Universität (TH) Karlsruhe, Karlsruhe, F.R.G.

ABSTRACT

The paper presents the simulation model DYNEMO which has been designed
for the development, evaluation and optimization of traffic control
systems for motorway networks. A new traffic flow model included with
the simulation package combines the advantages of a macroscopic model
(computational simplicity) with the advantages of a microscopic model
(output statistics relating to individual vehicles). For each stretch
in the network, the model needs as input a relationship between traffic
density and mean speed and the distribution of free flow speeds. The
new traffic flow model is validated by use of an example.
The simulation package is implemented on a 16-Bit microcomputer. A real
network with a traffic control system has been simulated with the model.

1. INTRODUCTION

The simulation model DYNEMO is presented. It has been designed as a tool for the development, evaluation and optimization of traffic control systems for motorway networks. For off-line applications a traffic control model can be easily included with the simulation model.

All input parameters of the model can be changed interactively by the user or by the program itself (for example by an implemented traffic control model). The following outputs are available from the simulation model:

- patterns over time of traffic volume and mean speed at any point in the network
- journey times of individual vehicles or a travel time distribution on any route in the network
- fuel consumption of vehicles (work still in progress).

2. MICRO- AND MACROSCOPIC CONCEPTS IN TRAFFIC FLOW SIMULATION

There are two basic concepts in traffic flow simulation, the microscopic concept based on driver behaviour and the interaction of individual vehicles and the macroscopic concept based on the hydrodynamic theory of Lighthill and Witham (1955).

Microscopic traffic flow simulation models describe the interaction of individual vehicles which in turn depends on the vehicle drivers. A large number of variables are used to describe driver behaviour and vehicle interaction. The microscopic models are based on the car-following theory which has been described by Gazis (1961). Additional perceptual factors in the car-following theory have been introduced by Michaels (1965) and Todosiev (1963). On this basis, Wiedemann (1974) presented algorithms for the microscopic simulation of traffic flow. These algorithms were included and further developed in various simulation packages (Brannolte, 1977, Brilon and Brannolte, 1977,

Leutzbach et al., 1977, Leutzbach et al., 1981) which were constructed for various problems.

In the USA, the microscopic NETSIM model (Lieberman, 1971) is used for the evaluation of signal control. The INTRAS model (Wicks, Lieberman, 1977) has been designed for studies of algorithms for incident detection. The behaviour of vehicles at isolated junctions has been investigated with the TEXAS simulation mode (Lee et al., 1977). Gibson (1981) and May (1981) give an overview over various simulation packages.

The advantage of microscopic simulation of traffic flow is that the user has complete information about the state of vehicle in the system over time and space. Nearly all microscopic simulation models are time step based, meaning that the system changes in a discrete time step which is normally about 1 s. It is evident that the large number of parameters and the small time step imply restrictions in the computer implementation of microscopic simulation models. The number of cars which can be simulated is restricted by the limits of the storage of the computer and acceptable computing time. A model described in Leutzbach (1981) can simulate 150 vehicles which are simultaneously in the system with the ratio 1:1 (computational time : real time).

The macroscopic simulation models are based on the hydrodynamic theory of traffic flow, originally presented by Lighthill and Witham (1955). Under the assumption that the traffic flow can be considered as a compressible medium, they applied the theorem of continuity to this medium. Applied to traffic flow, this theorem means that each vehicle which enters a time-space element of a stretch of road with no side entrances or exits will also leave this element.

Lighthill and Witham and others (Pipes, 1968, Payne, 1971) presented an extension of this macroscopic model by introducing additional terms describing for example 'diffusion' effects, 'inertia' effects, and 'relaxation' effects. Each of these continuum models needs a speed-density relationship as input. In the discrete form, the continuum theory is embodied in macroscopic simulation models. One example of such a model is the FREFLO simulation package (Payne, 1979). For a stretch of freeway with no side entrances or exits, this model is described as follows:

The freeway section is divided into N segments with length Δx_1, ...,
Δx_N. The following variables are used

Δt :   time step of the model

Δx_j :   length of section j

ρ_j^n :   density in section j at time $n\Delta t$

u_j^n :   mean speed of the vehicles in section j at time $n\Delta t$

λ_j^n :   volume which passes in the interval $[(n-1)\Delta t, n\Delta t]$ from
                     segment j-1 into segment j.

The equations of the model are:

$$\rho_j^{n+1} = \rho_j^n + \frac{\Delta t}{\Delta x_j} (\lambda_j^{n+1} - \lambda_{j+1}^{n+1}) \tag{a}$$

$$\lambda_{j+1}^{n+1} = \rho_j^n u_j^n \tag{b}$$

$$u_j^{n+1} = u_j^n - \Delta t \{u_j^n(u_j^n - u_{j-1}^n)/\Delta x_j\}$$
$$+ \frac{1}{T_j} (u_j^n - u_e(\rho_j^n) + \frac{v_j}{\rho_j^n}(\rho_{j+1}^n - \rho_j^n)/\Delta x_j) \tag{c}$$

Equation (a) expresses the conservation of vehicles. Equation (b) is the
relationship between volume, density and speed. Equation (c) is a
dynamic speed-density relationship, where $T_j = k_\tau \Delta x_j$ and $v_j = k_v \Delta x_j$.
Terms k_τ and k_v represent the relaxation time and anticipation
coefficients respectively. The three groups of terms in (c) express
three physical processes. The first of these is convection, namely that
the traffic condition in segment j-1 will influence the traffic
condition in segment j. The second term, $(\frac{1}{T_j}(u_j^n - u_e(\rho_j^n)))$, expresses the
fact that the traffic condition will relax to an equilibrium defined by
the relation $u_e(\rho)$. The third term is a model of anticipation of
changing traffic condition.

Cremer (1979) developed a macroscopic model in which additional noise terms describe the stochastic character of traffic flow.

It is evident that an advantage of a macroscopic model based on an algorithm such as that described above is the computational simplicity. But there are no output statistics available which relate to individual vehicles.

The simulation model described in this paper was developed for the investigation of traffic control systems in motorway networks. Since macroscopic models cannot produce output statistics like travel time distributions on certain routes or ratios of route choice, and since microscopic models are too time consuming, this new approach has been developed.

3. MODEL DESCRIPTION

The simulation model is based on a new traffic flow model. The model treats individual vehicles which are moved along a stretch of carriageway according to certain macroscopic parameters. These parameters relate to the traffic conditions in the neighbourhood of the vehicle. This new approach combines the advantages of the macroscopic model (namely computational simplicity) with the advantages of the microscopic model (namely output statistics relating to the characteristics of individual vehicles).

Microscopic models treat individual vehicles which are moved along a stretch of road in accordance with the states of the neighbouring vehicles. The state of a vehicle at time $t + \Delta t$ depends on the state of the vehicle and the neighbouring vehicles at time t (Fig. 1).

Macroscopic models treat segments of a stretch of road which are described by the traffic density ρ_i and the corresponding speed u_i. The traffic condition in a carriageway segment S_i at time $t + \Delta t$ depends on the traffic conditions which prevailed in the segments S_{i-1}, S_i and S_{i+1} at time t (Fig. 2).

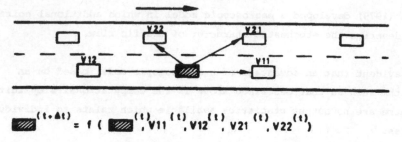

Fig. 1. The concept of vehicle behaviour in microscopic simulation.

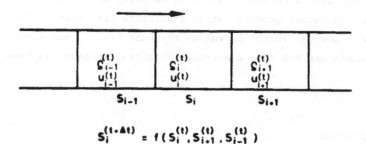

Fig. 2. The basic concepts in macroscopic traffic simulation.

The new approach presented in this paper treats both, individual vehicles and segments. The fundamental idea underlying this approach can be described in the following simplified form.

A stretch of carriageway is divided into segments S_i (i = 1, ..., N) of length 1. The traffic condition in a segment S_i at time t is described by the number of vehicles ρ_i in a segment. According to a relationship $u = u(\rho)$ at each time step a corresponding speed u_i is determined. The relationship $u = u(\rho)$ is input to the model. The values ρ_i are actualized by counting individual vehicles entering and leaving the segments. A vehicle which is at time t at the position $x^{(t)}$ in the segment S_i and drives with speed $v^{(t)}$ is moved in accordance with the speeds u_i and u_{i+1} and in accordance with its own state at time t. The speed of the vehicle at time $t + \Delta t$ is obtained from

$$v^{(t+\Delta t)} = (1 - \frac{x}{l}^{(t)})u_i + \frac{x}{l}^{(t)} u_{i+1} \qquad (1)$$

Eq. 1 models the anticipation of changing traffic conditions. The new position of the vehicle becomes

$$x^{(t+\Delta t)} = x^{(t)} + \frac{1}{2}(v^{(t)} + v^{(t+\Delta t)}) \cdot \Delta t \qquad (2)$$

Eq. 2 assumes a constant vehicle acceleration during the time step.
If $x^{(t+\Delta t)} > 1$ the vehicle is set into the next segment S_{i+1} and

$$x^{(t+\Delta t)} = x^{(t+\Delta t)} - 1, \ \rho_i = \rho_i - 1, \ \rho_{i+1} = \rho_{i+1} + 1 \qquad (3)$$

At time $t + \Delta t$ with the actual values ρ_i the corresponding speeds u_i
are determined from the relationship $u = u(\rho)$.

Fig. 3 shows the basis of the new approach.

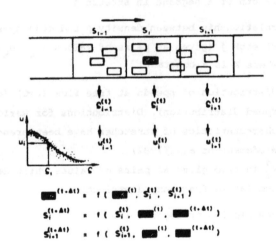

<u>Fig. 3.</u> The basis of the new approach.

In the following, the new model is described in more detail.

Not all of the following parameters are explicity treated in this paper.
However, it seems useful to define them, since they are necessary to
describe either the geometry of the network or the state of the vehicles
in the network.

The data requirements of the model DYNEMO can be divided into three
parts:

    a) description of the network

    b) description of the vehicles

    c) description of traffic demand.

a) Description of the network

The network is divided into stretches with similar characteristics
(number of lanes, etc.). The stretches are divided into segments S_i of
equal length. A stretch is defined by:

j	:	index identifying the stretch
B^j	:	stretch begin
E^j	:	stretch end
N^j	:	number of segments in stretch j
l^j	:	length of a segment in stretch j
$U(\rho)^j$	:	relationship between density ρ and mean speed u in stretch j (given as pairs of values (ρ_k, u_k), k = 1, ...,K, where K is a constant)
v_f^j	:	distribution of speeds at free flow ($\rho{\to}0$) (desired speed distribution). Distributions for various characteristics of stretches have been presented in Wiedemann et al. (1982). v_f^j is also given as pairs of values which define the cumulative frequency distribution.
s_1^j	:	description of segment 1 on stretch j
$\vdots$		
s_N^j	:	description of segment N^j on stretch j

A segment S_i^j is described by:

ρ_i^j	:	traffic density in s_i^j (number of individual vehicles in s_i^j)
u_i^j	:	mean speed at density ρ_i^j ($=f(\rho_i^j, U(\rho)^j)$)

b) Description of vehicles in the network

A vehicle is described by:

j	:	stretch in which the vehicle is driving
i	:	index of the segment in stretch j

x : position on s_i^j ($0 \le x \le 1^j$)

u : speed of the vehicle

δ : proportion of vehicles with a lower free flow speed
than this vehicle ($0 \le \delta \le 1$).

At vehicle generation, δ is computed as a uniform random
number. Fig. 4 shows how from a δ_o and any desired speed
distribution, the desired speed v_{δ_o} of the individual
vehicle is determined. v_o and v_1 are the minimum and
maximum desired speed respectively.

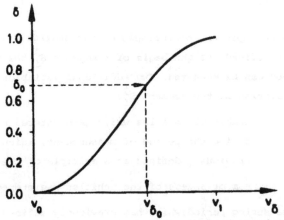

Fig. 4. Determination of the desired speed of a vehicle.

r : number of the route on which the car is driving

P_d : destination of the vehicle (there is at least one
stretch j with $E^j = P_d$)

c) Description of traffic demand

λ_k : input at origin k

$F_{(k, 1)}$: origin-destination matrix ($0 \le F_{kl} \le 1$)

$R_{(k, 1)}$: matrix to define route choice
R_{ij} is the end point of the next stretch if a vehicle
is at point i and its destination is j.

If there is more than one route from a point k to another point 1 in the
network, then each additional route is defined by

B   :   route begin

E   :   route end

p   :   probability that a vehicle which is at point B and has
               the destination E will drive on this route.
               Usually the values of p are not known, but it could be
               interesting to look at how traffic flow in the network
               changes when p varies.

S_1 :

.

.

.

S_N :   sequence of stretches lying on this route.

To obtain macroscopic and microscopic output statistics, further internal
variables are defined. At the begin of a segment S_i^j the traffic volume
and mean speed can be measured. Two additional variables are then needed
for the description of the segment S_i^j:

λ_i^j :   number of vehicles which have entered segment S_i^j
                    during the period of measurement, which has been
                    previously defined as a multiple of the time step.

U_i^j :   sum of speeds of the vehicles registered in λ_i^j.

After each measuring period which has previously defined by the user the
traffic volume and mean speed are evaluated. Then the variables λ_i^j and
U_i^j are reset to zero.

For each vehicle, the journey time in the network and the route chosen
can be determined by defining additional output variables for the
vehicles:

T_B :   the time at which the vehicle enters the network

T_E :   the time at which the vehicle leaves the network

R   :   number of the route on which the vehicle has driven
               through the network. If R = 0 the vehicle has not
               chosen an alternative route.

The following additional variables are also used:

v_δ^j :   desired speed of a vehicle at free flow in stretch j
                   ($=f(v_f^j,\delta)$)

                   (v_δ^j is determined in accordance with Fig. 4).

b_{max} :   maximum possible acceleration of a vehicle (a constant)

b_{min} :   maximum possible deceleration of a vehicle (a constant)

$\underline{u}_i^j$:   lower limit of assumed speeds in S_i^j

$\bar{u}_i^j$:   upper limit of assumed speeds in S_i^j

Δt :   duration of the time step.

The variables $\underline{u}_i^j$ and $\bar{u}_i^j$ represent the interval for individual speeds in the segment S_i^j at traffic density S_i^j.

In the following, the algorithm for determining the action of a vehicle which drives at time t with the speed $v^{(t)}$ at position $x^{(t)}$ in the segment S_i^j during the next time step Δt is described. For simplicity, it is assumed that $i < N^j$ (namely that there will be no interaction with the next stretch).

Therefore the superscript 'j' is omitted in the following presentation.

A basic assumption of the model is that the individual speeds of vehicles at a given traffic condition characterised by ρ_i vary within the interval defined by $[\underline{u}_i, \bar{u}_i]$. The assumed speed of a vehicle in this interval depends on its desired speed v_δ. For example this means that vehicles with the maximum desired speed ($v_\delta = v_1$) will assume the upper limit $\bar{u}_i$, whereas vehicles with the minimum desired speed ($v_\delta = v_0$) will assume the lower limit $\underline{u}_i$. The formulae for the definition of $\underline{u}_i$ and $\bar{u}_i$ are given later.

Under this assumption speeds $\hat{u}_i$ and $\hat{u}_{i+1}$ which can be considered as the relevant speeds of an individual vehicle (characterised by the value of δ) under the actual traffic conditions in S_i and S_{i+1} respectively are evaluated as follows:

$$\hat{u}_i = \underline{u}_i + \frac{v_\delta - v_0}{v_1 - v_0} \cdot (\bar{u}_i - \underline{u}_i)$$

The desired speed $\hat{u}_i^x$ for position x in S_i is set as follows:

$$\hat{u}_i^x = (1 - \frac{x}{l}) \cdot \hat{u}_i + \frac{x}{l} \cdot \hat{u}_{i+1} \tag{5}$$

where (5) takes the anticipation of traffic conditions by the motorists
into account. For example, if a vehicle is at the end of section S_i ($x = 1$),
it will be influenced only by the traffic conditions in section S_{i+1}.

It must now be assured that the expected value of the speeds $\hat{u}_i$ ($= \hat{u}_i(v_\delta)$)
equals the actual value of u_i:

$$\int_0^\infty \hat{u}_i \cdot f(v)\,dv = u_i \qquad (6)$$

where $f(v)$ is the density function of the cumulative distribution V_f.

To solve Eq. 6 it is necessary to define either the limit $\underline{u}_i$ or the
limit $\bar{u}_i$. In the model the lower limit is defined as follows:

$$\underline{u}_i = \begin{cases} v_o - \dfrac{v_o-u_{opt}}{\rho_{opt}} \cdot \rho_i, & \rho_i < \rho_{opt} \\[2ex] u_i, & \rho_i \geq \rho_{opt} \end{cases} \qquad (7)$$

where ρ_{opt} is the desity at which flow becomes a maximum and u_{opt} is
the corresponding mean speed (determined from the relationship $U(\rho)$). It
is assumed that in (7) $v_o \geq u_{opt}$. Since in reality the difference
between v_o and u_{opt} is small, the assumption of the linear decrease is
not a severe limitation.

From (4), (6) and (7) we obtain for the upper limit:

$$\bar{u}_i = \begin{cases} \underline{u}_i + \dfrac{v_i-v_o}{E(V_f)-v_o} \cdot (u_i-\underline{u}_i), & \rho_i < \rho_{opt} \\[2ex] u_i, & \rho_i \geq \rho_{opt} \end{cases} \qquad (8)$$

where $E(V_f)$ is the expected value of the cumulative distribution V_f.

The main hypothesis underlying this approach is that the variance of
individual speeds decreases when density increases. If $\rho_i > \rho_{opt}$ the
desired speed of a vehicle (at free flow) does not affect the possible
speed under the actual traffic conditions.

The speed of a vehicle at time $t+\Delta t$ is now set to:

$$\min\{v_\delta, \; u^{(t)} + b_{max} \cdot \frac{\hat{u}^x_i - u^{(t)}}{v_1} \cdot \Delta t\} \; , \; \hat{u}^x_i > u^{(t)}$$

$$u^{(t+\Delta t)} = \tag{9}$$

$$\max\{\hat{u}^x_i, \; u^{(t)} + b_{min} \cdot \frac{u^{(t)} - \hat{u}^x_i}{v_1} \cdot \Delta t\} \; , \; \hat{u}^x_i \leq u^{(t)}$$

In Eq. 9 a relaxation effect is considered. The terms $b_{max}(\hat{u}^x_i - u^{(t)})/v_1$ and $b_{min}(u^{(t)} - \hat{u}^x_i)/v_1$ are respectively the assumed acceleration and deceleration of a vehicle during the time step.

In the model the values of b_{min} and b_{max} are set as follows:

$$b_{min} = -7 \; m/s^2$$

$$b_{max} = 4 \; m/s^2$$

For the free flow distribution from Leutzbach et al. (1977) (Fig. 4) the relation between speed u and possible acceleration b, shown in Fig. 5 is obtained. The shaded area represents the possible range of values depending on the δ-value. The area between the discontinuous lines represents the corresponding relationship in a microscopic simulation model (Brannolte, 1977). This relationship has been calibrated from empirical data. Therefore the first part of Eq. 6 can be assumed to have been validated by empirical data.

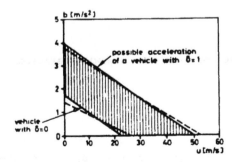

Fig. 5. The speed-acceleration relationship in the model.

With $b_{min} = -7 \text{ m/s}^2$ and a time step of $\Delta t = 10$ s (chosen in the following simulation) any diminuation of speed is realized in one time step. Therefore a validation of the second part of Fig. 9 is omitted.

The new position of the vehicle becomes:

$$x^{(t+\Delta t)} = x^{(t)} + \frac{1}{2} (u^{(t)} + u^{(t+\Delta t)})\Delta t \tag{10}$$

Eq. 10 assumes a constant acceleration during the time step.

If $x^{(t+\Delta t)} > 1$, the vehicle enters the next segment:

$$x^{(t+\Delta t)} > 1 \rightarrow x^{(t+\Delta t)} = x^{(t+\Delta t)} - 1,$$

$$i = i+1, \ \rho_i = \rho_i - 1, \ \rho_{i+1} = \rho_{i+1} + 1.$$

If $x^{(t+\Delta t)} > 1$, $i = N$ and $E \neq P_d$ a new stretch is determined. This is done according to the assignment matrix $R_{(i,j)}$ or according to the alternative routes.

When all vehicles in the system have been treated by the algorithm described, the new values $\underline{u}_i$ and $\bar{u}_i$ are determined by the new values of ρ_i and by (7) and (8). Since this is essentially a technical detail of the model, it will not be further described here.

4 MODEL VALIDATION

In the traffic flow model presented here the following parameters are to be set by the user:

$$U(\rho)^j, \ v_f^j, \ 1^j, \ \Delta t.$$

The model produces good results with $1^j = 500$ m and $\Delta t = 10$ s. These values have also been chosen by Cremer (1979) for his model. For 1^j, Δt

and v_1^j (maximum free flow speed in stretch j) the relation $\frac{l^j}{\Delta t} > v_1^j$ should be satisfied. Since the parameters $U(\rho)^j$ and v_f^j can be obtained from measurements there is no need of a model calibration since no parameters remain with unknown values. In the following, the traffic flow model is validated by presenting some results of a simulation in which a three-lane stretch runs into a two-lane stretch (Fig. 6).

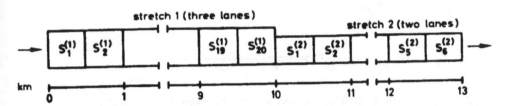

Fig. 6. The simulation stretch.

A similar case has also been used by Hauer and Hurdle (1979) for an examination of the validity of the FREFLO simulation package. They expected to observe the following general features:

1. When the demand exeeded the capacity of the two-lane stretch the begin of this stretch would become a bottleneck and begin to flow at capacity.
2. Once the bottleneck reached capacity, a congested region of high density and low speed would begin forming upstream of the bottleneck.
3. After demand dropped below the capacity of the bottleneck, the extent of the congested region would begin to diminish.
4. The flow in the bottleneck would remain at capacity until the congestion upstream had cleared.
5. The flow downstream of the bottleneck would never exceed the capacity of the bottleneck.

But without further calibration, FREFLO did not produce any of these features. In the following, it will be shown that these features are reproduced by DYNEMO.

The relations $U(\rho)^j$ as well as the corresponding relations $\underline{u}$ and $\bar{u}$ which were used in the following simulation are shown in Fig. 7a and 7b.

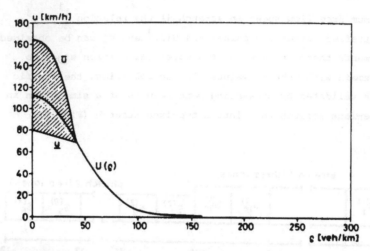

Fig. 7a. Relationship $U(\rho)$, $\underline{u}$ and $\bar{u}$ for the two-lane stretch.

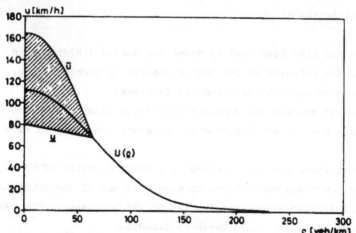

Fig. 7b. Relationship $U(\rho)$, $\underline{u}$ and $\bar{u}$ for the three-lane stretch.

The relationship $U(\rho)$ for the two-lane stretch has been calibrated by measurement in Cremer (1979). The relationship for a three-lane stretch has been assumed in this numerical example.

Fig. 8 shows the distribution of free flow speeds used in the simulation.

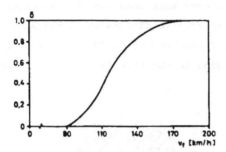

<u>Fig. 8.</u> Distribution of free flow speeds used in the simulation.

In the simulation, the pattern of the input flow over time temporarily exceeded the capacity of the bottleneck (**Fig. 9a** and **9b**).

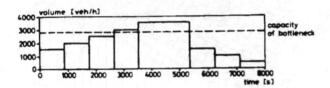

<u>Fig. 9a.</u> Pattern of input flow over time (model input).

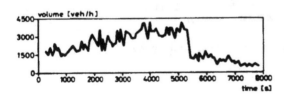

<u>Fig. 9b.</u> Pattern of input flow over time (produced by the model under the assumption of poisson arrivals).

Fig. 10 shows the pattern of speed over time and space. It can be seen from the figure that the performance of the model is consistent with the expected dynamic changes in traffic flow as listed earlier.

The obtained flow-density measurements at various points on the simulated stretch are shown in Fig. 11.

The relation $U(\rho)$ are represented by the curves in Fig. 11. As in reality, for $\rho > \rho_{opt}$ traffic conditions in the model are not in stationary

equilibrium. All measurements were obtained with a measuring period of
1 minute. The expected traffic conditions as listed above due to
capacity restraints are reproduced by DYNEMO. Fig. 12 shows the speed-
time patterns corresponding to the flow-density relationships in Fig. 11.

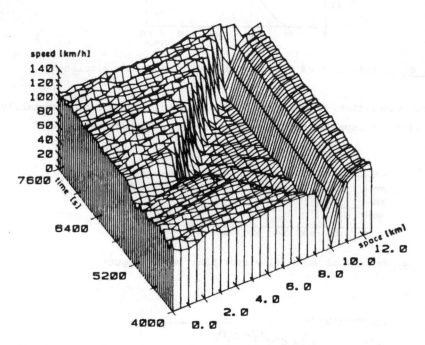

Fig. 10. Pattern of speed over time and space.

5. MODEL IMPLEMENTATION

The simulation model DYNEMO is implemented on a 16-Bit microcomputer in
the programming language PASCAL. For 6000 vehicles which are simultaneously
in the network, the relationship between simulation time and computing
time becomes one-to-one. Since the data for the description of the
vehicles are stored on disc, the maximum number of vehicles is limited
only by the available disc storage.

6. MODEL APPLICATION

The traffic flow in a German motorway network which consists of about
700 'lane kilometers' was simulated with DYNEMO. A traffic control system
is already installed in this network. This control system is also
implemented in the simulation model. For various situations, travel
times and fuel consumptions in the network can be compared.

7. SUMMARY

The paper presents the simulation model DYNEMO. This model is based on
a new traffic flow model which is described and validated by use of an
example. For each stretch in the network the model requires as input a
relationship between density ρ and mean speed u (or between volume λ
and speed u) in the form of pairs of (ρ_i, u_i) and a free flow distribution
of speeds. Also required are the input flows, an origin-destination
matrix and a matrix which defines the route choice of the vehicles.

Since DYNEMO treats vehicles individually, the model can generate
statistics which relate to individual vehicles, such as travel time
distributions on any route in the network. Furthermore the model
reproduces the stochastic character of traffic flow.

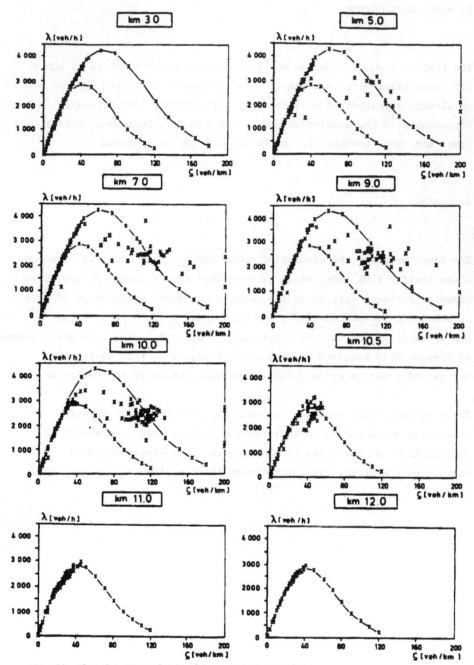

Fig. 11. Flow-density relationships at various points.

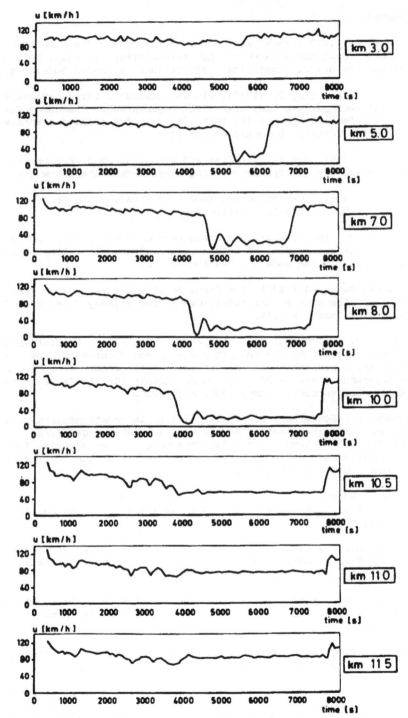

Fig. 12. Speed-time patterns corresponding to the flow-density
relationships.

REFERENCES

Brannolte, U. (1977). Verkehrsablauf an Steigungsstrecken von Richtungsfahrbahnen. Forschung Straßenbau und Straßenverkehrstechnik, Heft 318,
Schriftenreihe des Bundesministers für Verkehr, Bonn-Bad Godesberg.

Brilon, W., Brannolte, U. (1977). Simulationsmodell für den Verkehrsablauf auf zweispurigen Landstraßen mit Gegenverkehr. Forschung Straßenbau und Straßenverkehrstechnik, Heft 239, Schriftenreihe des Bundesministers für Verkehr, Bonn-Bad Godesberg.

Cremer, M. (1979). Der Verkehrsfluß auf Schnellstraßen. Fachberichte
Messen-Steuern-Regeln, Springer-Verlag, Berlin-Heidelberg-New York.

Gazis, D.C. (1961). Car Following Theory of Steady State Traffic Flow.
Op. Res. No. 9, pp. 545-567, Baltimore.

Gibson, D.R.P. (1981). Available Computer Models for Traffic Operations
Analysis. National Academy of Science, Special Rep. 194, "The Application
of Traffic Simulation Models", Washington D.C.

Lee, Rioux, Copeland (1977). The Texas Model for Intersection Traffic
Development. Research Report 184-1, Center for Highway Research,
University of Texas, Austin.

Leutzbach, W., Wiedemann, R., Hubschneider, H. (1977). Simulation des
Verkehrsablaufs auf Autobahnen mit zweispurigen Richtungsfahrbahnen im
Hinblick auf empirisch nicht ausreichend verifizierbare Situationen.
Forschungsauftrag Nr. 7450/12 der Bundesanstalt für Straßenwesen im
Rahmen des Großversuchs "Tempo 130", Karlsruhe.

Leutzbach, W., Schweizer, G., et al. (1981). Informationstechnische
Grundlagen für Leitsysteme im Straßenverkehr. Forschungsauftrag
Nr. TK 0073 des Bundesministers für Forschung und Technologie, Institut
für Verkehrswesen und Institut für Angewandte Informatik im Transport
und Verkehr, Universität Karlsruhe.

Liebermann, Worell, Bruggeman (1972). Logical Design and Demonstration
of UTC-1 Network Simulation Model. Highway Research Record No. 409,
HRB Washington.

Lighthill, M.J., Witham, G.B. (1955). On Kinematic Waves II. A Theory
of Traffic Flow on Long Crowded Roads. Proc. Roy. Soc., Series A,
No. 1178, Vol. 229, pp. 317-345, London.

May, A.D. (1981). Models for Freeway Corridor Analysis. National Academy
of Sciences, Spec. Rep. 194, "The Application of Traffic Simulation Models"
Washington D.C.

Michaels, R.M. (1965). Perceptual Factors in Car Following. Proc. 2nd.
Int. Symp. on the Theory of Traffic Flow, 1963, OECD, Paris.

Payne, H.J. (1979). FREFLO: A Macroscopic Simulation Model of Freeway
Traffic. Discussion by V.F. Hurdle and E. Hauer. Transp. Res. Rec. 722,
Urban System Operations, National Acad. of Sciences, Washington D.C.

Pipes, L.A. (1968). Topics in the Hydrodynamic Theory of Traffic Flow. Transp. Res., Vol. 3, pp. 229-234.

Todosiev, E.P. (1963). The Action Point Model of the Driver-Vehicle System. Engineering Experiment Station, The Ohio State University, Columbus, Ohio, Rep. No. 202 A-3.

Wicks, D.A., Liebermann, E.B. (1977). Developing and Testing of INTRAS, a Microscopic Freeway Simulation Model. Federal Highway Administration, Rep. FHWA-RD-76-76.

Wiedemann, R. (1974). Simulation des Straßenverkehrsflusses, Schriften-reihe des Instituts für Verkehrswesen, Universität Karlsruhe, Heft 8.

Wiedemann, R., Hubschneider, H. (1982). Zusammenhang zwischen der Wunschgeschwindigkeit und relevanten Kenngrößen des Verkehrsablaufs auf Autobahnen. Forschungsauftrag Nr. 01.071 G 80 H des Bundesministers für Verkehr, Karlsruhe.

Ninth International Symposium on
Transportation and Traffic Theory
© 1984 VNU Science Press, pp. 89–111

IMPROVED CONTINUUM MODELS OF FREEWAY FLOW

PANOS G. MICHALOPOULOS and
DIMITRIOS E. BESKOS*

*Department of Civil and Mineral Engineering, University of Minnesota,
Minneapolis, MN 55455, U.S.A.*

ABSTRACT

The problem of macroscopic modelling of freeway flow dynamics is addressed
in this paper. Existing simple and high order continuum models are examined,
modified and treated numerically. Subsequently they are implemented to a
number of situations representing uninterrupted and interrupted flow conditions in order to assess their effectiveness. This is accomplished by comparing
model results with a data base generated through a detailed microscopic simulation program recently developed by the FHWA. The problem of multilane
dynamics is also addressed. A simple continuum formulation is presented in
detail while two additional alternatives, a two dimensional one and a high
order continuum, are briefly discussed. Test results of all alternatives are
also presented.

*) Currently Department of Civil Engineering, University of Patras,
Patras, Greece.

1.   INTRODUCTION

Despite the need for improved macroscopic modelling and analysis of freeway
flow dynamics relatively little progress can be reported in more than a
decade. The most widely known dynamic formulations can be characterized as
either simple continuum (Lighthill and Whitham, 1955) or high order continuum
(Payne 1971, 1979; Phillips, 1978). The simple continuum formulation is based
only on the conservation equation which is supplemented by a quasi static
equation of state. In the high order continuum models a momentum equation is
added to the conservation in order to account for purely dynamic effects.
Although this should improve realism, due to the inclusion of acceleration and
inertia, recent experiments (Hauer and Hurdle, 1979; Derzko et. al. 1983)
did not reveal satisfactory performance of these models at realistic situations.

Since lack of satisfactory performance could be attributed to either the
particular method of solution of the governing equations or to the models
themselves, both possibilities were explored and the results are presented in
this paper. In particular, new solution algorithms were developed for both
the simple and high order models; in the latter case a comparison between the
existing and the proposed solution algorithm was made. Subsequently, the
performance of both modelling alternatives (simple and high order) was tested
under both uninterrupted and interrupted flow conditions by means of microscopic
simulation. Proceeding to more essential improvements three new continuum models
were developed which include the effects of lane changing. This should be of
interest due to the lack of a general macroscopic methodology allowing description of the lane changing process. The first model is based on the conservation
equation and the density oscillations principle proposed by Gazis et. al. (1962).
The second model includes the road width as an additional dimension, i.e., it is
two dimensional with respect to space. Finally, the third model is more complex
since it includes a momentum equation in order to take into account acceleration
and inertia effects. Due to space limitations only the simplest lane changing
model is presented here. However, comparison of the three alternatives based on
simulation are included along with a summary of the major findings. It should
be noted that the lane changing modelling allows a more realistic treatment of
merging, diverging and weaving areas. Model implementation to such situations
as well as inclusion of additional flow properties (such as friction) are also
discussed along with the practical implications of the developed models.

2. SIMPLE CONTINUUM TREATMENT

Macroscopic freeway models are needed not only for better understanding the collective behavior of traffic, but also for analyzing flow conditions in a dynamic fashion, devising efficient control strategies, simulation, assessing the effects of geometric or control strategy improvements, determining the adequacy of existing or proposed geometric configurations,etc. Existing macroscopic models fall into three main categories: a) Input-Output, b) simple continuum, and c) high order continuum. The models of the first category are rather simplistic in that they do not include space explicitly nor do they take compressibility into account. High order continuum models on the other hand are the most sophisticated but they have not, as yet, gained wide popularity or proved truly superior to the simple continuum alternative.

According to the simple continuum model, flow can be described by the conservation equation which has the general form:

$$\frac{\partial q}{\partial x} + \frac{\partial k}{\partial t} = g(x,t) \tag{1}$$

where $q = q(k) = ku$ is the flow rate of the traffic stream and $k = k(x,t)$, $u = u(k)$ are the density and speed respectively; t and x denote time and space while g is the generation rate. In freeway sections without entrances or exits $g(x,t) = 0$. The continuum model assumes that flow is only a function of density; this implies that Eq. (1) is a hyperbolic partial differential equation having density as the only unknown. Solution of this equation yields the value of k and therefore (q and u) at every point of the t-x domain. Analytical solutions are possible but only for continuous single regime equations of state (i.e., q-k models), simple initial and boundary conditions and zero generation terms. Thus, in order to improve realism (by relaxing simplifying assumptions) one must discretize in t and x and turn to numerical methods for solving the governing equation (Eq. (1)). Ordinarily, this would be a relatively simple task if solution of Eq. (1) were not discontinuous. However, such discontinuities (shocks) require employment of special solution algorithms. One shock fitting and three shock capturing algorithms were developed for this purpose by Michalopoulos et. al. (1983). The latter correspond to various orders of accuracy with respect to Δt and are much simpler (therefore faster) as they do not treat shocks explicitly; instead, shocks appear as part of the solution when and where they develop. The shock fitting scheme on the other hand is more accurate but also far more time consuming as it is iterative; in addition it is more complex since the shock represents a discontinuity in density and is treated by the appropriate jump conditions across its front. Due to space limitations only the simplest shock capturing solution method is presented here. It should be noted, however, that higher order of accuracy simply implies closer agreement to the analytical solution and not necessarily to field data. This is because the conservation equation alone is insufficient for complete description of flow. Thus, as the order of accuracy decreases shocks are smoothed out and this was found to agree better with reality (Michalopoulos et. al., 1983). Physically the lack of sharp shocks is explained by the deceleration of platoons as they approach areas of higher density.

Returning to the numerical solution of Eq. (1), the following first order (with respect to Δt) accuracy algorithm was found to be effective and performed only slightly worse than a similar second order one (which was also found effective) when compared with simulated data:

$$k_j^{n+1} = \frac{1}{2}(k_{j+1}^n + k_{j-1}^n) - \frac{\Delta t}{2\Delta x}(G_{j+1}^n - G_{j-1}^n) + \frac{\Delta t}{2}(g_{j+1}^n + g_{j-1}^n) \ v_j \qquad (2)$$

$$u_j^{n+1} = u_e (k_j^{n+1}) \qquad (3)$$

$$q_j^{n+1} = k_j^{n+1} u_j^{n+1} \qquad (4)$$

where

k_j^n, u_j^n, q_j^n = the density, speed or flow, respectively of node j at
                    $t = t_o + n\Delta t$

t_o = the initial time

$u_e (k_j^{n+1})$ = the equilibrium speed corresponding to the value of density k_j^{n+1}

$G_j^n = k_j^n u_j^n$

$g_j^n = g(x_j, t_n)$ = the generation rate of node j at $t = t_o + n\Delta t$

When the simplest equilibrium speed density model is assumed (Greenshields, 1934), then

$$u_e(K_j^{n+1}) = u_f(1 - \frac{K_j^{n+1}}{k_o}) \quad \text{and} \tag{5}$$

$$G_j^n = k_j^n u_f(1 - \frac{K_j^n}{k_o}) \tag{6}$$

where k_o and u_f represent the jam density and free flow speed, respectively. Figure 1 shows space discretization of a freeway section which consists of J. segments of length Δx. Naturally, for better accuracy the merging area can be divided into more segments δx such that $\delta x \leqslant \Delta x$. In order to save computation time the segment size can vary; this would only slightly affect Eq. 2.

It can be easily demonstrated that in order to keep the solution within reasonable bounds the time and space increments Δt, Δx must obey the rule: $(\Delta x/\Delta t) > u_f$. The numerical solution presented above allows employment of any speed-density model, including discontinuous ones; in such case Eqs. (5) and (6) can be altered accordingly. Further, initial and boundary conditions can be as complex as desired. Thus, arrivals and departures can follow a statistical distribution; this is particularly desirable in simulation. It should be noted

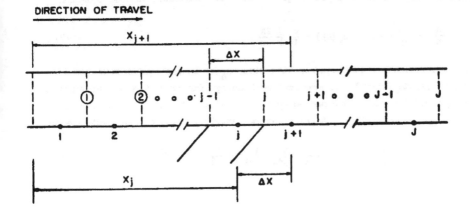

Fig. 1. Space Discretization of a Freeway Section.

that in merging or diverging areas the generation rate g is either given or it
can be derived dynamically from the ramp demands and the freeway flow and
density at the previous time step. The latter option is more attractive since
actual measurements are generally undesirable or even infeasible (i.e. in design,
real time control, simulation, etc.). In the simplest case where the entire ramp
equals one segment length Δx (i.e. at short ramps), an average value of g can
be assumed; however, as the ramp length increases or as Δx decreases this
assumption becomes unrealistic. In such case solution proceeds by considering
the conservation equation of the acceleration lane and solving it simultaneously
with the conservation equation of the freeway proper. This formulation must
take into account lane changing effects described in section 3. An algorithm
allowing numerical estimation of g(x,t) dynamically is presented in Michalopoulos
et. al. (1983).

3.   HIGH ORDER CONTINUUM TREATMENT

The simple continuum model does not include acceleration and inertia effects;
these can be taken into account by the momentum equation. Between the two
available high order models the simpler one (Payne, 1971) is presented here
although comparisons with the other alternative (Phillips, 1979) are also
included. Payne's model employs Eq. (1) and the following momentum equation:

$$\frac{\partial u}{\partial t} = -u\frac{\partial u}{\partial x} - \frac{1}{T}\{u - u_e(k)\} - \frac{\nu}{T}\frac{1}{k}\frac{\partial k}{\partial x} \tag{7}$$

where T represents a constant reaction time, $u_e(k)$ is the equilibrium speed
and $\nu = -1/2\{du_e(k)/dk\}$. A numerical methodology for simultaneous solution of
Eqs. (1) and (7) was also proposed by Payne (1971, 1979) and is also adaptable
to Phillips' model (1979). An alternative numerical solution proposed here is:

$$u_j^{n+1} = \frac{1}{2}(u_{j+1}^n + u_{j-1}^n) - \frac{\Delta t}{2\Delta x}(D_{j+1}^n - D_{j-1}^n) + \frac{\Delta t}{2}(s_{j+1}^n + s_{j-1}^n) \tag{8}$$

where

$$D_j^n = \frac{1}{2}(u_j^n)^2 + \{\frac{1}{T}\int\frac{\nu}{k}dk\}_j^n \tag{9}$$

$$s_j^n = -\frac{1}{T}\{u_j^n - u_e(k_j^n)\} \tag{10}$$

$$u_j^n = u(x_j, t_n) \neq u_e(k_j^n)$$

The remaining terms are as in Eq. (2). When the equilibrium u-k model is defined, the second term of the right hand side of Eq. (9) can be determined. For instance from Greenshields' (1934) model and the relationship

$$v = -\frac{1}{2}\frac{du_e(k)}{dk} = \frac{1}{2}\frac{u_f}{k_o}$$

it follows that

$$\frac{1}{T}\int\frac{v}{k}dk = \frac{1}{2}\frac{1}{T}\frac{u_f}{k_o}\ln(k) \tag{11}$$

Therefore Eq. (9) becomes

$$D_j^n = \frac{1}{2}(u_j^n)^2 + \frac{1}{2}\frac{1}{T}\frac{u_f}{k_o}\ln(k_j^n) \tag{12}$$

If the equilibrium relationship $u_e(k)$ is discontinuous or other than Greenshields' (1934) then Eq. (11) changes; if derivation becomes too involved, numerical calculation of v and the left hand side of Eq. (11) can easily be performed. Following computation of u_j^{n+1} from Eq. (8), the second dependent variable k_j^{n+1} is found from Eq. (2). Finally, flow is computed from $q_j^{n+1} = k_j^{n+1}u_j^{n+1}$.

Since experimentation with the model described by Eqs. (1) and (7) failed to demonstrate its superiority over the simple continuum model even after improving the solution algorithm, the possibility of modifying the momentum equation was considered. This was accomplished by comparing Eq. (7) with its equivalent developed by Phillips (1979). The latter was not considered suitable for further improvements because of its rigorousness, its complexity and the difficulties involved in estimating model parameters such as the equilibrium pressure. Comparison of the two modelling alternatives leads to the conclusion that if Phillips' model is correct, Payne's model is valid only at low densities. Since the experiments of section 5.2 tend to support this view, an improved practical treatment would be to drop the momentum equation at high densities, while at low densities both equations can be employed. This alternative is

referred to as hybrid model in the remaining of this text. A suggested threshold
value of density beyond which the momentum equation can be dropped is $k = 1/3k_o$;
this value was found experimentally by simulation and could conceivably change
for freeways having characteristics different to those assumed.

In addition to the above simple modification it can be easily realized that the
coefficient of the anticipation term $\lambda = 1/T$ of Eq. (7) should be a function
of density. This implies that $T = T(k)$ rather than $T = const.$; Phillips (1979)
suggests that the reaction time T should increase with density. This as well
as the opposite assumption were tested at the four situations of section 5.2;
however, no significant differences with the constant T alternative were detected.
Slight improvements were noticed only when $T(k)$ was assumed concave; a simple
piecewise linear $T(k)$ function was derived experimentally and used in the comparisons of section 5.2; however, employment of this relationship is not
recommended without further experimentation.

4.   LANE CHANGING

In the modelling presented to this point lane changing effects were not considered,
i.e., the previous models are aggregate. Although a few works on lane changing
are available, a sufficiently complete mathematical description and a practically
implementable method for treating this process macroscopically is still lacking.
Among the first pioneers Gazis et. al. (1962) proposed a model for describing
the interlane density oscillations and addressed the stability problem. The
major assumption of this model is that density oscillations occur about an equil-
brium density distribution. However, distance x as well as generation terms
were not included in the basic formulation. The former limitation was removed
by Munjal and Pipes (1971) and Munjal et. al. (1971) who incorporated the
Gazis et. al. (1962) model into the conservation equation. Even with this
improvement generation terms are not included but are imposed artificially.
Further, since analytical solutions were sought in both cases, oversimplications

had to be made. A stochastic approach describing the lane changing as a
Markov process was proposed by Rorberch (1976) and was later adopted by
Makigami (1981) for uncongested flow. In this latter work the Munjal and
Pipes model was adopted for congested flow. The major disadvantage of the
stochastic approach lies in the simplicity of the transition probability matrix
which is not time dependent. This modelling is more appropriate for steady
state conditions (rarely experienced in most real life problems) i.e. rapid
demand fluctuations cannot be accomodated.

The lack of a general macroscopic methodology for treating the lane changing
process led to the development of three continuum models. The first is discrete
and employs the conservation equation of each lane while the second is two
dimensional with respect to space, i.e., it employs only a single equation in
conservation form but includes the street width explicitly. Finally, the third
is high order continuum i.e. it includes an additional momentum equation per
lane which accounts for exchange of momenta between lanes (Michalopoulos et. al.
1983). Due to space limitations only the simplest continuum modelling alternative
is presented here; nevertheless comparative performance evaluation of all three
alternatives is included in subsequent sections.

A simple continuum model for describing flow along two or more homo-directional
lanes can be obtained by considering the conservation equation of each lane.
This is accomplished by observing that the exchange of flow between lanes
represents generation (or loss) of cars in the lane under consideration. Following
Gazis et. al. (1962), Munjal and Pipes (1972) and Michalopoulos et. al. (1983),
the simple continuum system of equations describing flow on a two-lane freeway is:

$$\frac{\partial q_1}{\partial x} + \frac{\partial k_1}{\partial t} = g + Q_1 \tag{14}$$

$$\frac{\partial q_2}{\partial x} + \frac{\partial k_2}{\partial t} = Q_2 \tag{15}$$

where:

$q_i(x,t)$ = the flow rate of the ith lane   (i=1,2)
$k_i(x,t)$ = the density of the ith lane   (i=1,2)
$Q_i(x,t)$ = the lane changing rate   (i=1,2)

$g(x,t)$ = the generation rate of lane 1 (right lane) due to merging or diverging; at exits $g < 0$

$$Q_1 = \alpha\{(k_2(x,t-\tau)-k_1(x,t-\tau))-(k_{20}-k_{10})\} \qquad (16)$$

$$Q_2 = \alpha\{(k_1(x,t-\tau)-k_2(x,t-\tau))-(k_{10}-k_{20})\} \qquad (17)$$

α = a sensitivity coefficient describing the intensity of interaction between lanes. In the simplest case α can be assumed
constant; alternatively:

$$\alpha = \begin{cases} 0 & , \quad |k_2(x,t-\tau)-k_1(x,t-\tau)| \leq k_A \\ \dfrac{\alpha_{max}}{k_o-k_A}(|k_2(s,t-\tau)-k_1(x,t-\tau)|-k_A) & , \quad |k_2(x,t-\tau)-k_1(x,t-\tau)| > k_A \end{cases} \qquad (18)$$

k_A = a constant value below which no exchange of flow occurs

τ = an interaction time lag; a value of zero could be assumed for
simplicity

k_{io} = an equilibrium density value which, if exceeded, will result in
lane changing; i = 1,2

k_o = the jam density

As before, the system of governing equations (Eqs. (14) and (15)) can be solved
numerically by discretizing in time and space. The numerical solution allowing
estimation of k, u, and q at each node and time increment is:

$$k_{1'j}^{n+1} = \frac{1}{2}(k_{1'j+1}^{n}+k_{1'j-1}^{n})-\frac{\Delta t}{2\Delta x}(q_{1'j+1}^{n}-q_{1'j-1}^{n})+\frac{\Delta t}{2}(g_{j+1}^{n}+g_{j-1}^{n})$$

$$+ \frac{\Delta t}{2}(Q_{1'j+1}^{n}+Q_{1'j-1}^{n}) \qquad (19)$$

$$k_{2'j}^{n+1} = \frac{1}{2}(k_{2'j+1}^{n}+k_{2'j-1}^{n})-\frac{\Delta t}{2\Delta x}(q_{2'j+1}^{n}-q_{2'j-1}^{n})$$

$$+ \frac{\Delta t}{2}(Q_{2'j+1}^{n}+Q_{2'j-1}^{n}) \qquad (20)$$

where

$k_{1'j}^{n}$: the density of the ith lane in the jth segment at t = n·Δt

$$Q_{1'j}^{n} = \alpha_{1'j}^{n-s}\{(k_{2'j}^{n-s} - k_{1'j}^{n-s}) - (k_{20}-k_{10})\}$$

$$q_{2'j}^{n} = \alpha_{2'j}^{n-s} \{ (k_{1'j}^{n-s} - k_{2'j}^{n-s}) - (k_{10} - k_{20}) \}$$

$s \cdot \Delta t$ = interaction time lag $(=\tau)$

$\alpha_{i'j}^{n-s} = f(k_{1'j}^{n-s} - k_{2'j}^{n-s})$ as suggested by Eq. (18); $\alpha_{i'j}^{n-s}$ could also be
assumed constant

$$G_{i'j}^{n} = k_{i'j}^{n} \cdot u_{i'j}^{n} = k_{i'j}^{n} \cdot u_e(k_{i'j}^{n}) \quad (i=1,2)$$

$u_e(k_{i'j}^{n})$ = the equilibrium speed corresponding to $k_{i'j}^{n}$; assuming

Greenshield's (1934) model $G_{i'j}^{n} = k_{i'j}^{n} u_f (1 - \dfrac{k_{i'j}^{n}}{k_0})$.

Following computation of density at each time step the flow rate $q_{i'j}^{n+1}$ and
speed $u_{i'j}^{n}$ are obtained from

$$u_{i'j}^{n+1} = u_e(k_{i'j}^{n+1}) \qquad \text{and} \qquad q_{i'j}^{n+1} = k_{i'j}^{n+1} u_{i'j}^{n+1}$$

Extension of the above treatment to more than two lanes is straightforward;
this extension as well as description of the remaining two modelling alternatives and further details concerning complete treatment of merging and diverging
areas are presented in Michalopoulos et. al. (1983). Naturally, employment of
models that take into account lane changing is particularly important at
such areas.

5.   TESTING AND VALIDATION

5.1   Preliminaries

Budgetary and manpower limitations did not allow model comparisons with field
data. However, the models were implemented to a number of situations which
covered a wide range of flow conditions and included both entrance, exit and
pairs of ramps as well as multiple lanes. Model effectiveness was judged by
comparing the results against those obtained from microscopic simulation. A

recently developed well documented, tested and calibrated microscopic simulation
program called INTRAS (Wicks et. al., 1980) was employed for this purpose. Microscopic simulation for generating a data base was further justified by the need
to allow demands to fluctuate sufficiently over a controlled and often wide
range in relatively short time intervals. Further, there was clearly a need
to impose tractable initial and boundary conditions in order to allow intuitive
inspection of the results. Incidentally, in most of the cases presented next
the step size was assumed to be $\Delta x = 30.5$ m (100 ft) and $\Delta t = 1s$ respectively.
Model accuracy improved by 40-45% when the step size was reduced by 1/2. Both
simple (Greenshields, 1934) and discontinuous equilibrium u-k relationships
were assumed in the calculations. In addition to visual inspections, which
were performed by plotting q, k or u vs. x and t, the deviations of the basic
flow variables from the simulated data during each 10s increment were found and
subsequently the mean square error (MSE) and the mean absolute error (MAE) were
computed. In this manner model (and/or solution algorithm) effectiveness was
assessed quantitatively. Finally, from k, total travel time was computed and its
percentage difference (at the end of simulation) from the INTRAS estimate was
calculated.

5.2 Aggregate Models

Initially four situations were considered for testing the effectiveness of the
various modelling alternatives and solution algorithms. Three of these cases
are shown in Figs. 2 and 3. In case 1 (Fig. 2a) demands are nearly constant
approaching maximum flow. In case 2 (Fig. 2b) arrival flow starts at about 1/3
of capacity and increases gradually to the maximum flow rate (capacity), where
it remains for some time, and then it gradually decreases to its initial level.
In case 3 merging flows are introduced for a short time resulting in light
congestion while the demands drop substantially after congestion sets in for
quick dissipation. Finally case 4 (Fig. 3) is similar to case 3 but it represents
a longer freeway section and higher ramp demands for heavier congestion. The four
cases represent single lane flow since it was thought that initial testing under
the simplest flow conditions should prevent distortion of true model performance;
such distortion could be introduced from lane changing especially in merging
areas. Unsatisfactory model (or algorithm) performance even under these simple
conditions would imply that there is little reason to expect better results as
the number of lanes increase. It is worthwhile noting that subsequent testing

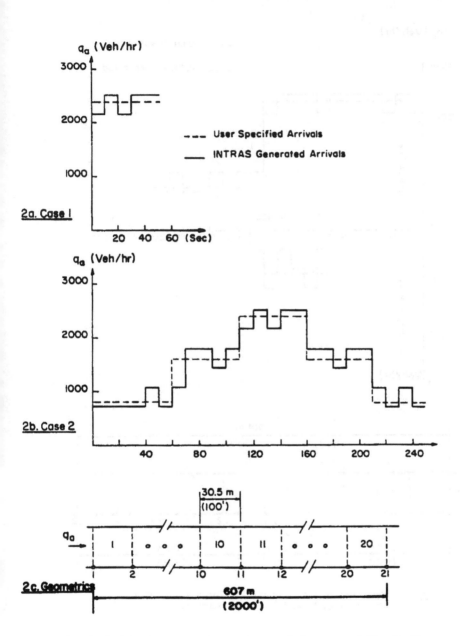

Fig. 2. Arrival Patterns and Geometrics for Uninterrupted Flow Testing (Case 1 and 2)

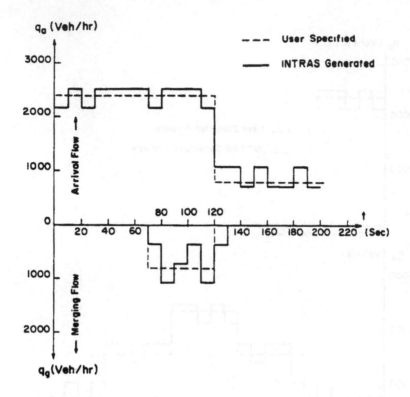

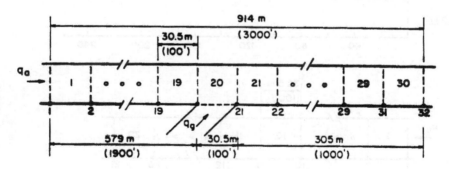

<u>Fig. 3.</u> Input Patterns and Generation Rates for Case 4
         (Interrupted flow, heavier congestion)

with multiple lanes did not alter the basic initial findings. In summary, the following conclusions can be drawn from the comparisons with the INTRAS generated data base:

1. Initial testing of both high order models revealed that their performance deteriorated as complexity increased (introduction of ramps).

2. Little quantitative difference in performance was detected between the two high order models (i.e. their MSE and MAE were about the same). Qualitatively it was found that in Phillip's model congestion did not propagate upstream of entrance ramps when demands in the merging area were excessively high; further, congestion dissipated rapidly when demand dropped below capacity. Payne's model on the other hand not only exhibited more severe congestion than the data indicated, but also this congestion was prolonged, i.e. it persisted much longer than simulation suggested. Some of the above findings can be visualized in Fig. 4 which presents the density distribution resulting from both high order models at a few selected time intervals corresponding to case 4 (Fig. 3).

3. The proposed numerical treatment improved the performance of the simplest high order model (Payne, 1971) substantially at interrupted flow (i.e. MSE of k was reduced by as much as 70% and MAE by 28% while similar reductions in u and q estimates were observed). This improved performance in case 4 can be seen partly in Fig. 4.

4. Existing high order models are only slightly better than the simple continuum model at uncongested flows; however at congested flow conditions the simple continuum model performs better even when the improved solution algorithm is employed in the high order models.

5. Payne's modified model and the hybrid model complemented by the improved solution algorithm surpassed the simple continuum model at congested flows (Fig. 5).

6. When discontinuous u-k relationships (derived from the INTRAS data) were employed, the simple continuum model performed consistently better than any other alternative regardless of flow conditions (i.e. MSE and MAE of k, u and q were 4-50% lower than the next best alternative). As expected, this finding was more pronounced at congested flows. The hybrid alternative combined with the new solution algorithm turned out to be the second best.

5.3 Lane Changing Models

So far the lane changing models were implemented in several two lane freeway sections with satisfactory results, i.e. no stability problems were encountered while the model results appeared to be reasonably close to those generated by INTRAS. Table 1 presents the quantitative comparisons of two additional situations representing uninterrupted and interrupted flow conditions. As

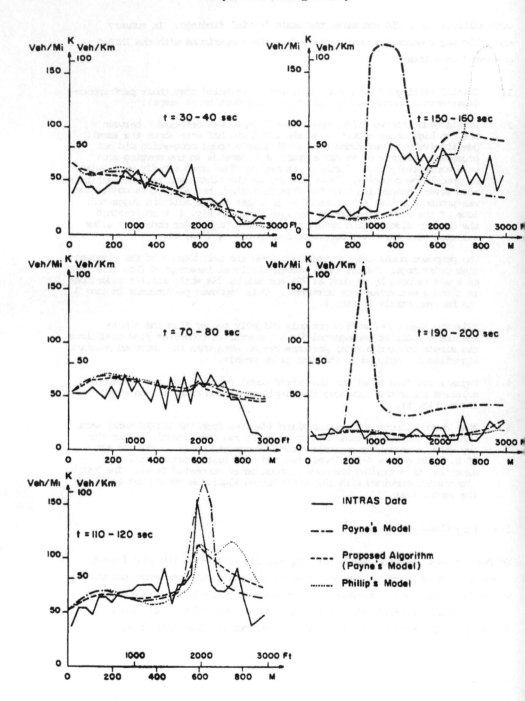

Fig. 4. High Order Model Comparisons at Interrupted Flow (Case 4)

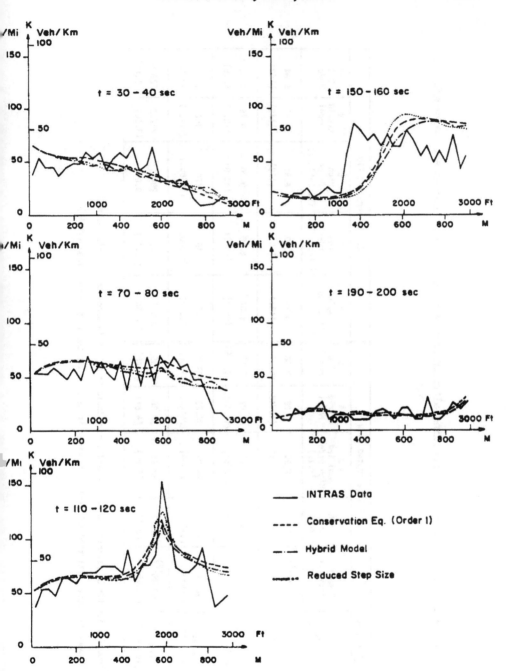

Fig. 5. Comparison of Simple Continuum and Hybrid Models (Case 4)

Table 1. Error indices of modelling alternatives when per lane estimates are averaged.

Situation →	1. Uninterrupted Flow					2. Interrupted Flow				
Model → MOE ↓	1 Simple Continuum	2 High Order T=cons	3 Modified High Order T=T(k)	4 Hybrid Model	5 2-D Model	6 Simple Continuum	7 High Order (Const. T)	8 High Order (Variable T)	9 Hybrid Model	10 2-D Model
Density (veh/mi)	2.06[1] (14.50)	2.07 (14.74)	2.21 (16.45)	2.22 (16.48)	2.38 (21.48)	6.76 (181)	7.30 (207)	6.91 (179)	6.81 (179)	6.84 (162)
Speed (mi/hr)	1.69 (8.88)	1.74 (9.46)	2.08 (12.63)	2.07 (12.59)	2.17 (13.65)	4.02 (55.3)	4.19 (60.0)	3.70 (48.8)	3.76 (48.5)	3.75 (47.5)
Flow Rate (veh/hr)	92.47 (28643)	93.37 (29157)	94.01 (29334)	94.23 (29448)	101.1 (36850)	143 (75353)	148 (80580)	174 (106850)	162 (92435)	164 (93388)
TTT (veh-min)	40.88[2] (-0.15)	40.53 (0.72)	38.86 (4.81)	38.88 (4.81)	36.95 (9.48)	230.1 (-2.56)	229.2 (-2.17)	204.1 (9.03)	217.3 (3.10)	228.5 (-1.85)

[1] Numbers outside of parenthesis in k, u and q rows indicate MAE; # in parenthesis indicates MSE.

[2] Number indicates estimated TTT while the number in parenthesis the % difference from INTRAS.

before, demands varied substantially while the volumes entering from the entrance ramp resulted in congested flow in lane 1 but only for a short time. The first column of the table presents the error indices resulting from the simple continuum model while the second to the high order one based on Payne (1971); the third column corresponds to the modified Payne model (i.e. one that employs variable T). In the fourth column the results of the hybrid model are presented. Naturally the last three models were appropriately changed to include lane changing effects (Michalopoulos et. al. 1983). Column 5 of Table 1 corresponds to the two dimensional model that includes the street width explicitly. Finally, columns 6-10 correspond to the case of interrupted flow and are similar to the first five. As Table 1 suggests the error indices are small while the differences among the various alternatives are not substantial. It is noted, however, that the 2-D model's performance should be better if a u-k relationship along the road width were available. Figure 6 presents a comparison of the results obtained from the simple continuum model in the uninterrupted flow case and the INTRAS generated data. The latter correspond to the average of both lanes since INTRAS does not output lane specific information. For this reason the comparisons of the multilane models against stimulated results are somewhat crude (i.e., the per lane estimates were averaged to accomodate comparison with INTRAS). Assuming, however, that these comparisons are at least partially indicative of actual model performance and from the experience gained by experimenting with a number of situations the following conclusions can be drawn:

1. Employment of variable sensitivity coefficient α (Eqs. (16-18)) is highly beneficial especially when time delay τ is assumed.

2. The introduction of time delay τ presented stability problems; therefore, its use is not recommended without further experimentation.

3. The two dimensional (2-D) model should be superior once an equilibrium u-k relationship along the road width is established. However, even in its simplified form the 2-D model has the advantage that it employs only one rather than I to 2I equations (where I represents the number of lanes) while its accuracy is about the same as the remaining models.

4. Due to the relatively insignificant performance differences among the various alternatives the simple continuum model (either in 1-D or 2-D formulation) is preferrable.

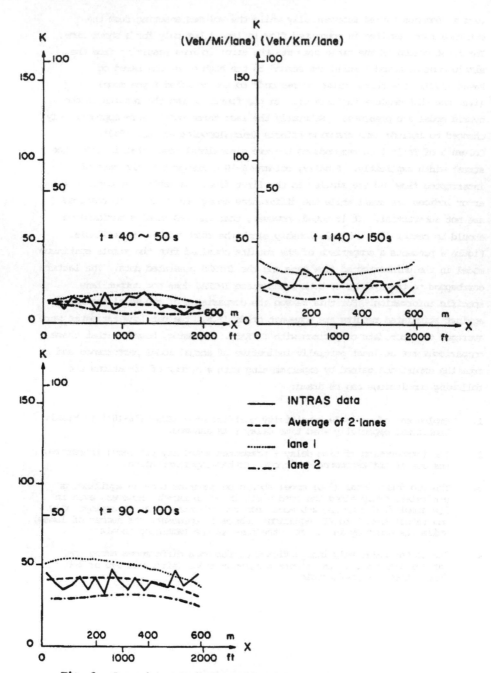

Fig. 6. Comparison of simple continuum and INTRAS results in a two-lane situation (uninterrupted flow).

6.   CONCLUDING REMARKS

In summary, the results presented earlier suggest that the effectiveness of the
simple continuum models cannot be easily surpassed especially at heavy flow
conditions. It also appears that the advantages of the high order models are
mainly realized at light flows where acceleration and inertia effects are more
pronounced.   The proposed numerical treatment improved the performance of the
simplest high order model substantially at interrupted flow conditions. This
along with the proposed modifications gave a minor advantage to the high order
modelling even at heavy interrupted flows. However, when the equilibrium u-k
relationships were adjusted according to the data the simple continuum model
performed considerably better regardless of flow conditions.

In addition to better understanding the inner workings of traffic the proposed
models can be used in simulation, control and analysis of situations where simple
demand-capacity considerations are insufficient. The relative simplicity of
these models combined with their macroscopic nature allow their implementation
in microcomputers. Recently, an interractive menu driven macroscopic freeway
simulation program with graphic capabilities was developed (Michalopoulos, 1983).
The program is written in UCSD Pascal language and runs on the IBM personal
computer. Input include conventional geometric and traffic parameters such as
freeway and ramp characteristics, demands and origin-destination information.
Output include dynamic description of k, u and q, estimation of the most common
measures of effectiveness and graphic presentation of flow conditions and congestion levels. The user can select one of the modelling alternatives presented
earlier. Naturally, better treatment of merging, diverging and weaving areas
can be expected when the models that take into account lane changing are selected.
This option, however, increases execution time substantially. The ramp proper,
acceleration, deceleration and auxiliary lanes are treated separately by the
program, i.e. their governing equations are added to those of the freeway
proper and solved simultaneously as a system. Currently this program is compared
to field data.
It is worthy to note that the finite difference methods employed in the solution
of the governing equations are the simplest and therefore the most efficient
computationally. Other alternatives include finite element (FEM) and perturbation
methods; the former is more complex but allows treatment of non-orthogonal systems
i.e. it is capable of treating freeways with more complex geometric characteristics

such as sharp curves. Since, however, these are rather uncommon in most facilities, analysis by FEM only adds complexities. Recent experimentation with such methods did not reveal higher accuracy, at least for segments where free flow speed is not affected by restricted geometrics. Presently perturbation techniques are being tested but it is expected that they should result in higher computation time.

Before concluding it should be recognized that the experimentation performed to this point was not very extensive in the sense that freeway sections up to 2 miles long with at most 2 lanes and up to a few ramps were tested. Most importantly, comparisons with real data were not made. Despite this some conclusions related to the performance of high order models are consistent with those of earlier independent studies (Hauer et. al. 1979, Derzko et. al. 1983). Further, it should be remembered that the initial experiments were kept deliberately simple in order to be able to easily inspect the results, avoid noise that could mask true model performance, generate the flow conditions of interest and assess the qualitative differences of the alternative formulations. Finally, it is noted that just recently a Finite Element (FEM) and a perturbation method were also developed by the authors for solving the simple and high order continuum models.

Acknowledgements

Financial support for this research was provided by NSF (grant #NSF/CEE-8210189).

REFERENCES

Derzko, N.A., A.J. Ugge, E.R. Case (1983). Evaluation of a dynamic freeway
    model using field data. Paper presented at the 62nd annual meeting of
    the Transpn. Res. Board. To be published at Transpn. Res. Rec.
Gazis, D.C., R. Herman, G.H. Weiss (1962). Density oscillations between lanes
    of a multilane highway. Ops. Res. 10, pp. 658-667.
Greenshields, B.D. (1934). A study of traffic capacity. In: Proc. Highway.
    Res. Board, Vol. 14, pp. 448-477.
Hauer, E., V.F. Hurdle (1979). Discussion on FREFLO. Transpn. Res. Rec. 722,
    pp. 75-76.

Lighthill, M.H., G.B. Witham (1955). On kinematic waves: II A theory of traffic flow on long crowded roads. In: Proc. R. Soc. Ser. A 229 No. 1178, pp. 448-477.

Makigami, Y., T. Makanishi, M. Toyama, R. Mizote (1981). On a simulation model for the traffic stream on freeway merging area. In: Proc. of 8th Intntl. Symp. on Transpn. and Traf. Theory, pp. 63-72.

Michalopoulos. P.G. (1983). A dynamic freeway simulation program for personal computers. Paper submitted to the Transpn. Res. Board for the 63rd annual meeting.

Michalopoulos. P.G., D.E. Beskos (1983). Dynamic modelling and analysis of traffic stream flow in complex transportation networks. University of Minnesota, Minneapolis, Min. Report NSF/CEE-8210189.

Munjal, P.K., L.A. Pipes (1971). Propogation of on-ramp density perturbations on unidirectional two and three lane freeways. Transpn. Res. 5, pp. 241-255.

Munjal, P.K., Y.S. Hsu, R.L. Lawrence (1971). Analysis and validation of lane drop effects on multilane freeways. Transpn. Res. 5, pp. 257-266.

Payne, H.J. (1971). Models of freeway traffic and control. In. Proc. Mathem. of Publ. Syst., published by Simul. Council, Vol. 1, No. 1, pp. 51-61.

Payne, H.J. (1979). FREFLO: A macroscopic simulation model of freeway traffic. Transpn. Res. Rec. 772, pp. 68-75.

Phillips, W.F. (1979). A new continuum model for traffic flow. Utah State Univ., Logan, Utah, Report DOT-RC-82018.

Rorberch, J. (1976). Multilane traffic flow process: Evaluation of queueing and lane changing patterns. Transpn. Res. Board 596, pp. 22-29.

Wicks, E.A., E.B. Lieberman (1980). Development and testing of INTRAS, a microscopic freeway simulation model. FHWA report No. DOT-FH-11-8502.

Ninth International Symposium on
Transportation and Traffic Theory
© 1984 VNU Science Press, pp. 113-132

DELAY AT A JUNCTION WHERE THERE IS PRIORITY FOR BUSES

BENJAMIN G. HEYDECKER*

Transport Studies Group, University College London, London, U.K.

ABSTRACT

Buses can experience considerable savings in delay at junctions where priority is provided for them by selective vehicle detection. However, these priority methods necessarily disrupt the normal operation of the traffic signals. This paper investigates the consequences of this for non-priority traffic. The methods developed here can be used to investigate proposed priority schemes in advance of their implementation.

Formulae are identified which are appropriate to estimate the mean overflow and delay in a stream where the durations of the effective red and green periods are random variables. The streams at a junction can be divided into 7 classes, each of which can be analysed individually to estimate the statistics of service which are required to use the delay formula. A direct analysis of 3 of these classes provides sufficient information to deduce the required statistics for any stream.

An embedded Markov chain is identified in the control process. This enables the analysis to proceed with the estimation of certain conditional expectations. One of the classes of stream is analysed in detail to estimate these quantities. This provides sufficient information to investigate a range of junctions. Example calculations are included for a junction based upon a real priority experiment.

*)   Present Address: Institute for Transport Studies,
                       University of Leeds

LIST OF SYMBOLS*

c   is the duration of an uninterrupted cycle,

m   is the number of stages in the normal sequence,

k   is the number of stages for which priority vehicles have right of way,

$\lambda_i c$ is the uninterrupted duration of stage i   ($1 \leq i \leq m$)

$\gamma_i c$ is the minimum green time for stage i ($1 \leq i \leq m$)

$\lambda_{i0} c$ is the duration of the transition period following stage i   in an uninterrupted cycle ($1 \leq i \leq m$),

$\lambda_{i1} c$ is the duration of the transition period following stage i   when stage 1 is recalled to give priority to a bus ($k < i \leq m$),

δc is the end lag less the start lag for an effective green period,

n_1 is the number of the stage which forms the first part of the effective green period for a stream,

n_2 is the number of the stage which forms the last part of the effective green period for a stream,

q   is the mean arrival rate of vehicles in a stream,

s   is the saturation departure rate for a stream,

Q_n is the number of vehicles in the queue at the start of the effective red period numbered n from some arbitrary origin,

r_n is the duration of the effective red period numbered n,

g_n is the duration of the effective green period which follows immediately on the effective red period numbered n,

d   is the mean delay incurred by vehicles in a stream,

D   is the total delay incurred in a stream during an effective red period and the sequent effective green period,

A   is the number of vehicular arrivals during an effective red period and the sequent effective green period,

τc is the length of time for which stage k is held in order to allow a bus to pass when it is given priority by extension,

β/c is the mean arrival rate of priority vehicles,

$p_0(\zeta)$ is the probability that no priority vehicles arrive at the detector in an interval of duration ζc,

υc, $\upsilon_2 c^2$ are the mean and variance of the time for which stage k exceeds $\lambda_i c$ because of priority extensions,

$\eta_i c$, $\eta_{2i} c^2$ are the conditional values of the first two moments of the duration of stage i given that it is interrupted by a priority recall of stage 1,

I   denotes the state "inhibition is in effect",

U   denotes the state "inhibition is not in effect",

$\underline{S}$ denotes the state space $\{U, I\}$,

$T_{PS}^{(n)}$ denotes the n-step transition probability from state P to state S ($n \geq 1$, $P, S \in \underline{S}$),

T_S denotes the equilibrium probability for state S ($S \in \underline{S}$).

*)   Greek letters are used throughout this paper to denote quantities which have their time components measured in units of c .

1.   INTRODUCTION

The provision of priority for buses is a readily available means to improve the quality and image of urban public transport services. Priority methods which use selective vehicle detection at signal-controlled road junctions have been used increasingly for this purpose. These methods work by enabling the signal controller to detect individual buses as they approach the junction in a stream of mixed traffic. This advance information is then used to alter the changing of the traffic signals accordingly. By being given priority in this way, buses can experience worthwhile reductions in average delay and in variability of delay.

Against the various benefits to bus passengers and operators of providing this kind of priority, the consequences for other traffic must be considered. The mean delay incurred by vehicles passing through a junction is the indicator most commonly used for this. This paper shows how analytical methods can be used to estimate changes in delay consequent upon the implementation of priority by selective vehicle detection. Using this analysis, the likely effects of a proposed priority scheme can be assessed in advance of implementation.

Amongst previous investigations of this topic, Gallivan, Young and Pierce (1980) noted that in many cases any changes in the mean delay incurred by non-priority vehicles are likely to be small compared with the standard deviation. Because of this, any experiment using field observations would need to be very large before it could lead to estimates of changes which are sufficiently accurate to be of any value. Indeed, in the case of the experiment reported by Gallivan et al, the least changes in delay which would have been stastically significant at the 95 per cent level were between 13 and 27 per cent of the mean delay before implementation: this is far larger than the changes which would be expected in most cases.

Vincent, Cooper and Wood (1978) used a simulation model to investigate the consequences of providing priority by selective vehicle detection at a simple cross-roads junction. Jacobson and Sheffi (1981) used numerical and approximate analytical methods for the same purpose. In neither case has the analysis been extended to more general junction configurations. However, Cottinet, Amy de la Breteque, Henry and Gabard (1980) have used a similar simulation model to assess the effects of providing priority in a network of simple junctions.

The main topic of this paper is to estimate various statistics of the durations of the effective red and green periods for the streams of traffic at a junction where priority is implemented by selective vehicle detection. This analysis leads to formulae to estimate the statistics which are required to estimate mean delay from an appropriate method. The method used considers the relationship between the effective green periods for the stream containing priority vehicles and those for other streams. At a simple cross-roads junction, there are just two possibilities: for each stream, either the effective red or the effective green periods are essentially concurrent with the effective green periods for the stream containing priority vehicles. Thus the statistics of the duration of these periods for one stream provide sufficient information to deduce those for all the others. As junctions with more complicated configurations are considered, the number of different possibilities increases. Rather than try to elaborate on an analysis of the simplest possible case, this paper presents an analysis which is appropriate to quite general junction configurations. When applied to more straightforward junctions, the formulae derived here can be simplified considerably.

2.   PRIORITY FOR BUSES BY SELECTIVE VEHICLE DETECTION

Many different rules have been devised to determine how the signal controller should respond to the detection of an approaching bus. The combination of rules investigated here is that recommended by the Department of Transport (1977) in Britain. While some variations on the individual rules and on their combination have been reported in the literature, this combination is the one used in the majority of theoretical and practical applications.

The action of the first priority rule is quite straightforward. This rule is effective towards the last part of the period during which the stream containing priority vehicles has right of way. If a bus arrives at the detector when the stream in which it travels is about to lose right of way, then the green indication can be held for a sufficient time to allow the bus to pass. This is called priority by extension.

The second priority rule is effective throughout the period during which the stream containing priority vehicles does not have right of way. If a bus is detected during this period, then right of way is given to the stream in which the bus travels as soon as is safely possible. Once this has been done, the control process continues as usual. This is called priority by recall.

In some cases, priority by recall can change the order of the stages in the control sequence. In order to limit any disruption to non-priority traffic caused by this, an additional rule is often used to prevent recalls from being granted too frequently. The rule to be analysed here ensures that each stream of traffic receives a full complement of green time between successive priority recalls. This is called the inhibition rule.

3. REPRESENTATION OF A JUNCTION

3.1 Introduction, definitions and assumptions

In this section, the representation of a junction which is used in the analysis is introduced, several terms are defined and various assumptions used in the analysis are set out. The assumptions fall into two distinct groups: those which are often adopted in analyses of delay at traffic signals and those which are specific to this analysis of priority by selective vehicle detection. The action of the priority rules to be investigated is then defined in precise terms.

As traffic approaches a signal-controlled junction, it divides itself between the lanes available according to the manoeuvres desired and those permitted in each lane. The lanes can be collected together into sets which receive identical signal indications and within which the vehicles behave like customers in a single queue independently of that in the others. The traffic using any such set of lanes is called a stream. The analysis developed here is appropriate to a single stream: it can be applied in turn to each of the streams at a junction.

The traffic signals controlling each stream display either red or green or some combination including amber to indicate that they are changing. A maximal interval during which all streams at a junction receive a constant indication of either red or green is called a stage. Between two consecutive stages occurs a transition period when some signal indications change. The durations of the stages are the main control variables at the junction. Associated with each stage is a minimum permissible duration, called the minimum green time. The durations of the transition periods and the minimum green times are governed by safety considerations.

From the point of view of the controller, time can be divided up into an alternating series of stages and transition periods. However, from the point of

view of a stream of traffic, time can be divided up into intervals when it has
right of way separated by intervals when it does not. Applying suitable end
corrections by way of start and end lags, traffic in each stream is supposed to
experience alternating periods when it can enter the junction at a constant
rate, called the <u>saturation departure rate</u>, and periods when it cannot enter the
junction at all. A period during which traffic can enter the junction in this
way is called an <u>effective green period</u> and the intervening periods are called
<u>effective red periods</u>.

The assumptions used in the analysis which follows can now be stated
conveniently. The first three assumptions are often adopted in the analysis of
normal signal-controlled junctions while the last two are specific to this
analysis of priority.

A1) Other than when priority is granted to buses, the signal-controller runs a
fixed sequence of stages, each of which has some fixed duration.

A2) Each stream of traffic has right of way during a single contiguous set of
stages during each repetition of the normal sequence of stages. If a
stream has right of way during any two consecutive stages, then it also
has right of way throughout the intervening transition period. Each stream
has right of way during at least one stage of the sequence but not during
all of them.

A3) Vehicles in each stream arrive according to the Poisson law, as do the
buses for which priority is provided.

A4) If priority is provided for buses in more than one stream, then all of the
streams in which they travel have right of way at the same time.
Furthermore, the signal-controller responds to the detection of a priority
vehicle in a manner which is independent of the stream in which it
travels.

A5) The time taken by an unimpeded priority vehicle to travel from the
selective vehicle detector to the stop-line is less than the duration of a
normal green period for the stream in which it travels.

Two conventions of notation are adopted for convenience. These are as follows.

C1) The first stage during which priority vehicles have right of way is stage
number 1 in the sequence.

C2) Any summation of the form $\sum_{n=N}^{M} X_n$ takes the value 0 whenever $M < N$,
whatever values X_n have.

Assumption A1 is equivalent to assuming fixed-time operation of the signal-
controller. Assumption A3 that vehicles arrive at random is particularly convenient to analyse: in the case of buses, it will model arrivals in urban areas
quite well, especially when buses on several routes are given priority at a
single junction. Assumption A5 ensures that any bus which is given priority by
recall will not require an extension as well.

3.2   The priority rules

The priority rules to be analysed can now be specified in precise terms. The
response of the extension and recall rules depends on the time after the start
of stage 1 at which the bus is detected as follows.

Arrival before τc before the end of stage k : no effect.

Arrival after this but before the time at which stage k would otherwise end:
stage k is extended until τc after the time at which the bus was detected.

Arrival during the transition period preceding a later stage or during the
minimum green time for such a stage: the duration of that stage is
reduced to the minimum which is permissible and a transition is then made
directly to stage 1.

Arrival during the remainder of a later stage: that stage is terminated
immediately and a transition is made directly to stage 1.

The inhibition rule to be analysed here overrides the recall priority rule but
has no effect on the extension rule. It can be specified as follows.

A recall may be granted only if the previous occurrence of stage 1
did not result from a priority recall.

Inhibition is said to be in effect from the start of an occurrence of stage 1
which resulted from a priority recall until the start of the next occurrence of
stage 1 . The way in which the controller responds to the detection of a bus
when the stream in which it travels does not have right of way depends upon
whether or not inhibition is in effect. This gives rise to different conditional distributions for the durations of the effective red and green periods
for the streams depending upon the inhibition state at the start of the period.
This state plays a crucial role in the analysis presented in Section 5.

3.3   Classification of streams

The streams at a junction can be classified according to the relationship
between the times at which they have right of way and those at which priority
vehicles do. If assumptions A1 and A2 are satisfied, then 7 classes of
stream can be identified so that streams in each class behave in a similar
manner. In the first instance, a different analysis is required for each class
of stream. However, some preliminary consideration shows how the behaviour of
streams in four of the classes can be deduced from that of streams in the other
three.

As a consequence of assumptions A1 and A2, in the absence of priority, the durations of the effective red and green periods for a stream can be calculated from the signal settings, the start and end lags and the numbers of the stages which run during the first and last parts of an effective green period, n_1 and n_2. The values of these last two numbers can be used to classify the streams according to the following scheme.

Class 1 $1 \leq n_1 \leq n_2 \leq k$

Class 2 $1 < n_1 \leq k < n_2 \leq m$

Class 3 $1 \leq n_2 < n_1 \leq k$

Class 4 $k < n_1 \leq n_2 \leq m$

Class 5 $1 \leq n_2 \leq k < n_1 \leq m$

Class 6 $k < n_2 < n_1 \leq m$

Class 7 $1 = n_1 \leq k < n_2 < m$

If assumptions A1 and A2 are satisfied, then these 7 classes are mutually exclusive and exhaustive. As a consequence of assumption A4 and convention C1, the streams containing priority vehicles are always in Class 1. At a simple cross-roads junction for which $k=1$ and $m=2$, only streams in Classes 1 and 4 can occur. As the number of stages in the sequence increases, so does the number of Classes to which a stream can belong.

Inspection of the classification scheme shows that if one stream has right of way at exactly those times when another does not, then the two streams will always belong to different Classes. For complementary pairs of streams like this, the statistics of the duration of the effective red and green periods for one can be deduced from those of the other. Using this observation, if formulae are available for the statistics for any stream in Classes 1, 2 and 4, then they can be deduced for any stream. This follows because for any stream, either it is in one of these three Classes, in which case formulae are available, or a notional complementary stream can be constructed which will be in one of these Classes. In the latter case, the statistics can be calculated for the notional stream and those for the real stream can then be deduced. The worked example given in section 6 illustrates the use of this classification scheme and of complementary streams.

4.   AN APPROPRIATE DELAY FORMULA

Most of the formulae to estimate mean delay which appear in the literature (see, for example, McNeil and Weiss, 1974) are appropriate to traffic signals where the durations of the effective red and green periods for the streams are constant. When responsive priority is provided for buses, the durations of the stages, and hence of the effective red and green periods, are random variables. In this section a brief review is given of methods which have been devised to estimate mean delay in cases such as this. A suitable formula is identified and the statistics which are required for its use are listed.

McNeil (1968) gave formulae for $E(D)$ and $E(A)$ which are appropriate to fixed-time traffic signals. The formula for $E(D)$ was conditional on the value of the mean overflow, $E(Q_n)$. An estimate was derived for this quantity using a bulk-service approximation. The value of the mean delay was then estimated from the fundamental relationship

$$d = \frac{E(D)}{E(A)} \tag{1}$$

Griffiths (1981) developed a bulk-service model of a queue of traffic with intermittent service to investigate the overflow at uncontrolled pedestrian crossings. However, this analysis was specific to the problem and the method used depended upon the independence of the durations of all effective red and green periods.

Jacobson and Sheffi (1981) used McNeil's formulae to estimate the conditional mean values of D and A given particular values of r_n and g_n. They then used numerical methods to take expectations of the ratio of these quantities and interpreted the result as the mean delay at traffic signals where the durations of the effective red and green periods are random variables. This estimate for the mean delay can be expressed as

$$d = E\left\{ \frac{E(D|r_n, g_n)}{E(A|r_n, g_n)} \right\} \tag{2}$$

Equation (2) can be interpreted as the mean delay at fixed-time traffic signals where there is some uncertainty as to the exact values of r_n and g_n. When r_n and g_n are random variables, eq.(1) remains appropriate when expectation is taken over the distributions of all random variables. Heydecker (1982) proceeded like McNeil (1968) but without using the assumption of fixed durations for the effective red and green periods. Using eq.(1), a formula was derived which in the present context reduces to

$$d = \frac{s\{E(r^2) + 2[Cov(Q_n,r_n)+E(Q_n)E(r_n)]/q - qE(r_n)/[s(s-q)]\}}{2(s-q)[E(r_n) + E(g_n)]} \qquad (3)$$

An estimate for $E(Q_n)$ was derived after Miller (1963) under the same assumptions. This reduces to

$$E(Q_n) = Max\{0, \{[q^2+sq(E(r)-E(g))/E(g)]Var(g)-2(s-q)[qCov(r_n,g_n) + \\ + Cov(Q_n,g_n)]+2qCov(Q_n,r_n)+q^2Var(r)+[2qE(r)+(2q-s)E(g)]\} / \\ / 2[(s-q)E(g) - qE(r)] \} \qquad (4)$$

Finally, the covariances of Q_n with r_n and g_n were estimated from the formula

$$Cov(Q_n,X) = \sum_{i=1}^{\infty} [q\,Cov(r_{n-i},X) - (s-q)Cov(g_{n-i},X)][P(q_n\neq 0)]^i \qquad (5)$$
$$(X = r_n,g_n)$$

where the probability of a non-zero overflow is approximated from the steady-state formula given by Miller (1968):

$$P(Q_n\neq 0) = exp\{-1.58\sqrt{sE(g)}\,[(s-q)E(g)-qE(r)]/[q(E(g)+E(r))]\} \qquad (6)$$

These formulae will be used here to estimate the mean delay incurred by non-priority vehicles at a junction where priority is provided for buses by selective vehicle detection.

In order to make use of eqs(3-6), various statistics of the durations of the effective red and green periods are required. These are $E(r)$, $E(r^2)$, $E(g)$, $E(g^2)$, $Cov(r_{n-i},g_n)$ ($\forall i$), $Cov(r_{n-i},r_n)$ ($i\geq 1$) and $Cov(g_{n-i},g_n)$ ($i\geq 1$). Appropriate formulae are derived for these statistics in the next section.

5. THE STATISTICS OF SERVICE

5.1 Introduction and preliminaries

In this section, formulae are derived for the statistics of the durations of the effective red and green periods which are required to estimate delay from eqs(3-6). The formulae given here are specific to the particular combination of priority rules analysed: other combinations can be analysed in a similar manner. Formulae are derived only for streams in Class 4. The analysis of an arbitrary stream in this Class is rather more intricate than that for streams in Classes 1 and 2 because of the possibility that stage 1 is recalled before the start of an effective green period. If this happens, then two occurrences of stage 1 will be included in a single effective red period for these streams. Classes 1 and 2 can be analysed in a similar but rather more straightforward manner.

Before proceeding to consider effective red and green periods, some preliminary results are given in this sub-section. In the next sub-section, an embedded Markov chain is identified which is then used to reduce the extent of the analysis required. The transition and equilibrium probabilities for the chain are found. Expressions are given in terms of these probabilities which enable the required statistics of the durations of the effective red and green periods to be found from certain conditional expectations. In sub-section 5.3 explicit formulae are derived for these conditional expectations, thus completing the analysis.

Because of the assumption of Poisson arrivals for priority vehicles,

$$p_0(\zeta) = \exp(-\beta\zeta) \tag{7}$$

Furthermore, the values of υ and υ_2 can be found from formulae given by Tanner (1951) and Allsop (1977) to be

$$\left. \begin{array}{l} \upsilon = [\exp(\beta\tau) - (1+\beta\tau)] \ / \ \beta \\[2mm] \upsilon_2 = [\exp(2\beta\tau) - 2\beta\exp(\beta\tau) - 1] \ / \ \beta^2 \end{array} \right\} \tag{8}$$

Finally, given that stage i is interrupted by a priority recall of stage 1, a priority vehicle is detected before the minimum green time for stage i has elapsed with conditional probability $[1 - p_0(\lambda_{i-1,0}+\gamma_i)] \ / \ [1 - p_0(\lambda_{i-1,0}+\lambda_i)]$; in this case the duration the stage i is exactly $\gamma_i c$. Otherwise the duration of the stage has the truncated negative exponential distribution on $[\gamma_i c, \lambda_i c]$. Thus

$$\left. \begin{array}{l} n_i = \gamma_i + \dfrac{\{1 - [1+\beta(\lambda_i-\gamma_i)]\exp[-\beta(\lambda_i-\gamma_i)]\}}{\{1 - \exp[-\beta(\lambda_i-\gamma_i)]\}} \qquad (k<i\leq m) \\[6mm] n_{2i} = \gamma_i^2 + 2\gamma_i \dfrac{\{1 - [1+\beta(\lambda_i-\gamma_i)]\exp[-\beta(\lambda_i-\gamma_i)]\}}{\{1 - \exp[-\beta(\lambda_i-\gamma_i)]\}} + \\[6mm] \qquad + \dfrac{2\{1 - [1+\beta(\lambda_i-\gamma_i) + (\beta(\lambda_i-\gamma_i))^2/2)]\exp[-\beta(\lambda_i-\gamma_i)]\}}{\beta^2\{1 - \exp[-\beta(\lambda_i-\gamma_i)]\}} \qquad (k<i\leq m) \end{array} \right\} \tag{9}$$

5.2   The Markov chain

There are two possibilities for the inhibition state of the controller when
stage n_1 is called: either inhibition is in effect or it is not. The state
of the controller when stage n_1 is next called is determined by its current
state and by arrivals of priority vehicles at the detector during some parts of
the interim period. Since the arrivals of priority vehicles form a Poisson
process, the arrivals during this period are independent of those at all other
times. Thus the inhibition state of the controller at these instants forms a
Markov chain in a space with two states.

During some transitions into state I , stage 1 will be called twice. Thus
there is not always a one to one correspondence between transitions in the
Markov chain and occurrences of any particular stage. However, there is by
definition a one to one correspondence between these transitions and
occurrences of stage n_1 . As a result of this, there are similar
correspondences between transitions in the chain and each of effective red and
green periods for the stream under consideration. The transition and
equilibrium probabilities for the chain can be derived as follows.

Suppose that inhibition is not in effect when stage n_1 is called. Inhibition
will not be in effect when stage n_1 is next called provided that no priority
vehicles arrive at the detector between the time at which stage n_1 was called
and the time at which stage m completes its normal running time and that none
arrive between the next call of stage k+1 and the following call of stage n_1.
Otherwise a priority recall will be granted before stage n_1 is next called and
inhibition will be in effect throughout a complete sequence of stages, including
stage n_1 .

Now suppose that inhibition is in effect when stage n_1 is called. This will
prevent any priority recalls of stage 1 from being granted before the next
normal call of stage 1. Thus inhibition will not be in effect when stage n_1
is next called provided that no priority vehicles arrive at the detector between
the next call of stage k+1 and the following call of stage n_1 . Otherwise
inhibition will be in effect at the next call of stage n_1 . Thus the 1-step
transition probabilities (abbreviated to T_{PS}) are given by

$$
\left.
\begin{aligned}
T_{UU} &= P_0\left(\sum_{i=1}^{m-1}\lambda_{i0} + \sum_{i=k}^{m}\lambda_i\right) \\
T_{IU} &= P_0\left(\sum_{i=k}^{n-2}\lambda_{i0} + \sum_{i=k+1}^{n-1}\lambda_i\right) \\
T_{SI} &= 1 - T_{SU} \qquad (S \in \underline{S})
\end{aligned}
\right\}
\qquad (10)
$$

The n-step transition probabilities can be calculated recursively from eq.(10) and the relationship

$$T_{S_1 S_2}^{(n+1)} = \sum_{S \epsilon \underline{S}} T_{S_1 S} T_{SS_2}^{(n)} \qquad (S_1, S_2 \epsilon \underline{S}) \tag{11}$$

The equilibrium probabilities for the chain can be found in the usual manner by solving the two equations $T_U = \sum_{S \epsilon \underline{S}} T_S T_{SU}$ and $\sum_{S \epsilon \underline{S}} T_S = 1$ for T_U and T_I. Thus

$$\left.\begin{aligned} T_U &= T_{IU}[1 + T_{IU} - T_{UU}]^{-1} \\ T_I &= T_{UI}[1 + T_{UI} - T_{II}]^{-1} \end{aligned}\right\} \tag{12}$$

Consider the conditional expectations $E(X|S_1 S_2)$ for each X in the set $\{r_n, r_n^2, g_n, g_n^2, r_n g_n, g_{n-1} r_n\}$ where the conditioning event $S_1 S_2$ denotes a transition in the Markov chain from state S_1 occurring at the call of stage n_1 immediately preceeding the start of the interval under consideration to state S_2 at the next call of stage n_1. These conditional expectations can be used together with the transition and equilibrium probabilities given by eqs(10-2) to calculate the statistics required for the use of eqs(3-6) form the following formulae.

$$\left.\begin{aligned} E(X|S_1) &= \sum_{S \epsilon \underline{S}} T_{S_1 S} E(X|S_1 S) \qquad (S_1 \epsilon \underline{S}) \\ E(X) &= \sum_{S \epsilon \underline{S}} T_S E(X|S) \\ E(X_{N-n} Y_N) &= \sum_{S_1 S_2 S_3 \epsilon \underline{S}} T_{S_1} T_{S_1 S_2} T_{S_2 S_3}^{(n-1)} E(X|S_1 S_2) E(Y|S_3) \\ &\quad (X=r, \; Y=r, \; n{\geq}1), \; (X=g, \; Y=g, \; n{\geq}1), \; (X=g, \; Y=r, \; \forall n) \end{aligned}\right\} \tag{13}$$

5.3 The conditional expectations

In this sub-section, formulae are given for the conditional expectations used in eq.(13). These formulae are derived by considering in turn the consequences of each stage being interrupted by a priority recall of stage 1.

Consider a transition of the Markov chain from state U to state I. This arises as a result of a priority recall of stage 1. The recall could be granted during stage n $(n_1{\leq}n{\leq}m)$ or, if no such recall is granted, during stage n $(k{<}n{<}n_1)$ of the next sequence. The conditional probability of each of these events is given by P_n $(k{<}n{\leq}m)$ in the following formulae, where P_n^a denotes the conditional probability that stage n is interrupted by a priority recall of stage 1 given that inhibition is not in effect.

$$P_n^a = P_0\left(\sum_{i=k}^{n-2}\lambda_{10} + \sum_{i=k+1}^{n-1}\lambda_i\right)\left[1 - P_0(\lambda_{n-1,0} + \lambda_n)\right] \qquad (k<n\leq m)$$

$$P_0 = P_0\left(\sum_{i=k}^{m-1}\lambda_{10} + \sum_{i=k+1}^{m}\lambda_i\right) = 1 - \sum_{i=k+1}^{m}P_n^a$$

$$P_n = \begin{cases} P_n^a / \left(\sum_{i=n_1}^{m}P_i^a + P_0\sum_{i=k+1}^{n-1}P_i^a\right) & (n_1\leq n\leq m) \\ P_0 P_n^a / \left(\sum_{i=n_1}^{m}P_i^a + P_0\sum_{i=k+1}^{n-1}P_i^a\right) & (k<n<n_1) \end{cases} \qquad (14)$$

If the recall of stage 1 is granted while stage n $(n_1\leq n\leq n_2)$ is running, then the effective green period for the stream will be truncated. The conditional values of the first two moments of the duration of the effective green period in this case are given by

$$\begin{aligned} E(g|n) &= c\,(\sigma_n + \eta_n) \\ \text{and}\quad E(g^2|n) &= c^2(\sigma_n^2 + 2\sigma_n\eta_n + \eta_{2n}) \end{aligned} \qquad (n_1\leq n\leq n_2) \qquad (15)$$

$$\text{where}\quad \sigma_n = \sum_{i=n_1}^{n-1}(\lambda_i + \lambda_{10}) + \delta$$

If the recall of stage 1 is granted while any of the other stages is running, then the duration of the effective green period will be exactly

$c\left(\sum_{i=n_1}^{n_2}\lambda_i + \sum_{i=n_1}^{n_2-1}\lambda_{10} + \delta\right)$. Using this with eqs(15) and the conditioning probabilities of eq.(14) gives

$$E(g|UI) = c\left[\sum_{n=n_1}^{n_2}P_n(\sigma_n+\eta_n) + \left(\sum_{i=n_1+1}^{m}P_n + \sum_{n=k+1}^{n_1-1}P_n\right)\sigma_m\right]$$

$$E(g^2|UI) = c^2\left[\sum_{n=n_1}^{n_2}P_n(\sigma_n^2+2\sigma_n\eta_n+\eta_{2n}) + \left(\sum_{n=n_2+1}^{m}P_n + \sum_{n=k+1}^{n_1-1}P_n\right)\sigma_m^2\right] \qquad (16)$$

$$\text{where}\quad \sigma_n = \begin{cases} \sum_{i=n_1}^{n}(\lambda_i + \lambda_{10}) + \delta & (n_1\leq n\leq n_2) \\ \sum_{i=n_1}^{n_2}\lambda_i + \sum_{i=n_1}^{n-1}\lambda_{10} + \delta & (n_2<n\leq m) \end{cases}$$

If the recall of stage 1 is granted while stage n $(n_1\leq n\leq n_2)$ is running, then the mean duration of the effective red period thus initiated is equal to

$c\left[\lambda_{n1} + \sum_{i=1}^{n_1-1}(\lambda_i + \lambda_{10}) + \upsilon - \delta\right]$. If the recall is granted while stage n

$(n_2<n\leq m)$ is running, then the conditional values of the first two moments of the duration of the effective red period during which the recall is granted are given by

$$\begin{aligned} E(r|n) &= c\,(\rho_n + \eta_n) \\ E(r^2|n) &= c^2(\rho_n^2 + 2\rho_n\eta_n + \eta_{2n}) \end{aligned} \qquad (17)$$

$$\text{where}\quad \rho_n = \sum_{i=n_2}^{n-1}\lambda_{10} + \sum_{i=n_1+1}^{n-1}\lambda_i + \lambda_{n1} + \sum_{i=1}^{n_1-1}(\lambda_i + \lambda_{10}) + \upsilon - \delta \qquad (n_2<n\leq m)$$

If the recall is granted while stage n $(k < n < n_1)$ is running, then the effective red period will continue through a second occurrence of stage 1. Thus the moments of r can be calculated from eqs(17) together with

$$\rho_n = \sum_{i=n_2}^{m} \lambda_{i0} + \sum_{i=n_2+1}^{m} \lambda_i + \sum_{i=1}^{n-1} (\lambda_i + \lambda_{10}) + \lambda_{n1} + \sum_{i=1}^{n_1-1} (\lambda_i + \lambda_{10}) + \upsilon - \delta \quad (k < n < n_1).$$

Combining these conditional expectations gives

$$E(r|UI) = c \left[\sum_{n=n_1}^{n_2} P_n \rho_n + \sum_{n=n_1+1}^{m} P_n(\rho_n + \eta_n) + \sum_{n=k+1}^{n_1-1} P_n(\rho_n + \eta_n) \right]$$

$$E(r^2|UI) = c^2 \left[\sum_{n=n_1}^{n_2} P_n(\rho_n^2 + \eta_2) + \sum_{n=n_2+1}^{m} P_n(\rho_n^2 + 2\rho_n \eta_n + \eta_{2n} + \upsilon_2) + \right.$$

$$\left. + \sum_{n=k+1}^{n_1-1} P_n(\rho_n^2 + 2\rho_n \eta_n + \eta_{2n} + 2\upsilon_2) \right]$$

where

$$\rho_n = \begin{cases} \sum_{i=n_2}^{m} \lambda_{i0} + \sum_{i=n_2+1}^{m} \lambda_i + \sum_{i=1}^{n-1} (\lambda_i + \lambda_{10}) + \lambda_{n1} + \sum_{i=1}^{n_1-1} (\lambda_i + \lambda_{10}) + 2\upsilon - \delta & (k < n \le n_1) \\ \lambda_{n1} + \sum_{i=1}^{n-1}(\lambda_i + \lambda_{10}) + \upsilon - \delta & (n_1 \le n \le n_2) \\ \sum_{i=n_2}^{n-1} \lambda_{i0} + \sum_{i=n_2+1}^{n-1} \lambda_i + \lambda_{n1} + \sum_{i=1}^{n_1-1} (\lambda_i + \lambda_{10}) + \upsilon - \delta & (n_2 < n \le m) \end{cases} \quad (18)$$

Consideration of the expected value of the product of the durations of the effective green period with that of the next effective red period in each case gives

$$E(g_{N-1} r_N | UI) = c^2 \left\{ \sum_{n=n_1}^{n_2} P_n(\sigma_n + \eta_n)\rho_n + \right.$$

$$\left. + \sigma_m \left[\sum_{n=n_2+1}^{m} P_n(\rho_n + \eta_n) + \sum_{n=k+1}^{n_1-1} P_n(\rho_n + \eta_n) \right] \right\} \quad (19)$$

where σ_n and ρ_n $(k < n \le m)$ are as given in eqs(16) and (18).

Applying similar arguments to the other three state transitions yields the formulae

$$E(g|UU) = c\,\sigma_m$$

$$E(g^2|UU) = c^2 \sigma_m^2$$

$$\left. \begin{aligned} E(g|IS) &= c\,\sigma_m \\ E(g^2|IS) &= c^2 \sigma_m^2 \end{aligned} \right\} \quad (S \in \underline{S})$$

$$E(r|II) = c \sum_{n=k+1}^{n_1-1} P_n^a (\rho_n + \eta_n) \Big/ \sum_{i=k+1}^{n_1-1} P_i^a$$

$$E(r^2|II) = c^2 \sum_{n=k+1}^{n_1-1} P_n^a (\rho_n^2 + 2\rho_n \eta_n + \eta_{2n} + 2\upsilon_2) \Big/ \sum_{i=k+1}^{n_1-1} P_i^a$$

$$\left. \right\} \quad (20)$$

$$E(r|SU) = c\,\rho_0$$

$$E(r^2|SU) = c^2(\rho_0^2 + \upsilon_2)$$

$$\left.\vphantom{\begin{array}{c}1\\1\\1\\1\end{array}}\right\}\quad (20)$$

$$\text{where}\quad \rho_0 = \sum_{i=n_2}^{m} \lambda_{10} + \sum_{i=n_2}^{m} \lambda_1 + \sum_{i=1}^{n_1-1} (\lambda_1 + \lambda_{10}) + \upsilon - \delta$$

Finally, the conditional expectation of the product of the duration of an effective red period with that of the next effective green period is just the product of the appropriate conditional expected values. Thus

$$E(r_n g_n | S_1 S_2) = E(r|S_1 S_2)\, E(g|S_1 S_2) \qquad (S_1,\, S_2 \,\epsilon\, \underline{S}) \qquad (21)$$

which completes the analysis of an arbitrary stream in class 4.

6.   EXAMPLE CALCULATIONS

In order to illustrate the use of the analysis presented here, it is applied to an example junction which is based upon a real priority experiment in Derby, England (Department of the Environment (DoE), 1973). A diagram of the junction is given in Fig. 1. Buses enter the junction from the west and leave to the south. They have right of way only during the first of the three stages.

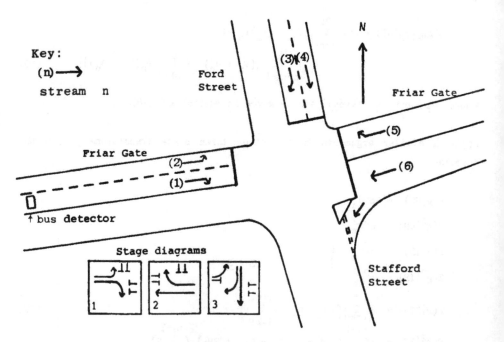

Figure 1.   The experimental junction at Derby.

Table 1. Stage durations used in the example (seconds)

Stage number	minimum green time	uninterrupted stage durations		transition period following
		original	adjusted	
1	7	16	15.0	5
2	7	24	24.6	5
3	7	28	33.9	5
	$\tau c = 8.00$			

Table 2. Stream data used in the example

Stream number	Arrival rate (veh/h)	Saturation departure rate (veh/h)	Class of stream	Class of complementary stream
1	481	2778	1	4
2	1481	3462	5	4
3	853	2941	4	2
4	808	3462	4	2
5	623	2830	4	5
6	941	3529	4	5

The original stage durations given in Table 1 are those published in the DoE report. Details of the arrival and departure rates were not published: those in Table 2 were devised to be consistent, in the absence of priority, with the signal-settings used. The Class to which each stream belongs is given in Table 2: streams 2-6 all belong to Class 4, so a direct analysis is possible for them. Streams 1 and 2 belong to Classes 1 and 5 respectively: the notional streams to which they are complementary both belong to Class 4. Thus each of the 6 streams is amenable, either directly or indirectly, to the analysis given in Section 5.

Table 3. Statistics of service when priority is implemented with the original stage durations (mean bus arrival rate = 25 per hour)

Stream		1	2	3	4	5	6
E(r)	(s)	57.98	31.34	60.11	60.11	51.87	51.87
Var(r)	(s^2)	258.8	21.9	195.7	195.7	142.4	142.4
E(g)	(s)	17.23	43.87	26.81	26.81	23.24	23.34
Var(g)	(s^2)	1.3	147.4	37.6	37.6	21.9	21.9
$Cov(r_n, g_n)$	(s^2)	0.0	44.4	13.6	13.6	-10.0	-10.0
$Cov(Q_n, g_n)$	(veh.s)	0.0	0.9	2.8	0.5	0.0	-0.2
$Cov(Q_n, r_n)$	(veh.s)	-1.3	0.1	7.2	1.3	-2.1	-12.1

Values of the statistics of service were calculated according to the methods given in Section 5 for a mean bus arrival rate of 25 per hour. Those in Table 3 result from the use of the original stage durations. The adjusted stage durations in Table 1 were calculated according to a method devised by Allsop (1977) and developed by Heydecker (1983). They are calculated to compensate for any changes in capacity for each stream which result from the implementation of priority. Estimates of the mean delays are given in Table 4 for each set of stage durations.

Table 4a. Estimates of mean overflow and delay when priority is implemented with the original stage durations and changes compared to operation without priority.

	Stream	1	2	3	4	5	6
$E(Q_n)$	(veh)	1.74	0.34	10.12	2.37	1.43	5.00
Change in $E(Q_n)$	(veh)	-0.49	-0.38	8.43	1.86	0.57	1.67
d	(s)	40.97	11.99	73.01	38.06	31.08	42.65
Change in d	(s)	-6.71	-0.43	42.17	13.37	0.81	2.95

Table 4b. Estimates of mean overflow and delay when priority is implemented with the adjusted stage durations and changes compared to operation without priority.

	Stream	1	2	3	4	5	6
$E(Q_n)$	(veh)	3.53	0.11	3.26	1.37	1.83	6.64
Change in $E(Q_n)$	(veh)	1.30	-0.61	1.59	0.86	0.97	3.28
d	(s)	57.64	11.43	41.38	31.80	35.37	50.77
Change in d	(s)	9.96	-0.99	10.54	7.11	5.10	11.07

When the original stage durations are used, the mean time saving for each bus, estimated from methods similar to those of DoE is 35.4 seconds. This corresponds to a reduction in the rate of delay of .246 bus hours per hour. Against this saving, there is an increase of 12.57 vehicle hours per hour for non-priority traffic. Thus in order for the priority scheme to give rise to a net reduction in passenger delay, the ratio of the mean occupancy of priority vehicles to that of non-priority ones would need to exceed 51.1 .

Inspection of the changes in mean delay in Table 4a shows that the increase in delay to non-priority traffic is dominated by that in streams 3 and 4. This is as would be expected since these streams have right of way only during stage 3 which is sometimes omitted from the sequence. The adjusted stage durations given in Table 1 allow rather more time for stage 3 than do the original ones. As a result of this, the changes in mean delay given in Table 4b are more evenly distributed between the streams than are those in table 4a.

The change in rate of delay for non-priority traffic consequent upon the implementation of priority with the adjusted stage durations is 8.63 vehicle hours per hour. The estimated mean time saving for each bus in this case is 31.4 seconds, corresponding to a reduction in the rate of delay of .218 bus hours per hour. Accordingly, when priority is implemented with these stage durations, the ratio of the mean vehicle occupancies required for a net saving in passenger time is 39.7 . In this case then, buses benefit less than they did before the stage durations were adjusted. However, the total increase in delay incurred by non-priority traffic is smaller and the increases are distributed between streams more equitably. Furthermore, a net reduction in total passenger delay is achieved at a lower mean bus occupancy.

7.   SUMMARY AND CONCLUSIONS
When priority is provided for buses by selective vehicle detection, the durations of the stages in the sequence are random variables. This in turn causes variability in the durations of the effective red and green periods experienced by the streams of traffic. A number of statistics of the durations of these periods are required in order to estimate the mean delay incurred by non-priority traffic.

The consequences for any particular stream of traffic of granting priority to a bus depends on the relationship between the parts of an uninterrupted sequence during which it has right of way and those at which priority vehicles do. The streams at a junction can be classified accordingly. At junctions with fairly general configurations, 7 different classes of streams can be identified, each of which can be analysed separately. However, a direct analysis is required only for 3 of these Classes: the statistics for streams in the other 4 can be deduced from this. Indeed, a variety of junctions, including simple crossroads, can be treated using a direct analysis of just one of these 3 classes.

An embedded Markov chain can be identified in the control process. This enables the statistics required for an analysis of delay to be deduced from certain conditional expectations. These conditional expectations can be found from a detailed analysis of the consequences of the detection of a priority vehicle during each stage of the sequence.

The resulting estimates of overflow and delay in each stream can be used to assess in detail the effects of a proposed priority scheme in advance of implementation. As well as changes in the total rate of delay at the junction, the changes in individual streams can be examined. This enables the evaluation of possible variants on the design of the scheme in a manner which might otherwise be impractical.

ACKNOWLEDGEMENTS

This paper arose from work financed by the British Science and Engineering Research Council. The author wishes to acknowledge the discussions and helpful suggestions of Professor Richard Allsop and of an anonymous referee.

REFERENCES

Allsop, R.E. (1977) Priority for buses at signal-controlled junctions: some implications for signal timings. Proceedings of the 7th International Symposium on Transportation and Traffic Theory, eds T. Sasaki and T. Yamaoka. (The Institute of Systems Science Research, Kyoto), 247-70.
Cottinet, M., L. Amy de la Breteque, J.J. Henry, F. Gabard (1980) Assessment by observation and simulation studies of the interest of different methods of bus preemption at traffic lights. Proceedings of the International Symposium on Traffic Systems, Berkeley, California, August 1979. Vol 2a, 92-105.
Department of the Environment (1973) Bus detection: bus priorities at traffic control signals. Bus Demonstration Project Summary Report No. 1. (London)
Department of Transport (1977) Bus priority at traffic signals using selective vehicle detection. Technical Memorandum H2/77. (HMSO, London).
Gallivan, S., C.P. Young, J.R. Peirce (1980) Bus priority in a network of traffic signals. Paper presented to the ATEC Conference, Paris, April 1980. (Transport and Road Research Laboratory, Crowthorne, unpublished)
Griffiths, J.D. (1981) A mathematical model of a nonsignalized pedestrian crossing. Transportation Science, 15(3), 222-32.
Heydecker, B.G. (1982) A delay formula for traffic signals with variable red and green times. Proceedings of the 14th Conference of the Universities Transport Study Group, Bristol University (Unpublished).
Heydecker, B.G. (1983) Capacity at a signal-controlled junction where there is priority for buses. Transportation Research, 17B(5), 341-57.
Jacobson, J., Y. Sheffi (1981) Analytical model of traffic delays under bus signal preemption: theory and applications. Transportation Research, 15B(2), 127-38.
McNeil, D.R. (1968) A solution to the fixed-cycle traffic light problem for compound Poisson arrivals. Journal of Applied Probability, 5(3), 624-35.
McNeil, D.R., G.H. Weiss (1974) Delay problems for isolated intersections. In: Traffic Science, ed. D.C. Gazis. (Wiley, London), 109-74.
Miller, A.J. (1963) Settings for fixed-cycle traffic signals. Operational Research Quarterly, 14(4), 373-86.
Miller, A.J. (1968) Australian road capacity guide: provisional introduction and signalized intersections. Australian Road Research Board Bulletin No. 4.
Tanner, J.C. (1951) The delay to pedestrians crossing a road. Biometrika, 38(3/4), 383-92.

Ninth International Symposium on
Transportation and Traffic Theory
© 1984 VNU Science Press, pp. 133–154

STUDY AND NUMERICAL MODELLING OF NON-STATIONARY TRAFFIC FLOW DEMANDS AT SIGNALIZED INTERSECTIONS

JANUSZ CHODUR and MARIAN TRACZ

Cracow Technical University, Cracow, Poland

ABSTRACT

Empirical studies of traffic flow variations and theoretical analyses
have been conducted in order to derive characteristics of traffic flow
variability, particularly in relation to peak periods. The length of
flow counting interval, t_c, as well as profiles of flow rate variations
for peak periods have been determined. The flow profiles are based on
a time scale divided into units t_p, within which demand flow is assumed
as stationary. A computer simulation program has been developed for
modelling of traffic flow with varying flow rates accordingly to an
assumed curve. This program simulates the vehicle arrival process at
an approach to fixed-time traffic signals and calculates delays, stops,
queue lengths, saturation ratios and their probabilistic statistics,
which characterize level of service in the considered period of time.
The primary application of this program lies in a simulation study of
the effect of flow rate variability on the effectiveness of signal
settings and estimation of traffic conditions during peak periods.

1.   INTRODUCTION

A large fraction of the total delay and stops, as well as accidents
experienced by traffic, arises from limitations and interruptions of
flow at single level junctions. Such junctions will continue to form
essential elements of the street network. Both delays and accidents
can be considerably reduced, as has been shown in the recent research
(Tracz, 1979) by the efficient design of junction geometry and choice
of traffic control methods, and also by the improvement of signal
settings algorithms. This improvement is mostly related to elimination
of the simplified assumptions used in these methods and algorithms,
which do not adequately represent the real effects of junctions. One
of major simplified assumptions to be found both in mathematical
models of traffic and in capacity estimation methods and one used also
in signal settings design among others by Webster (1966) and Miller
(1963), is considering demand flow and its composition as a time-
-stationary stochastic process within one hour, which is the basic
period for signal settings and capacity calculations. However, traffic
observations show that traffic arrival rates and its composition are
variable in time; they vary from one period of day to another and also
within one hour.

In order to identify a range of non-stationary effect on the efficiency
of traffic signal control, the research programme was undertaken
(Chodur, 1982) to investigate first of all the variability of flow
rates - particularly during peak periods, with quantitative
characteristics of variations in time, and then to analyse the effect
of the determined flow variability on the efficiency of a single
approach. As the next step, the comparative analysis of some well-known
formulae for junction delays were also undertaken. In the present
paper, empirical studies of the vehicle arrival process and its
characteristics, as well as numerical modelling, including flow
variations, have been emphasized.

2. CHARACTERISTIC OF TRAFFIC FLOW VARIATIONS

The research programme developed by the authors was designed primarily
to determine the characteristics and range of short-term flow
variations. Observational studies have been conducted at 19 junctions.
Detailed analyses of peak period data (usually from 6.00 to 9.00 and
13.30 to 16.30) have been done. The basic problems which have been
investigated on the basis of these empirical data were:

(i)   determination of an appropriate interval t_r for modelling of
        time-dependent traffic flow,

(ii)   development of the traffic flow variation characteristics based
        on various mathematical statistics,

(iii)   determination of the typical demand flow profiles for peak
        periods,

(iv)   evaluation of the stability of the flow variation trend.

2.1. Statistics describing flow variability

Figure 1a shows a time-series of flow rates $\{q(t)\}_{t=1,..,n}$, i.e. rates
of flow corresponding to counting intervals t_r, and also it illustrates
statistical parameters related to the time-series. The probability
approach allowed the set of vehicles observed within interval TO to be
characterized by known statistical variables which are defined in
Table 1. These statistics were used for the comparative evaluation of
various time-series, for selecting the time interval t_r appropriate for
identification of flow variability and for evaluation of the
sensitivity of flow rate variations in time in the computer simulation
model.

Relative measures of flow variations in the observation period TO such
as: z, z_n, ε, are particularly useful in the comparative analysis of
various traffic flows, as well as in observing changes in short-term
variations during peak periods. Values of the statistics given in
Table 1 are calculated by computer program EMPIR, on the basis of time-
-series in numerical form. This program draws a graph of flow time-
series and repeats the sequence of calculations for different lengths

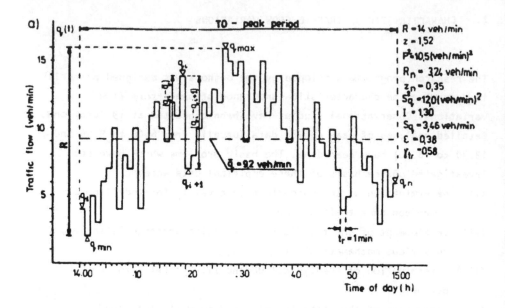

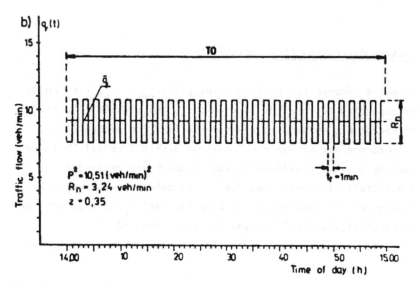

Fig. 1.   Part (a): Example of traffic flow time-series $\{q(t)\}_{t=1,\ldots,n}$, part (b): Theoretical normalized time-series corresponding to flow rate notation in time - in relation to statistic P^2

of time interval t_r; t_r being a multiple of an interval used in the observations. Figure 2 shows the flow profile during one day at one of the approaches to a junction, as well as its variability characteristic represented by statistics z_n and ε. It can be seen that, in general,

Table 1. Statistical parameters (statistics) used in paper

Statistic	Formula	Statistic	Formula
Mean number of vehicles in interval t_r	$\bar{q} = \dfrac{t_r}{TO} \cdot \displaystyle\sum_{i=1}^{n} q_i$	Relative intensity	$RI_i = \dfrac{q_i}{\bar{q}}$
Range	$R = q_{max} - q_{min}$	Relative range of variations	$z = \dfrac{R}{\bar{q}}$
Mean quadratic flow increment[1]	$p^2 = \dfrac{1}{n-1} \cdot \displaystyle\sum_{i=1}^{n-1} (q_i - q_{i+1})^2$		
Range of normalized variations[2]	$R_n = \sqrt{p^2}$	Relative range of normalized variations	$z_n = \dfrac{R_n}{\bar{q}}$
Variance	$S_q^2 = \dfrac{1}{n} \cdot \displaystyle\sum_{i=1}^{n} (q_i - \bar{q})^2$	I-ratio[3]	$I = \dfrac{S_q^2}{\bar{q}}$
Standard deviation	$S_q = \sqrt{S_q^2}$	Variation coefficient	$\varepsilon = \dfrac{S_q}{\bar{q}}$
Peak-hour factor	$Y_{t_r} = \dfrac{t_r}{TO \cdot q_{max}} \cdot \displaystyle\sum_{i=1}^{n} q_i$		

[1] Mean quadratic flow increment describes traffic variability between subsequent intervals,

[2] Range of normalized variations is the quadratic mean of differences between numbers of vehicles in subsequent intervals t_r of time--series (graphical interpretation of this statistic is shown in Fig. 1b),

[3] I-ratio represents mean quadratic deviation of flow rate in interval t_r in relation to mean rate $\bar{q}$.

larger flow rates N correspond to smaller values of the statistics, although maximum flow rates do not correspond to extreme values of the statistics. Using regression analyses, mutual relations between various statistics were derived and also relationships of their values to the mean flow rate, with regard to hourly analysis periods TO for short-term variations. Figures 3 and 4 show the most important relationships obtained for interval $t_r = 5$ min. There is low correlation ($r = 0.58$) between the variation coefficient ε and the mean number of vehicles $\bar{q}$ passing a given point within 5 minutes (Fig. 3). On the other hand correlations between statistics z and ε (Fig. 4a) as well as between z_n and ε (Fig. 4b) are strong (respectively $r = 0.968$ and $r = 0.866$). Consequently corresponding correlations between statistics z and z_n

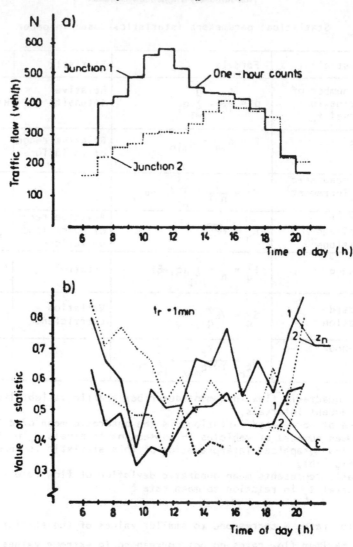

Fig. 2. Volume characteristics. Part (a): hourly traffic volumes,
part (b) : characteristic of one-minute rate variations during
successive one-hour periods.

on one hand and flow rate $\bar{q}$ on the other are low. Thus, when correlations
between individual statistics are concerned, there is no need to
calculate each considered statistic for a given time-series, because on
the basis of the value of one statistic and established regression
relationships, it is possible to derive approximately the other
statistics. It is apparent that low correlation between these statistics
and mean flow rates $\bar{q}$ in successive short intervals confirms the need
of providing mean flow rates together with the value of a chosen

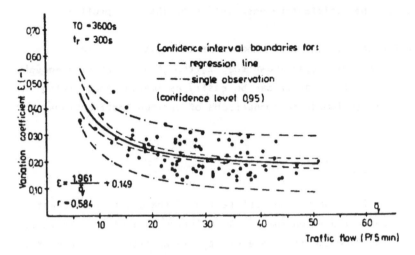

Fig. 3. Variation coefficient as a function of the mean flow rate q̄ for counting interval t_r= 5 min.

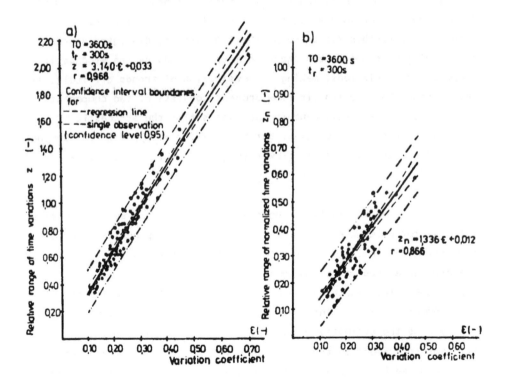

Fig. 4. Relative range of variations (part a) and relative range of normalized variations (part b) as a function of variation coefficient for counting interval t_r= 5 min.

statistic for the satisfactory description of the flow profile.

The variation coefficient ϵ was chosen as the most appropriate
statistic. It has also been shown that z, though not sensitive enough,
can be useful statistic as it can be easily estimated, thus giving
the possibility of immediate comparisons of various traffic flows.

2.2. Counting interval

Time-series of flow, being the realisation of the discrete stochastic
process, depend not only on the characteristics of a given traffic flow
but also on the flow counting interval t_r. In earlier researches, Kimber
and Hollis (1978, 1979) used a time interval of 5 min., Branston (1978)
analysed variations in intervals of 6, 12, and 18 min., Retzko and Tonke
(1979) analysed flow profiles with an interval of 1 min., while Catling
(1977) used 5 to 10 min. counting intervals for description of demand
flow. Adopting too long a counting interval (i.e. 15 minutes) brings
about an undesirable smoothing of the time-series, disregarding real
trends of flow changes. Too short intervals emphasise the stochastic
component of variations, making the evaluation of trends difficult. It
was assumed that the traffic flow process consists of two components:
w(t) - the constant component (trend) and s(t) - the stochastic
component, which are equivalent but have different properties. On such
assumptions the value of the flow rate q(t) in the discrete time
$t = 0,1,2,3,....$ can be expressed by the sum of w(t) and q(t).

The objective of the analysis undertaken was to select such a counting
interval for which the effect of the stochastic component s(t) on the
time-series is low, while the trend of flow variations w(t) is preserved.
The procedure used was as follows:
(I)   Exponential smoothing has been applied for filtering the stochastic
      component s(t) of the time-series (obtained for t_r= 1 min.) which
      allows the estimation of the real trend of changes w(t).
(II) For such smoothed time-series and for the set of time-series
      obtained by using a greater and greater time interval t_r (from 1
      to 10 min. with 1 min. step) the values of the statistics shown
      in Table 1 were calculated.

(iii) The differences between the statistics of the time-series after the exponential smoothing and the statistics obtained for the time-series with various time intervals t_r were analysed.

The exponential smoothing procedure in step (i) is based on calculating the smoothed value of the time-series $\hat{q}(t)$ in time t by interpolation between the current value of the series $q(t)$ and the previously smoothed one $\hat{q}(t-1)$:

$$\hat{q}(t) = a \cdot q(t) + (1-a) \cdot \hat{q}(t-1) \qquad\qquad 0<a<1 \qquad\qquad (1)$$

where a - smoothing constant.
Considering the fact that the smoothed value in time (t-1) comprises component $\hat{q}(t-2)$, which in turn comprises $\hat{q}(t-3)$ and so on, Eq.(1) may be given as follows:

$$\hat{q}(t) = a \cdot \left[\sum_{i=0}^{T} (1-a)^i \cdot q(t-i) \right] + (1-a)^{T+1} \cdot \hat{q}(t-T-1) \qquad\qquad a \neq 1 \qquad\qquad (2)$$

where T is the number of time intervals from the beginning of the analysed time-series to the moment t. For large values of T the last component in the Eq.(2) can be omitted, because:

$$\lim_{T \to \infty} (1-a)^{T+1} \cdot \hat{q}(t-T-1) = 0$$

It is evident from Eq.(2) that $\hat{q}(t)$ is a linear combination of all previous values of the time-series and the weight related to information from the previous intervals t_r decreases geometrically with the increase of the time range. The weight attributed to information from the (t-i)-th interval is given by:

$$w_i = a \cdot (1-a)^i \qquad\qquad (3)$$

so it is an exponential function of information age i. For smaller values of a, historical information has a greater effect on the smoothed value $\hat{q}(t)$ of the time-series. Sensitivity of exponential smoothing is defined by a so-called "mean age of information" $\bar{t}_{wi}$, which is a weighted mean time elapsing between consecutive historical pieces of information. It is given by the formula:

$$\bar{t}_{wi} = t_r \cdot (a^{-1} - 0,5) \qquad\qquad (4)$$

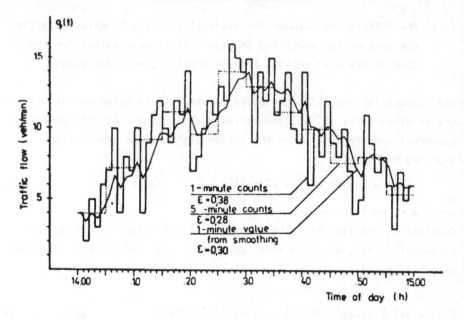

Fig. 5. The effect of exponential smoothing and increasing length of counting interval.

The mean age of information and therefore the sensitivity of smoothing depends on the value of the parameter a. Thus, it is possible to estimate the value of the smoothing constant a by specifying the required $\bar{t}_{wi}$ in advance. The necessity of providing the proper characteristics of the process {s(t)} can be assumed as the criterion for chosing the correct parameter a. It can be shown that for a = 0.3 the autocorrelation coefficient of the process {s(t)} takes zero value practically for the whole range of flow rates at a junction approach; moreover the distribution of the stochastic component s(t) is normal with mean value equal to zero. For a = 0.3, the mean age of information is $2.8 \cdot t_r$ and the corresponding weight of information is 0.11. Pieces of information that originated earlier than 10 intervals t_r before are omitted in practice ($w_i < 0.01$). The time-series from Fig. 1a, after application of the smoothing process with parameter a = 0.3, is shown in Fig. 5.

Considering that the length of the counting interval t_r affects the self-contained smoothing of short-term flow variations, comparisons of smoothing products related to various lengths of intervals t_r and flow variability curves obtained from exponential smoothing (steps (II) and

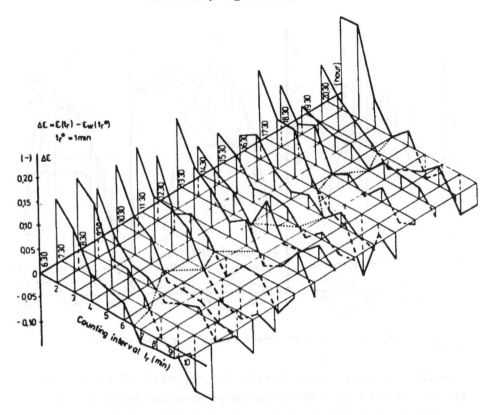

Fig. 6. Relationship between differences $\Delta\varepsilon$ and length of time intervals for consecutive counting hours.

(iii)) have been drawn. One such comparison, which has been carried out using statistics from Table 1, is presented in Fig. 6. It shows the differences $\Delta\varepsilon$ as functions of the counting interval t_r for flow rate time-series in consecutive hours of the day. $\Delta\varepsilon$ expresses the differences between the values of the variation coefficient $\varepsilon(t_r)$, calculated for a flow profile in a given hour with counting interval $t_r = 2$ to 10 min., and values of variation coefficient $\varepsilon_w(t_r^0)$ for a smoothed flow profile obtained for $t_r^0 = 1$ min. The zero line (dotted line) which stands for the equalization of the products of exponential smoothing and the products of the time interval length, runs within the range $t_r = 5$ to 7 min. It can be seen that for intervals $t_r = 1$ to 5 min., a rapid smoothing of flow variations occurs as the length of the interval increases, while for the range over 5 min. further smoothing is proceeding slowly. This means that at the interval value $t_r = 5$ min., the effect of short-term random variations of flow rates on the

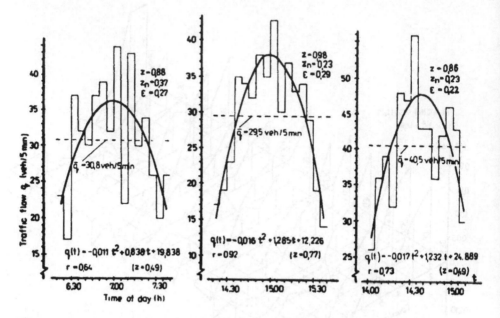

Fig. 7. Parabolic fit of the demand flows from observations.

variability trend is limited. Considering the practical application, on the basis of the results of the conducted measurements and analyses, it was possible to specify t_r= 5 min., as the optimum value of counting interval for representation of determined variations of flow rate in time. This takes into account a compromise between the demands of accuracy of traffic model identification and traffic flow evaluation. For this value of the interval t_r, close to zero values of the statistic differences are reached, characterizing the time-series obtained after the exponential smoothing and with use t_r= 5 min.

2.3. Typical traffic flow profiles for peak-periods

Traffic flow, especially during peak periods, is characterized by an evident dependence of 5-minute flow rates on current time. An analysis of the dispersion of time-series statistics recorded in Cracow during peak hours and the visual evaluation of traffic flow variations for these periods made it possible to specify three types of profiles to incorporate into non-stationary numerical models. We define them as symmetrical time-series (LS) and asymmetrical time-series (LL and LP)

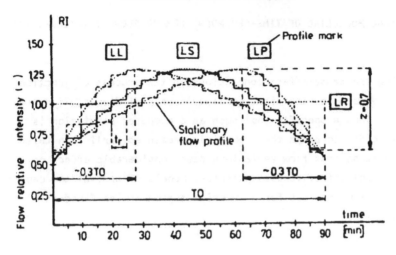

Fig. 8. Normalized profiles of flow variations in time.

with the top of the curve shifted in relation to the middle of the
considered peak period and a disproportion in the rate of its formation
and decline. Further analysis of each of the above groups of flow
profiles included characteristic features of their shape, that is,
length of time base TO, placement of the top in the period TO and
range of flow rate R.

A parabolic curve (Fig. 7.) appears to be the optimal function for the
approximation of a symmetrical flow profile. Average values of
parameters for a typical shape of flow profile are as follows:

- length of profile time base TO = 90 min.,
- relative range of 5 min. flow rate variations z = 0.7,
- tops of curves are within the distance -0.3·TO from the beginning
  and the end of the time base in the case of asymmetrical profiles.

Figure 8 shows normalized, typical peak profiles in the relative form.
These selected profiles of determined flow variations are theoretical
idealizations of the flow variation trend. In practice, flow time-
-series, as well as its trend, fluctuate, which was proved in
comprehensive studies on the stability of trend. To enlarge the range
of traffic conditions which are represented by normalized flow profiles
and for the needs of numerical modelling of traffic, flow profiles LS
type with z varying in the range 0.25 to 1.25 (for which ε = 0.08 to 0.43)
and TO = 45 to 135 min. were assumed in addition to the profiles shown
in Fig. 8.

3. NUMERICAL MODELLING OF TIME-DEPENDENT DEMAND FLOW AT INTERSECTIONS

Determining the cause-effect connections in the process of junction operation by means of empirical studies is very difficult, if not impossible. Thus a specific tool such as a computer simulation is needed. In order to prove the hypothesis presented earlier in this paper suggesting that flow variations have considerable effect on traffic control effectiveness, numerical models of arrival and departure processes were worked out for a junction approach with fixed-time traffic signals.

3.1. <u>Model of non-stationary flow</u>

Vehicle demand flow, considered at a location sufficiently distant from the junction as to avoid difficulties with queueing effects, is a stochastic process, determined in the traffic model by the distribution of headways between successive vehicles. Bearing in mind that this process is non-stationary, its parameters are dependent on the real time in which the process occurs. According to recent studies (Chodur, 1982), it was assumed that the demand process is a stochastic stationary process within a time interval t_s with the length determined earlier in this paper, that is $t_s = t_r = 5$ min. The core of the demand process in the model is the generating of the demand stream accordingly to the predicted distribution, in this case a shifted-exponential distribution of headways t_o between successive vehicles, with distribution function given by:

$$F(t_o) = 1 - \exp\{-(t_o - t_{om})/[\bar{t}_o(\tau_1, t_s) - t_{om}]\} \qquad t_o \geqslant t_{om} \qquad (5)$$

where: t_{om} - minimum headway between successive vehicles in seconds, $\bar{t}_o(\tau_1, t_s)$ - mean headway in the traffic stream, corresponding to the interval t_s (the equilibrium value) beginning at the moment τ_1, given by the equation:

$$\bar{t}_o(\tau_1, t_s) = 3600/N \cdot RI(\tau_1, t_s) \qquad (6)$$

where: N - traffic volume (veh/h), $RI(\tau_1, t_s)$ - relative intensity of the demand stream (see Table 1) within the interval t_s, beginning at the moment τ_1.

According to the typical variability of flow assumed on the basis of empirical studies, moments τ_i beginning intervals t_s, are determined by a linear function:

$$\tau_i = (i-1)\ t_s \qquad\qquad\qquad i = 1,\ldots,n \qquad\qquad (7)$$

where: i - number of successive intervals t_s with constant intensity of stream, T0 - length of the peak period (see Fig. 8), which is equal to the observation time in the simulation runs, $n = T0/t_s$. The inverse--function to Eq. (5) is:

$$(t_o - t_{om})/\left[\bar{t}_o(\tau_i,t_s) - t_{om}\right] = -\ln\left[1 - F(t_o)\right] \qquad\qquad (8)$$

Let $F(t_o)$ be the random fraction r_n between $[0,1]$. Then from Eq. (8) we obtain the series of headways of the shifted exponential distribution. If the random fraction is generated and substituted for $(1-r_n)$, the arrival headway can be computed by using the equation:

$$t_o = -\left[\bar{t}_o(\tau_i,t_s) - t_{om}\right]\cdot\ln(r_n) + t_{om} \qquad\qquad (9)$$

In the general, the basic problem in the generating procedure — when t_s can theoretically take very small values - is the determination of the mean intensity of the demand stream within the interval $(\theta,\theta+t_o)$, where θ is the moment of arrival of the last j-th vehicle and t_o is the time headway after which the currently generated vehicle will arrive. In the situation when both vehicles (j and $j+1$) are in the same k-th interval t_s, the problem lies in assuming the stream intensity RI_k characteristic for the k-th interval t_s and putting this value into Eq. (6). But when the j-th vehicle arrived in the k-th interval and the end of the generated headway is to be in interval $k+1$ as shown in Fig.9, then the mean stream intensity in the constructed interval $(\theta,\theta+t_o)$ will be:

$$\overline{RI}(\theta,\theta+t_o) = \{d\cdot RI_k + t_s\cdot\sum_{i=1}^{\ell-1} RI_{k+i} + [t_o - d - (\ell-1)\cdot t_s]RI_{k+\ell}\}/t_o \qquad (10)$$

The mean stream intensity, being a parameter of the distribution, is calculated accordingly to Eq. (6) and is dependent on the time headway t_o, which has to be estimated. Generation of time headways according to the distribution with parameters from Eq. (6) and Eq. (10) can easily be solved in the case of the ordinary exponential distribution; thus

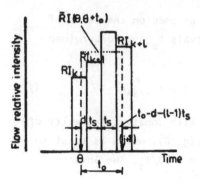

Fig. 9. Schema for determinating mean intensity of demand stream in the interval $\bar{RI}(\theta, \theta + t_o)$

$$t_o = -\bar{t}_o(\theta, \theta + t_o) \, \ln(r_n) \tag{11}$$

Transforming Eq. (10) and substituting into Eq. (6), and then into Eq. (11) gives:

$$t_o = -\frac{1}{RI_{k+\ell}} \cdot \{d \cdot (RI_k - RI_{k+\ell}) + t_s \cdot [\sum_{i=1}^{\ell-1} RI_{k+i} - (\ell-1) \cdot RI_{k+\ell}] + \frac{3600}{N} \cdot \ln(r_n)\} \tag{12}$$

This equation has two unknown parameters: t_o and ℓ. Taking into account their mutual relation (Fig. 9) the equation can be solved by an iterative method, examining the condition:

$$t \cdot RI_{k+\ell} + \{d \cdot (RI_k - RI_{k+\ell}) + t_s \cdot [\sum_{i=1}^{\ell-1} RI_{k+i} - (\ell-1) \cdot RI_{k+\ell}]\} > -\frac{3600}{N} \cdot \ln(r_n) \tag{13}$$

where ℓ takes consecutive values of $1, 2, 3, \ldots$, while $t = d + \ell \cdot t_s$.

Because, as a result of an earlier assumption, the basis of generation is the ordinary exponential distribution, in order to obtain time headways not shorter than t_{om}, a correction procedure for the generated headways has been applied. This procedure works when a generated headway shorter than t_{om} is recorded. This correction is applied to one or more time headways according to their length. So, correction of time headways is done on the expense of local shifts, without changes of flow rate in the whole observation period.

The described procedure was named SEK. In the situation when flow rates are high and intervals of constant intensity are long (for example $t_s = 5$ min.) the probability of occurrence of a time headway much longer

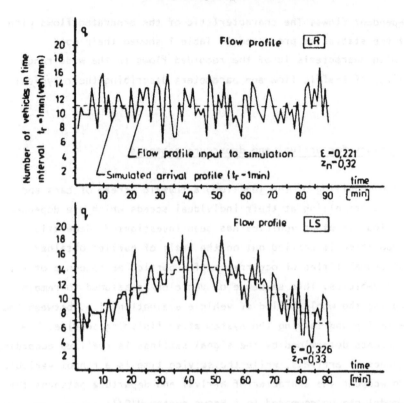

Fig. 10. Flow profiles obtained from simulation for volume 700 veh/h.

than interval t_s is very low. In such a situation a simplified
procedure MIN, using relationships (9) and (6) directly has been applied
for generating headways. Flow intensity RI (τ_I, t_s) is taken from that
interval $(\tau_I, \tau_I + t_s)$ in which the last generated vehicle occurred.
A comparison showed that this simplification does not induce a
significant distortion of the flow profile in relation to the one
generated by procedure SEK.

The first part of simulation program SYGNAL-2 consists of the above
described procedures for generating traffic streams and other procedures
for making graphs of generated streams as well as for calculating
values of statistics, so enabling quantitative characteristics of flow
to be determined. This part of SYGNAL-2 can be used independently.
Examples of generated streams (stationary and non-stationary) are shown
in Fig. 10. Verification of the results obtained from generation versus
field observations confirmed the correctness of the numerical modelling

of time-dependent flows. The characteristic of the generated flows with the use of the statistics presented in Table 1 showed their good agreement with characteristic of the recorded flows in the wide range of variability of traffic flow and parameters describing their typical profiles.

3.2. Assumptions for arrival and departure patterns

In simulation modelling of the junction, a traffic stream of cars and heavy vehicles travelling at their individual speeds which are dependent on the behaviour of other vehicles has been investigated. Generation of traffic composition is carried out on the basis of earlier obtained conditional probabilities of occurrence of a particular sequence of the two types of vehicles. This sequence of vehicles is assumed to remain constant during the whole period of vehicle evaluation, i.e. between the vehicles entering and leaving the system after finishing service. The departure process determined by the signal settings is realized according to the sequence of arrivals, while the service time is a random variable. Taking into account the character of arrival and departure patterns the developed model can be compared to a known system M/G/1.

Studies on the simulation system in which the data of its states are recorded is realized by a discrete method. The obtained data make it possible to determine the basic measures of traffic progress quality on the approach to the junction, such as: delays, queue lengths, queues at the end of green period, stops, saturation ratios, speed changes, etc. The obtained results are statistically analysed in a scale of individual cycles and for the whole period of calculations TO.

The model of traffic progress on the junction approach presented here only in the shortened form is described in full in the earlier Chodur's paper (1982) and this model enters into the SYGNAL-2 computer program. These two parts of the model, i.e. the first part including demand stream generation and the second including simulation, are connected by the matrix containing the recorded times of vehicle arrivals - related to tne cross-section situated just upstream of the tail of the queue.

Due to shortage of time and means, the validation of the model is rather insufficient. In spite of this the validation results convey optimism about the usefulness of the model. Validation of the numerical model and the simulation program (the second part of the SYGNAL-2) was done by comparing the most important final results of simulation (i.e. delays, queue lengths) with the data obtained from field observations at junctions. This comparison included flow variability in time and distribution functions of the analysed parameters, and showed statistically good agreement between empirical data and the simulation results.

4. FINAL REMARKS

Empirical studies and analysis and a numerical model of traffic made it possible to investigate the essence and effects of flow non-stationarity at isolated traffic signals. Figure 11 shows relationships between average delays of vehicles $\bar{d}$ and saturation ratio $\bar{x}$ for different types of flow variation patterns. These relationships show what errors can be made while estimating delays with the assumption that traffic flow within the peak period is stationary and they also reveal how the delays are affected by the shape of flow rate variations. Charts of average delays calculated according to the known formulae of Clayton, Webster, Miller, Pollaczek-Khintchine, Doherty and Akcelik in Fig. 12 are presented together with the results of simulation. As the comparison shows, these formulae, except perhaps that by Clayton, can be used in practice for estimation of delays in peak period in the situation when the saturation ratio is lower than ~0.7, that is when the effect of the determined flow variability on the junction effectiveness is limited. Neglecting non-stationarity of flow for high saturation ratios ($\bar{x}>0.7$) leads to underestimating of such measures of junction efficiency as delays, queues, stops, etc., in comparison with the real values. Therefore the correct evaluation of signal setting in such situations is impossible. Runs of the delay curves for time-dependent flows, characteristic of peak periods, showed that in most cases, when $\bar{x} = 0,7$

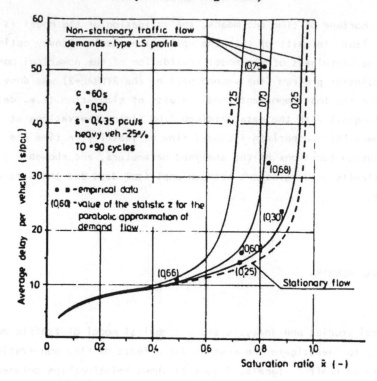

Fig. 11. Delays as a function of the saturation ratio for different flow variation profiles.

to 0.8 traffic conditions at a junction deteriorate. Hence the conclusion: signal settings used for peak periods should be designed for geometry providing the practical capacity, which is 20 to 30 % lower than possible capacity. The method of delay calculation described by Catling (1977) gives a better estimation of delays when flow demands are variable in comparison to other analysed formulae (Fig. 13). The procedure, however, is very time-consuming with manual calculation.

The developed numerical model, including time-dependent arrival patterns, is a good tool for investigation of signal setting efficiency. A comparative study of the optimum signal timings obtained using the model given in this paper and those by Webster, Miller and Akcelik (1980) is in progress. Analysis of junction traffic conducted with the application of this model makes it possible to estimate average values of traffic control efficiency measures and their time variability and distributions. This is particularly important for determination of those

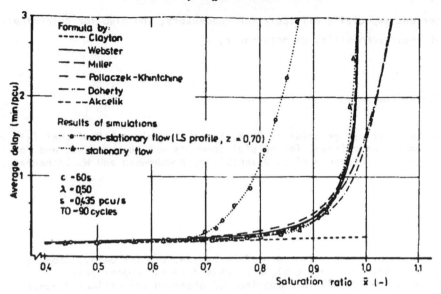

Fig. 12. Delays predicted by various formulae compared with values obtained from simulation.

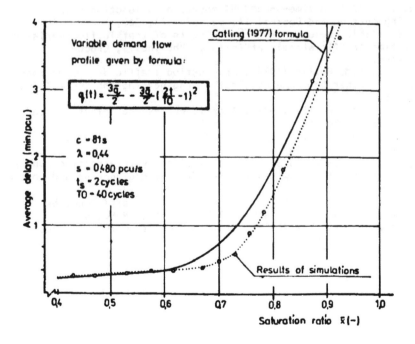

Fig. 13. Delays predicted by Catling's method compared with values obtained from simulation.

periods especially difficult for road-users, and also for determination
of their probability of occurrence.

Acknowledgements

The authors are grateful to Dr R. Ashworth, University of Sheffield and
to anonymous referees for helpful comments and suggestions, and for the
technical assistance of J. Jeleński, M. Nowakowska and W. Sonnenberg.

REFERENCES

Akcelik, R. (1980). Time-dependent expressions for delay, stop rate and
queue length at traffic signals, ARRB Internal Report 367-1.
Branston, D. (1978). A comparison of observed and estimated queue
lengths at oversaturated traffic signals, Traffic Engineering and
Control 7, pp. 322-327.
Burrow, I.J., R.M. Kimber, N. Hoffmann, D. Wills (1983). The prediction
of traffic peak shapes from hourly flow counts, TRRL Supplementary
Report, SR 765.
Catling, I. (1977). Development of junction delay formulae, LTR1 Working
Paper 2, Department of Transport.
Catling, I. (1977). A time-dependent approach to junction delays,
Traffic Engineering and Control 11, pp. 520-523, 526.
Chodur, J. (1982). Research into the effects of traffic flow variations
on fixed-time traffic signals efficiency, PhD Thesis, Cracow,
(unpublished).
Kimber, R.M., E.M. Hollis (1978). Peak-period traffic delays at road
junctions and other bottlenecks, Traffic Engineering and Control 10,
pp. 442-446.
Kimber, R.M., E.M. Hollis (1979). Traffic queues and delays at road
junctions, TRRL Laboratory Report LR 909.
Miller, A.J. (1969). Settings for fixed-cycle traffic signals,
Opnl Research Quarterly 14, pp. 373-386.
Retzko, H.G., F. Tonke (1979). Delays at priority intersections under
non-stationary traffic flow conditions, Traffic Engineering and Control
12, pp. 596-598.
Tracz, M. (1979). Evaluation of at-grade junctions efficiency by
computer simulation, Cracow Technical University, ZN BL, 9.
Webster, F.V., B.M. Cobbe (1966). Traffic Signals, HMSO, Road Research
Technical Paper No. 56, London.

Ninth International Symposium on
Transportation and Traffic Theory
© 1984 VNU Science Press, pp. 155-178

A TRAFFIC FLOW MODEL WITH
TIME DEPENDENT O–D PATTERNS

RODNEY VAUGHAN,* V. F. HURDLE and EZRA HAUER
Department of Civil Engineering, University of Toronto, Ontario, Canada

ABSTRACT

A myriad of trips, each with its own origin, destination, and departure
time, interact on a network to produce an intricate pattern of traffic
flows. The task of traffic flow theory is to weave a web of cause and
effect which - given the triad of origin, destination, and departure
time - can replicate the major features of what one would observe in
reality. Traditionally, this theory has dealt with exceedingly simple
origin-destination structures, a fact which may have limited its applicability. The aim of this paper is to remove this limitation.

The building blocks of the broader theory are: a road which traffic may
enter and leave at any point, a speed-flow relationship which may vary
along the road and depends on the local capacity, and a description of
the pattern of trip making by origin, destination, and departure time.
We show that the reconstitution of the traffic flow pattern from these
building blocks requires two equations: a 'local equation' which governs
what happens in a small space-time neighbourhood just as in the traditional theories, and a 'history equation' which supplies all the per-
tinent information about the evolution of the flow pattern which is
needed to specify its future course. The use of these two equations is
illustrated by a numerical example.

* Permanent Address: Mathematics, University of Newcastle, Australia

1.   INTRODUCTION

The landmark 1955 paper by Lighthill and Whitham set the tone for two
generations of researchers' investigations into the theory of traffic
flow. Because of the fertility of the basic ideas, it was easy to find
many intriguing and unresolved problems within the scope of traffic flow
theory as originally defined. The present paper is an attempt to encom-
pass within the purview of traffic flow modelling some features which
are intrinsic to most real problems, but which are not present in the
original formulation and its derivatives.

Like the authors of many previous papers, we consider the flow of traffic on a single, long, and rather idealized road, but we have tried to
give our road the general characteristics of an arterial street in a
large metropolitan area. The road itself may or may not be infinitely
long, but the people who travel it make trips of finite length at
definite times. In fact, it is the incorporation of the origin-destination structure of trip-making that distinguishes this work from the rest
of the body of traffic flow theory.

The problem is initially set in a two-dimensional space defined by an x-
axis which measures distances and locations along the road and a t-axis
which measures times of day. We suppose that the characteristics of the
road and the origin-destination pattern of trips on the road are known,
and ask what the resulting flow, speed, and concentration patterns
$Q(x,t)$, $V(x,t)$, and $K(x,t)$, will be.

In the simplest version of the model, each traveller is going home in
the evening from a workplace located at point x_w (w for work) to a
destination at x_h (h for home). The traveller enters the x-axis street
at time t_w, but the time at which his or her trip is completed depends
on the traffic conditions enroute, hence on the origins, destinations,
and departure times of other motorists. The travel pattern of the
entire group of travellers using the street is described by a trip
density, $g_{wh}(x_w,x_h,t_w)$, defined in such a way that, for sufficiently
small δx_w, δx_h, and δt_w,

$g_{wh}(x_w, x_h, t_w) \delta x_w \delta x_h \delta t_w$ = the number of persons who leave workplaces on road segment $(x_w, x_w + \delta x_w)$ headed for home destinations in road segment $(x_h, x_h + \delta x_h)$ during time interval $(t_w, t_w + \delta t_w)$.

Since we consider the two directions of travel as separate, independent problems, the trip density is defined only for $x_w < x_h$. For notational convenience, we also assume that there is no travel at negative values of either t or x, but allow both t and x to extend to infinity. Thus, $g_{wh}(x_w, x_h, t_w)$ is defined for $0 < x_w < x_h < \infty$ and $0 < t_w < \infty$, but may well be zero beyond some boundary points marking the ends of the city and the day.

The trip density function $g_{wh}(x_w, x_h, t_w)$ conveys exactly the same information as an origin-destination matrix. It is unconventional only in that it is assumed to be continuous rather than discrete, and varies as a function of time as well as location.

The other major element of the problem is the standard assumption that traffic slows down when the flow increases. However, we depart from convention in a number of ways in making this assumption, so it must be discussed in detail. The first step is to define a function

$\Lambda(x,q)$ = the travel time per unit distance at
location x if the flow past x is q. (1)

We assume that $\Lambda(x,q)$ is a known function which does not vary with time and call it the tardity. This name was settled upon only after considerable difficulty, research, and even dissension. The problem is that the English language does not have a word for the reciprocal of speed - a shortcoming apparently shared by most languages, if our none too systematic search for a suitable word is any indication. Yet surely modern man needs to know how long it will take to get somewhere, not how far he or she can travel in an hour.

Perhaps the requirements of our ancestors were different. One can conjecture that they may have been more interested in horse-racing, or perhaps escape, than cost-benefit analyses. Or perhaps they foresaw, not our needs, but those of the Canadian railway planners of yesteryear

who laid out prairie railroads a day's wagon trip apart, so that every
farmer could take his grain to the trackside elevator and be home in
time for supper. In any case, and for whatever reason, our predecessors
left us no word, so we must either invent one or do without.

In searching for the inventions of others, we found *lenteur* used by
Tournerie (1969) and *Langsamkeit* by Leutzbach (1972); both translate as
slowness. Not quite satisfied with this, we persevered and eventually
added to our list: ρυθμoσ (rhythmos), suggested by Constantin Lalenis;
sloth, Rodney's suggestion and still a favorite; and tardity, the fruit
of a search begun after receiving a suggestion from a helpful linguist
who remembered another archaic word that meant the opposite of acceleration rather than the opposite of velocity. Tardity was selected primarily because the Oxford English Dictionary indicated that it had been
used before in technical writing, and because its very unfamiliarity
discourages the idea that slowness must necessarily be slow (and also,
perhaps, because a collaborator 13 000 km away is easily outvoted, no
matter how loudly he may argue). We assume that the reason for choosing
a capital lambda to represent tardity requires no explanation.

In the analysis that follows, only three assumptions are made about the
tardity function $\Lambda(x,q)$: that it is at least piecewise continuous, that
it is defined only for flows less than the capacity of the roadway, and
that the tardity remains finite even when the flow approaches the roadway capacity.

The second and third of these assumptions require some explanation.
Both have to do with the fact that all roadways have fixed capacities
which cannot be exceeded unless a physical change such as widening the
road or retiming the signals is made:

$$Q(x,t) \leq c(x,t), \text{ for every } (x,t), \tag{2}$$

where the functions $Q(x,t)$ and $c(x,t)$ describe the flow and capacity,
respectively, at all locations and times. The capacity is a function of
the roadway width and the signal timing, both of which can, of course,
be changed, but are regarded as fixed for the purposes of this paper.

Real roads do reach capacity and long queues form as a result, but the whole system does not grind to a halt. To be both complete and correct, any traffic flow model must reflect these basic facts. It is our contention that this can be done only by modelling the transient queue behaviour, something we are not prepared to do at this time. Therefore, since we prefer incompleteness to incorrectness, the model in this paper is limited to applications where Constraint (2) is satisfied with strict inequality. We hope to report a solution to the version of the problem with oversaturated intersections at a later date.

The tardity function used in a numerical example later in the paper is

$$\Lambda(x,q) = a(x) + b(x)R^2(x)/[2C(x)(1-q/s(x))], \quad 0<q<c<s(x) \tag{3}$$

where

    $a(x)$ = tardity in the vicinity of x if all signals were to be
        always green and $q = 0$,

    $b(x)$ = number of intersections per unit distance in the
        vicinity of x,

    $R(x)$ = length of red interval at the signal in the
        vicinity of x,

    $C(x)$ = cycle length of the signal in the vicinity of x,

and

    $s(x)$ = saturation flow in the vicinity of x.

This tardity function gives a constant travel time plus the delay indicated by the first term (often called the uniform delay) of Webster's (1958) delay equation, reflecting an assumption on our part that all delay on the road is caused by traffic signals and that this delay can be modelled adequately in a deterministic manner. Since Equation (3) neglects stochastic effects, it will underestimate the travel time at high flows, but the error will be substantially less than if a steady-state queueing model such as Webster's full equation were to be used, since peak traffic flows at real intersections never last long enough to permit formation of the long steady-state queues predicted by the models.

A compromise would have been to use a delay estimator such as the one included in TRANSYT (Robertson and Gower, 1977; Kimber and Hollis, 1979). We use Eq. (3) only for illustration, however, and users of our method are free to use any tardity function they consider appropriate. This discussion of Eq. (3) is not intended to justify our choice, but simply to bring out some of the issues involved in choosing a tardity function.

2. MODELLING FRAMEWORK - AN ABSTRACT TRAJECTORY SURFACE

To begin the analysis, we consider the journey of a vehicle which starts at time zero from a workplace at $x_w = z$ and travels to a destination at the far edge of the city. Fig. 1 shows the trajectory of such a vehicle in the xt-plane; for identification purposes we assign to its trajectory the label z. The shape of trajectory z is specified by the function

$$T(x,z) = \text{the time at which trajectory z passes point x.} \qquad (4)$$

$T(x,z)$ is, of course, an unknown function - in fact, a very important unknown function. Once we discover a means to evaluate $T(x,z)$, the flows, speeds, travel times, and densities will all be easy to derive.

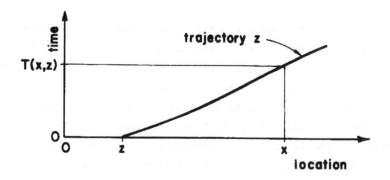

Fig. 1. Trajectory of a Single Vehicle

Fig. 2. Trajectory Surface

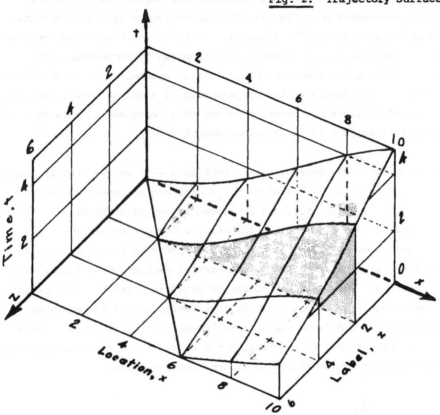

In introducing T(x,z), we assumed that the vehicle whose trajectory is shown in Fig. 1 travelled the full length of the city, something few vehicles do. If, however, this vehicle were to end its journey at some point (x,t) and another vehicle were to enter at (x,t), then that second vehicle would simply continue along the trajectory of the first. Thus a particular trajectory, z, does not usually represent the path of a single vehicle, but describes the movement of some group of vehicles. One could think of what we call trajectory z as a bundle of trajectories of vehicles which share a path across the xt-plane. A more fruitful approach, however, is to think of the set of all trajectories as a surface in a three-dimensional space with axes that measure location, x, time, t, and trajectory label, z. Such a trajectory surface is shown in Fig. 2. The numerical values of x and z in the figure are arbitrary, but could be distances in hundreds of metres.

Figure 1 is a cross-section cut through this surface at z = 2, the
shaded cross-section in Fig. 2. The curves formed by the intersection
of the T(x,z) surface with planes at z = 0, 4, and 6 are also shown in
Fig. 2, and Fig. 3 is a z-contour map produced by projecting these
curves onto the location-time plane. It is convenient to think of the
contours thus produced as the trajectories of a set of marker vehicles.
The dashed contours for negative values of z represent an extension of
the surface which will be explained shortly.

Examining the trajectory of marker vehicle z = 0, we see that it passes
x = 2 at time t_2 and x = 4 at time t_4. Its average speed during this
interval is $(4-2)/(t_4-t_2)$, and its speed at any point is dx/dt, the
reciprocal of the slope of the contour. Using our three-dimensional
notation, however, we say instead that marker vehicle z's tardity at any
location x is equal to $T_x(x,z)$, the partial derivative of T(x,z) with
respect to x. Note that the number of other vehicles between any pair
of marker vehicles is not constant, but increases and decreases as
vehicles enter and leave the stream. These changes, in turn, induce
tardity changes, and thus determine the slope of the z-contours, hence
the shape of the three-dimensional trajectory surface.

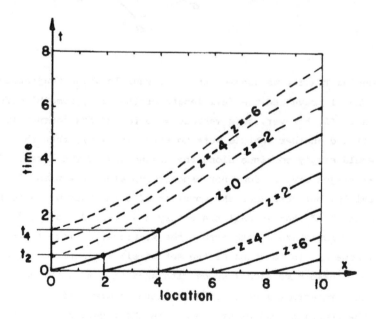

Fig. 3. Trajectories or z-Contours on the x-t Plane

We have used the function T(x,z) to describe the surface in Fig. 2, but this is not always the most convenient description, so we also define a second, and completely equivalent, descriptive function

$$Z(x,t) = \text{the label of the trajectory which passes through}$$
$$\text{point } (x,t) \quad (5)$$

and note that any vehicle which passes through point (x,t) must travel along trajectory z = Z(x,t).

It should also be noted that the numerical values of z(x,t) are established by an arbitrary labelling convention. In introducing the idea of trajectory labels, we implicitly established a relationship between the z and x variables by labelling each trajectory with the value of x at which it crossed the positive x-axis at time zero. Thus, by our definition,

$$Z(x,0) = x, \quad x \geq 0. \quad (6)$$

Other labelling conventions such as $Z(x,0) = 12.7x^2 + 13.4x + 374.5$ or $Z(x,3) = 12\log x$ would also be possible, as would conventions tying z to t instead of x. The use of a different convention would change the shape of the trajectory surface by transforming the scale on the z-axis, but such a change would not be very important because it would leave the shape of the z-contours on the xt plane unchanged. After the transformation, the z-contour running through any point (x,t) would have a new name, but its shape would remain unchanged.

The z-labelling convention $Z(x,0) = x$ provides labels only for that portion of the T(x,z) surface for which x, t, and z are all positive, the octant shown in Fig. 2. As can be seen in Fig. 3, this part of the surface describes only the history of those vehicles which are downstream from the z = 0 marker vehicle, the vehicle that started its trip at location x = 0 and time t = 0. Naturally, there are also vehicles upstream from the z = 0 vehicle; their trajectories are labelled with negative values of z according to the convention

$$Z(0,t) = -t/\Lambda \left(x{=}0, q{=}Q(0,t) \right). \quad (7)$$

In other words, the negative labels in Fig. 3 were obtained by extending
each of the dashed contours leftward at a constant slope until it
crossed the negative x-axis. The complete trajectory surface is shown
in Fig. 4; the fact that the surface intersects both the tz and xz
planes along straight lines is a result of the z-labelling convention
and the assumption of zero flow at x=0.

3.   CONTOURS AND SLOPES

Until this point, the discussion has focused on projections of the
trajectory surface onto the xt-plane and on one particular partial
derivative, $T_x(x,z)$, associated with this plane. Other projections and
other partial derivatives will, however, play a role in the development
that follows, so it is useful to dwell on what they are and what they
represent.

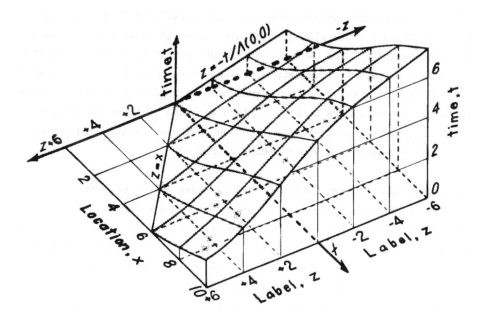

Fig. 4. Trajectory Surface, including Octant with Negative z-Labels

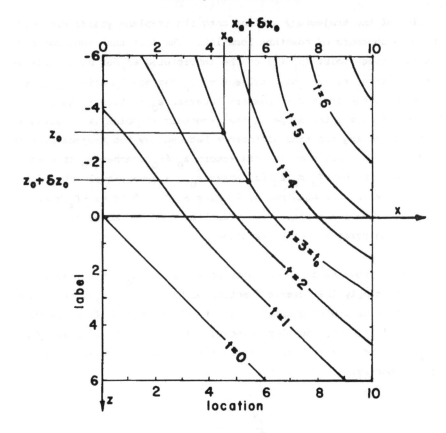

<u>Fig. 5.</u>   t-Contours (Isochrones) on the xz-Plane

Figure 5 shows the projection of the trajectory surface onto the xz-plane. The time contours may be interpreted in two ways. If one concentrates on a location x_0 and reads vertically down the graph, time contour t_0 is encountered, indicating that trajectory z_0 passes location x_0 at time t_0. Alternatively, if one selects some trajectory z_0 and moves horizontally to time contour t_0, the position reached by trajectory z_0 at time t_0 is read off the x-axis as x_0. The slope of a given time contour, say t_0, is given by $Z_x(x_0,t_0)$, so at a given point in time the trajectories passing through two close locations x_0 and $x_0+\delta x_0$ will be z_0 and $z_0+\delta z_0 \simeq z_0+Z_x(x_0,t_0)\delta x_0$. The importance of this observation in the calculation of concentration will become apparent below.

Projection of the trajectory surface onto the tz-plane yields Fig. 6,
which shows contours of constant location. The fact that some contour
x_0 passes through point (z_0, t_0) can be interpreted either as an indication of the time, t_0, at which trajectory z_0 reaches location x_0 or as a
statement that the trajectory passing location x_0 at time t_0 is z_0.
Notice that $T_z(x,z)$, the slope of the x-contour at point (x,z) gives a
means of estimating the time difference between any two trajectories at
a fixed location. For example, trajectory $z_0 - \delta z_0$ reaches x_0 at time
$t_0 + \delta t_0$, approximately $-T_z(x_0, z_0)\delta z_0$ after t_0. A more visual interpretation of $T_z(x,z)$ can be obtained by looking at Fig. 2 or 4: $-T_z(x,z)$ is
the rate at which the trajectory surface rises as one moves to the right
in a direction parallel to the tz-plane.

Figure 6 also provides an opportunity to interpret $Z_t(x,t)$. For fixed
x, $Z(x,t)$ is simply the inverse function of $T(x,z)$, and $Z_t(x,t)$ is the
rate of change of z with respect to t; it follows that $Z_t(x,t)$ is the
reciprocal of $T_z(x,z)$, the slope discussed in the previous paragraph.
Both $T_z(x,z)$ and $Z_t(x,t)$ bear a critical relation to flow which is
discussed shortly.

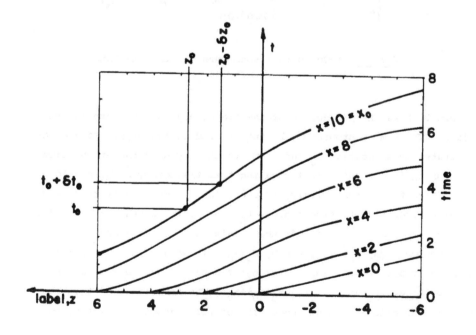

Fig. 6.   Contours of Constant Location on the tz-Plane

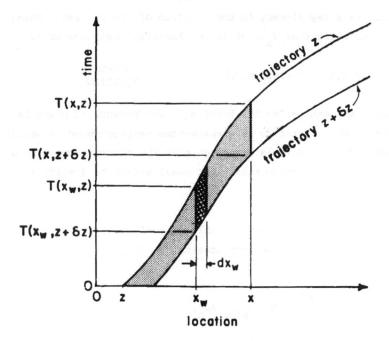

Fig. 7. Illustration for the Derivation of the Local and History Equations

4. KEY EQUATIONS

To continue the analysis, we return to the xt-plane in Fig. 7, which shows trajectory z and a near neighbour, z+δz. If the effect of over-taking is neglected, all the vehicles passing point x during time inter-val $\left(T(x,z+\delta z),\ T(x,z)\right)$ must have destinations beyond x and origins and departure times within the darker portion of the shaded band bounded by trajectories z and z+δz. Suppose that there are H(x,z)δz such vehic-les. Then the average flow during time interval $\left(T(x,z+\delta z),\ T(x,z)\right)$ is

$$H(x,z)\delta z\ /\ [T(x,z)-T(x,z+\delta z)]. \tag{8}$$

In the limit as δz approaches zero, the denominator of this fraction is equal to $-T_z(x,z)\delta z$, so the flow past x at time T(x,z) is

$$Q\left(x,T(x,z)\right)\ =\ -H(x,z)/T_z(x,z). \tag{9}$$

This equation is a key element in the solution of the problem. Using Eq. (9) and recalling that $T_x(x,z)$ is the tardity, we may now write

$$T_x(x,z) = \Lambda\left(x, -H(x,z)/T_z(x,z)\right). \qquad \begin{array}{l}\text{LOCAL}\\\text{EQUATION}\end{array} \qquad (10)$$

This equation has been called the local equation because it describes the local behaviour of traffic in a space-time neighbourhood, a neighbourhood which we assume is large enough that the fluctuations caused by individual vehicles are unimportant, but small enough to justify a continuous model.

To make further progress, we must be able to evaluate $H(x,z)$. To do this, we first obtain from the trip density, $g_{wh}(x_w, x_h, t_w)$, a cumulative trip density:

$$G_{wh}(x_w, x, t_w) = \int_x^\infty g_{wh}(x_w, x_h, t_w)\,dx_h. \qquad (11)$$

$G_{wh}(x_w, x, t_w)$ is the density of trips that begin in the vicinity of x_w at time t_w and go to destinations beyond x. Since $H(x,z)\delta z$ is, by definition, the total number of such trips originating within the darker area in Fig. 7, it can be obtained by integrating $G_{wh}(x_w, x, t_w)$ over this region. To carry out the integration, we note that, for sufficiently small δz, the height of the cross-hatched element in Fig. 7 is

$$T(x_w, z) - T(x_w, z+\delta z) = -T_z(x_w, z)\delta z. \qquad (12)$$

Therefore,

$$H(x,z) = -\int_z^x G_{wh}\left(x_w, x, T(x_w, z)\right)\, T_z(x_w, z)\,dx_w. \qquad \begin{array}{l}\text{HISTORY}\\\text{EQUATION}\end{array} \qquad (13)$$

This equation has been called the history equation because it contains all relevant traffic information at (x,z) that can be determined from the past history of trajectory z. As far as local traffic behaviour is concerned, how the traffic arrived is not of concern; all that is important is how much traffic there is, which is conveyed to the local equation from the history equation by $H(x,z)$. The traffic analyst may, however, also be interested in the history equation for its own sake since it allows him to trace back along a trajectory band and determine the number of vehicles joining or leaving the stream. Looking at the local and history equations together, it can be seen that, since both the Λ and G functions are known, the two equations form a coupled set of partial integro-differential equations in $T(x,z)$ and $H(x,z)$; the local behaviour depends on the history, but the history in turn depends on the local behaviour at other times and places. In general, these equations must be solved numerically, though in very simple circumstances an analytical solution might be possible.

5. RELATION TO CLASSICAL THEORY

We have already seen that the flow can be expressed in terms of $H(x,z)$ and $T(x,z)$:

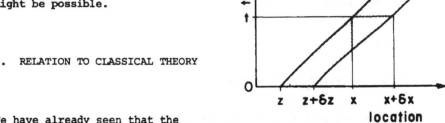

Fig. 8. Illustration for the Derivation of Eq. (15)

$$Q(x,t) = -H(x,z)/T_z(x,z).$$

Since $T_z(x,z) = 1/Z_t(x,t)$, this relationship, which appeared earlier as Eq. (9), can be expressed even more compactly as

$$Q(x,t) = -H(x,z)Z_t(x,t). \tag{14}$$

Similar relationships can be found between the partial derivatives of the trajectory surface and the concentration $K(x,t)$, which is defined by

$$K(x,t)\delta x = \text{number of vehicles lying between } x \text{ and } x+\delta x \text{ at time } t.$$

In Fig. 8, it is clear that the vehicles lying between x and x+δx at time t are those whose trajectories are in trajectory band $(z,z+\delta z)$ as it passes x. There are, by definition, $H(x,z)\delta z$ such vehicles, so

$$K(x,t)\delta x = H(x,z)\delta z. \tag{15}$$

With t held constant, $\delta z = Z_x(x,t)\delta x$, so the concentration is

$$K(x,t) = H(x,z)Z_x(x,t) . \tag{16}$$

To find concentration in terms of $H(x,z)$ and $T(x,z)$, a more circuitous argument must be adopted. Note that with t still held constant,

$$T_x(x,z)\delta x + T_z(x,z)\delta z = 0,$$

so

$$\delta z = -\delta x T_x(x,z)/T_z(x,z).$$

Therefore, from Eq. (15),

$$K(x,t) = -H(x,z)T_x(x,z)/T_z(x,z) . \tag{17}$$

Thus, both flow and concentration can be expressed as simple functions of $H(x,z)$ and either $T(x,z)$ or $Z(x,t)$. In every case, in fact, it is necessary only to multiply or divide $H(x,z)$ by an appropriate partial derivative of the trajectory surface.

With these relations established, both the fundamental relation of traffic flow and the conservation equation of classical theory are easily found. To obtain the fundamental relation, all that is required is to multiply the expression for flow from Eq. (9) by the tardity, $T_x(x,z)$. The resulting product is identical to the expression for concentration on the right hand side of (17), so

$$\text{Concentration} = \text{Flow} \times \text{Tardity.} \qquad \begin{array}{c}\text{FUNDAMENTAL}\\\text{RELATION}\end{array} \qquad (18)$$

The other equation which figures prominently in classical theory is the conservation equation in which $Q_x + K_t$ represents the net loss or gain of vehicles to the system per hour per kilometre. To see what value this expression takes in the present formulation, equations (14) and (16) are rewritten explicitly in terms of x,t:

$$Q(x,t) = -H\big(x,Z(x,t)\big)\ Z_t(x,t),$$

$$K(x,t) = H\big(x,Z(x,t)\big)\ Z_x(x,t).$$

Partial differentiation then yields

$$Q_x + K_t = -H_x(x,z)Z_t(x,t)$$
$$= -H_x(x,z)/T_z(x,z). \tag{19}$$

The classical theory does not yield the right hand side; it can only treat the case in which vehicles may enter the road at a variable rate $r(x_w)$ (vehicles per hour per km) and travel to the extreme right of the road section under study. As all vehicles travel to the extreme right,

$$G_{wh}(x_w,x,t_w) = r(x_w), \quad x_w < x. \tag{20}$$

In this special case, the history equation yields

$$H(x,z) = -\int_z^x r(x_w)T_z(x_w,z)dx_w$$

and

$$H_x(x,z) = -r(x)T_z(x,z). \tag{21}$$

Substituting this expression into Eq. (19) yields the classical equation

$$Q_x + K_t = r(x). \qquad \text{CONSERVATION EQUATION} \tag{22}$$

Note that the classical theory employs these two key equations, (18) and (22), and a speed flow relation. The conservation equation and the fundamental relation have been derived from the present theory, but it is not possible to derive the history equation or Eq. (19) from the classical theory.

6. NUMERICAL EXAMPLE

The beauty of working with the xz coordinate system is that the numerical solution of the coupled local and history equations is straightforward. The method is standard in that the solution space is divided into a grid, and the solution is generated by moving stepwise from known points. Initially the values of $H(x,z)$ and $T(x,z)$ are known on

$x=z$, $z>0$, and $x=0$, $z<0$, as shown in Fig. 4 and 9. The step algorithm
detailed below enables the determination of T and H at any point, provided the values to the right and below are known. Consequently, a
numerical solution may be generated by passing through the grid as
indicated in Fig. 9.

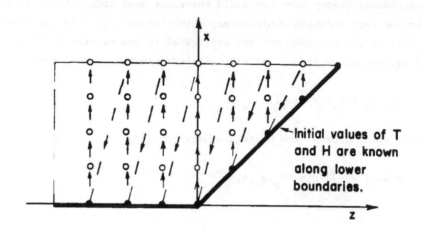

Fig. 9. Path of the Algorithm on the xz-Plane

6.1. The Step Algorithm

This algorithm will find the values of T and H at a target point if the
values to the right and below are known. (See Fig. 10.)

Step 1: Assume $T(x,z)$, $H(x,z)$, $T_x(x,z)$, and $T_z(x,z)$ are all known at the
points indicated by black dots in Fig. 10.

Step 2: Moving along arrow number 1 in Fig. 10, find $T(x+\delta x,z)$ using the
known values $T(x,z)$, $T_x(x,z)$ and

$$T(x+\delta x,z) = T(x,z) + \delta x T_x(x,z).$$

Step 3: Moving along arrow number 2 in Fig. 10, find $T_z(x+\delta x,z)$ by using
the known values $T(x+\delta x,z)$, $T(x+\delta x,z+\delta z)$ and

$$T_z(x+\delta x,z) = [1/\delta z][T(x+\delta x,z+\delta z) - T(x+\delta x,z)].$$

Step 4: Find H(x,z) by applying the history equation. This requires an integration over the known values within range 3 of Fig. 10:

$$H(x+\delta x,z) = -\int_{z}^{x+\delta x} G_{wh}\Big(x_w,x+\delta x,T(x_w,z)\Big)\, T_z(x_w,z)\,dx_w.$$

Step 5: Find $T_x(x+\delta x,z)$ by using the local equation and the now known values of $H(x+\delta x,z)$ and $T_z(x+\delta x,z)$:

$$T_x(x+\delta x,z) = \Lambda\Big(x,-H(x+\delta x,z)/T_z(x+\delta x,z)\Big).$$

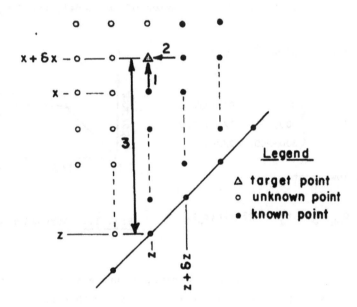

Fig. 10.   Steps in the Numerical Solution

The Calculation of the tardity at step 5 is the essential portion of this algorithm. Step 5 reflects the behaviour of the motorists: in order to advance, an appropriate tardity (i.e., speed) must be selected and this selection is based on the local traffic conditions. These local traffic conditions, in turn, depend on information about the past history of the system which is conveyed by $H(x+\delta x,z)$. The algorithm has been programmed in FORTRAN.

6.2. The Example

In order to avoid confusing the dynamic interpretation, a very simple
example in which neither tardity nor travel demand depends on the location x along the road was chosen. Travel demand is assumed to be uniform with respect to origins and destinations over the 5 km length of
the road, but has a trapezoidal time profile. This profile, illustrated
in Fig. 11, represents a typical buildup, stabilization and falloff in
demand over an hour. The total number of trips is 2500, ensuring that
flow levels remain below capacity, as is the assumption of the model.
Specifically, in terms of the notation of the model, for $0<x_w<x_h<5$,

$$g_{wh}(x_w,x_h,t_w) = f(t_w),$$

where

$$f(t) = \begin{cases} 900t, & 0<t<1/3 \\ 300, & 1/3<t<2/3 \\ 300-900t, & 2/3<t<1. \end{cases}$$

It follows that

$$G_{wh}(x_w,x,t_w) = (5-x)f(t_w).$$

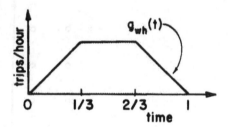

Fig. 11. Assumed Travel Pattern

The road has two lanes in each direction, hence a one-way saturation
flow of s = 3600 vehicles/h was assumed and, with a little more than
half the green time available to the major road, a capacity of c = 2000
vehicles/h. Other parameters appropriate to a street in a downtown area
were substituted into Eq. (3) to yield a tardity in minutes/km of

$$\Lambda(x,q) = 1.2 + 1.5/(1-q/s), \quad q<c.$$

6.3. Results

The results are shown in Fig. 12. The actual calculation used z increments of 500m; four of these trajectories are plotted on the diagram, as
are contours of equal traffic flow.

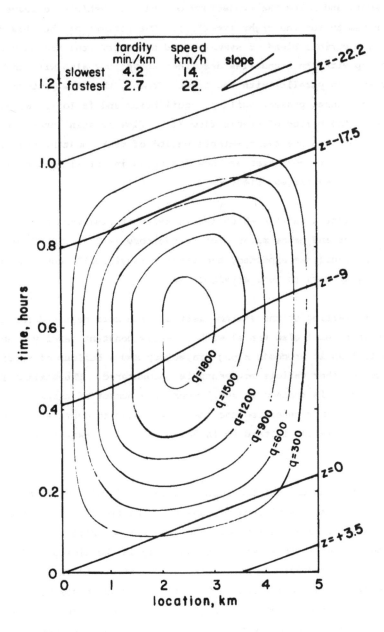

Fig. 12. Flows and Selected Trajectories which
Result from the Travel Pattern Shown in Fig. 11

Note that the region of traffic activity lies to the right of the time
axis, above the distance axis, to the left of the right hand end of the
road (x=5), and below the trajectory of the last vehicle to leave the
left extremity for the right (z=-22.2). The pattern of the flow contours is a curious blend of symmetry and asymmetry: the contours attempt
to line up with the boundaries described above, but also with the trajectories. In parallel with the travel demand, the flow pattern can be
broken into three phases: buildup, equilibrium and falloff. Note, however, that the period of stable flow (i.e. flow constant over time) is
much shorter than the twenty-minute period of peak constant travel
demand. These three phases are now discussed in detail by the examination of four key trajectories.

First, consider a vehicle which leaves x=3.5 at time zero for the
extreme right and hence moves along trajectory z=+3.5. The flow is very
small, and hence, as expected, the tardity is virtually constant and the
trajectory appears to be a straight line.

A vehicle starting at the extreme left at time zero destined for the
extreme right will also travel with a nearly constant tardity. But
notice that as it proceeds along trajectory z=0 a buildup of traffic
flow due to other vehicle movements is encountered. The maximum flow of
420 vehicles/h is reached at x=3.5 where the number of trips terminating
exceeds the number generated and the flow begins to decrease. Even
though the variation in tardity is very small, the trajectory has a
discernible ogive shape.

A vehicle leaving the extreme left at time t=0.33 will do so at the
beginning of peak travel demand. However, the traffic flow pattern has
not yet stabilized at this time. The only trajectories that can possibly be stable are those which lie entirely in the time period 0.33 to
0.66, i.e. -9<z<-7.33. In this example, the first and virtually only
trajectory with a stable flow pattern is z=-9.0 which starts at time
0.405. If the flow is stable along this trajectory, then the flow can
be calculated by the following argument. The number of vehicles passing
a point x will be in proportion to the length of road to the left
multiplied by the length of road to the right, i.e. x(5-x). At the
time of peak travel demand, trips are being generated at the rate of 300

vehicles per kilometre per kilometre per hour. Hence, the steady state flow along the trajectory is given by $300x(5-x)$, i.e.

x:	0	1	2	2.5	3	4	5
flow:	0	1200	1800	1875	1800	1200	0

It can be seen that these values are consistent with trajectory $z=-9.0$ in Fig. 12 and give it a distinct ogive shape.

A trajectory typical of the falloff phase ($z=-17.5$) has a different ogive shape. A vehicle leaving the extreme left at time $t=0.788$ experiences a rapid buildup of traffic flow to 600 vehicles/h at $x=1$, but until it reaches $x=3.5$, the flow remains relatively constant in the range 600 to 850 vehicles/h.

7. DISCUSSION

In its present form, the model deals with situations where no intersection is saturated. The consideration of oversaturation is the obvious next task. Once this difficulty is overcome, we hope that the model will prove to be a powerful tool for the examination of complex traffic phenomena in urban areas. Even in its current form, however, the model provides a limited capability for analyzing realistic situations which could not be handled by its predecessors. While we have only considered the journey from work to home, the journey from home to work can be dealt with in a very similar fashion.

Acknowledgements

This paper would not have been possible without the assistance of the Department of Civil Engineering, University of Toronto, and the Natural Sciences and Engineering Research Council of Canada.

REFERENCES

Kimber, R.M., and E.R. Hollis (1979). Traffic Queues and Delays at
    Road Junctions, Laboratory Report No 909 (Transport and Road
    Research Laboratory, Crowthorne).

Lighthill, M.J., and G.B. Whitham (1955). On Kinamatic Waves; II. A
    Theory of Traffic Flow on Long Crowded Roads, Proceedings of
    The Royal Society, Series A, 229, pp. 317-345.

Leutzbach, W. (1972). Einführung in die Theorie des Verkehrsflusses
    (Springer-Verlag, Berlin).

Oxford English Dictionary (1933). (Oxford University Press, Oxford),
    Vol. XI, pg. 91.

Robertson, D.I., and P. Gower (1977). User Guide to TRANSYT Version 6,
    Supplementary Report No 255 (Transport and Road Research
    Laboratory, Crowthorne).

Tournerie, G., (1969). Sur la définition des grandeurs caractéristiques
    d'une circulation, in Beiträge zur Theorie des Verkehrsflusses,
    edited by W. Leutzbach and P. Baron, Strassenbau und Strassenver-
    kehrstechnik, Heft 86.

Webster, F.V. (1958). Traffic Signal Settings, Road Research Technical
    Paper No. 39   (Her Majesty's Stationary Office, London).

PAPER 9

Ninth International Symposium on
Transportation and Traffic Theory
© 1984 VNU Science Press, pp. 179–195

EVALUATION OF DIMENSIONALITY REDUCTION ON NETWORK TRAFFIC PATTERN RECOGNITION

SHIH-MIAO CHIN[1] and AMIR EIGER[2]
[1] *Greenman-Pedersen, Inc., Albany, NY, U.S.A.*
[2] *Rensselaer Polytechnic Institute, Troy, NY, U.S.A.*

ABSTRACT

The traffic flow on a network at time t is defined by a vector V_t of dimensionality L comprising the L link volumes on the network. This paper examines four techniques for reducing the dimensionality of V_t and evaluates these techniques in terms of the resulting effectiveness of a network traffic pattern matching scheme. The results indicate that a significant reduction in dimensionality can be attained with little sacrifice in reliability.

1. INTRODUCTION

Given the spatial and temporal nature of urban activities, urban network
traffic flows are continuously changing throughout the day. The network
flow at time t is defined by a vector $V_t = [v_{1,t}, v_{2,t}, \ldots, v_{L,t}]$ of
volumes for the L links comprising the network. In many instances
arising in network traffic control, it is convenient to classify network
flows into one of several representative patterns in order to reduce the
complexity of the control system. Given a set of J predefined patterns
and   their   corresponding   pattern   signatures   (identifiers)
$\{S_j : S_j = [s_{1j}, s_{2j} \ldots, s_{Lj}]\}$,   any observed network flow vector V_t is
classified as pattern m at time t, denoted by m_t if the deviation
corresponding to the mth signature is minimal. That is,

$$D_{m,t} = \min_{j \in J} \{D_j, t\} \tag{1}$$

where

$$D_{j,t} = \sum_{\ell \in L} |v_{\ell,t} - s_{\ell j}| \tag{2}$$

is the jth signature deviation computed at time t and is the set of
links. The pattern matching scheme defined by Eqs. (1) and (2) can be
extended to include weighted occupancy factors in addition to traffic
volumes. However, in this paper we deal with the flow vectors in their
simplest form.

Urban traffic control systems acquire the link volume data through
area-wide detection systems which are costly to install and maintain
(FHWA, 1976; FHWA, 1979). In addition, system failure rates as high as
30% of detectors per year have been reported (Chandler, 1979).
Consequently, it is important to maximize the information collected
through the detection system at an acceptable cost. Equivalently, the
problem is to reduce the dimensionality of V_t without adversely affecting
the reliability of the pattern matching scheme.   This paper presents and
evaluates potential dimensionality reduction procedures.

The traffic data used in the study was made available from the computerized traffic control system in the City of Clearwater, Florida. The Clearwater system consists of 43 computer controlled intersections with 36 instrumented links (L = 36). Figure 1 illustrates the location of the links on the network diagram. The database consisted of 96 daily 15-minute volume counts at each detector for a period of 8 days.

For illustrative purposes in this study 4 independently determined network traffic patterns were used (i.e., J = 4). Their signatures correspond to the network flow vectors at 9:00, 17:45, 20:00 and 22:30 on October 14, 1981.

2. DIMENSIONALITY REDUCTION PROCEDURES

2.1 Zero-One and Linear Programming

Recalling the definitions of $D_{j,t}$ and $D_{m,t}$, let

$$\Delta D_{j,t} = D_{j,t} - D_{m,t} \geq 0$$

$$= \sum_{\ell \varepsilon} (|v_{\ell,t} - s_{\ell j}| - |v_{\ell,t} - s_{\ell m}|) \geq 0$$

$$= \sum_{\ell \varepsilon L} a_{\ell j,t} \geq 0 \tag{3}$$

where $a_{\ell j,t}$ is defined by the above expression.

Consider a subset $L' \subset L$ of L network links and let the minimum signature deviation in the reduced space be computed by

$$D_{m',t} = \min_{j \varepsilon J} \sum_{\ell \varepsilon L} |v_{\ell,t} - s_{\ell j}|. \tag{4}$$

It follows that if for all time intervals t, the indices m'_t and m_t of the pattern classification are the same, then the identified patterns using the volume data from the reduced set of links are identical

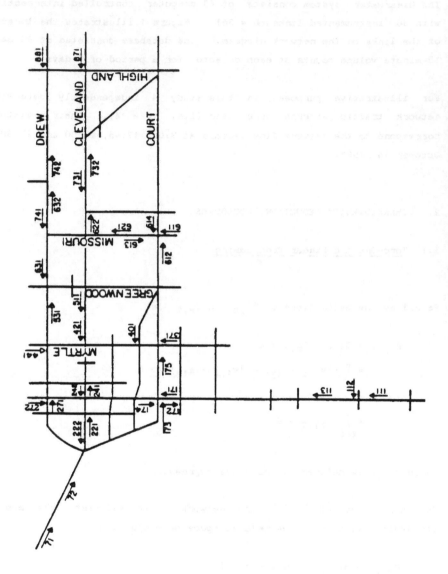

Fig. 1. Link diagram showing detector locations.

to those using the entire set. Equivalently, if

$$\sum_{\ell \in L'} a_{\ell j, t} \geq 0 \qquad \forall t, \; j \neq m_t \tag{5}$$

then $m'_t = m_t$ for all t.

The requirement imposed by Eq. (5) can be incorporated into the following zero-one programming formulation, in which one could also minimize the dimensionality of L':

$$\min \sum_{\ell \in L} \delta_\ell \tag{6}$$

$$\sum_{\ell \in L} a_{\ell j, t} \, \delta_\ell \geq 1 \qquad \forall t, j \neq m_t$$

$$\delta_\ell = 0, 1.$$

Solving the above problem using one day's network volume data (Wednesday: October 14, 1981) yields a subset of 9 links $L' = \{71, 112, 113, 172, 175, 222, 241, 271, 871\}$.

If the integer constraints are relaxed, Prob. (6) results in a linear programming (LP) formulation in which the final values of the basic variables can be considered as weights in the pattern matching algorithm. In this case

$$D_{m',t} = \min_{j \in J} \sum_{\ell \in L'} w_\ell \; | \; v_{\ell, t} - s_{\ell j} |, \tag{7}$$

where w_ℓ is the weight assigned to link ℓ.

The linear programming solution to Prob. (6) using the same data is $L' = \{71, 113, 171, 222, 242, 271, 421, 441, 511, 611, 612, 622, 731, 732, 741\}$ with corresponding weights: 0.01256, .00153, .00196, .00985, .00202, .01302, .0088, .00043, .00369, .00530, .00456, .00122, .06257,

.00189, .00875.   As can be seen, the LP formulation produces a rather
arbitrary set of weights which have no particular physical significance.
Consequently, the zero-one programming solution is preferred.

Note in the above discussion that the pattern matching is accomplished by
computing the distance metric in a <u>reduced</u> <u>space</u> as determined either by
the zero-one or LP problem solution.   In the following sections we
discuss another class of techniques in which the pattern matching is done
in the full space but some elements of the network flow vector V_t are
estimated (i.e., field data is only required for the subset L' of links).

2.2   Linear Model and Set Covering

Assume there exists a linear relationship between the volumes on links i
and k given by:

$$v_{i,t} = \alpha_{ik} + \beta_{ik} v_{k,t} + e_{ik,t} \tag{8}$$

where $v_{i,t}$ denotes the volume on link i at time t, α_{ik} and β_{ik} are
constants, and $e_{ik,t}$ is the error.   If one further assumes that the
pattern signatures have been derived through linear combination of
observed flow vectors, then

$$s_{ij} = \alpha_{ik} + \beta_{ik} s_{kj} + e_{ikj}. \tag{9}$$

Consequently,

$$D_{j,t} = \sum_{\substack{\ell \in L \\ \ell \neq i,k}} |v_{\ell,t} - s_{\ell j}| + |v_{i,t} - s_{ij}| + |v_{k,t} - s_{kj}|$$

$$\cong \sum_{\substack{\ell \in L \\ \ell \neq i,k}} |v_{\ell,t} - s_{\ell j}| + (1 + \beta_{ik}) |v_{k,t} - s_{kj}| + e_{ik} \tag{10}$$

where e_{ik} is a composite error term. If e_{ik} in Eq. (10) is small, the effectiveness of the pattern matching algorithm can be maintained. The acceptable magnitude of e_{ik} depends on the number of predefined patterns and the distance separation between them. The dimensionality reduction problem is to find a subset $L' \subset L$ of links so that the effectiveness of the pattern matching algorithm is not significantly reduced by using volume estimates for the links in the complementary set $\bar{L} = L - L'$.

In a network with L links, one can estimate L^2 linear functions of the form shown in Eq. (8). Suppose that the error sums of squares,

$$SSE_{ik} = \sum_t | v_{i,t} - \alpha_{ik} - \beta_{ik} v_{k,t} |^2, \tag{11}$$

is considered as a measure of distance from link k to i. Moreover, if SSE_{ik} is smaller than a prespecified tolerance, link i is said to be "covered" by link k. Then, one can use a set covering algorithm to determine the smallest subset of links which covers the entire set. Note, of course, that the size of the determined subset depends on the preselected error tolerance.

Let $T = [t_{ij}]$ be an LxL matrix where the elements are defined by:

$$t_{ij} = \begin{cases} 1 & SSE_{ij} < \text{error tolerance} \\ 0 & \text{otherwise.} \end{cases} \tag{12}$$

The set covering problem is to reduce the number of rows in T so that each column in the reduced matrix T' has at least one non zero element. The minimal link subset L' corresponding to the preselected tolerance is then composed of the row indices of T'.

Setting the error tolerance level at 52,900 (i.e., an average of 23.5 vehicles/15 minute time interval) and using the set covering algorithm described by Larson and Odoni (1981), the reduced set becomes $L' = \{72, 113, 171, 421, 611, 614, 731, 732, 742\}$.

Having obtained L' one must evaluate the effectiveness of the resulting pattern matching algorithm which, in this case, is based on the following minimum signature deviation:

$$D_{m',t} = \min_{j \epsilon J} \sum_{\ell \epsilon L'} (1 + \sum_{i \epsilon L} \beta_{i\ell}) \mid v_{\ell,t} - s_{\ell j} \mid . \qquad (13)$$

This can be done by computing the frequency of occurrence of pattern misclassification as will be described in subsequent sections.

2.3 Transfer Function Model and Set Covering

The development in section 2.2 was based on the assumption that the relationships existing between link traffic volumes can be adequately represented by linear models. Chin and Eiger (1984) have shown that for upstream-downstream link pairs it is more reasonable to assume a time series transfer function model of the form (see Box and Jenkins, 1976):

$$v_{i,t} = \nu(B)v_{k,t} + \psi(B) a_t \qquad (14)$$

where a_t is a white noise series, and $\nu(B)$ and $\psi(B)$ are polynomial operators. In term of the backward shift operator B, where $Bv_{i,t} = v_{i,t-1}$, $\nu(B)$ and $\psi(B)$ are given by:

$$\nu(B) = \nu_0 + \nu_1 B + \ldots + \nu_r B^r \qquad (15)$$

$$\psi(B) = \psi_0 + \psi_1 B + \ldots + \psi_q B^q . \qquad (16)$$

As in the previous section, one can consider the sums of squares of the white noise series (computed during the parameter estimation procedure) as a distance measure between link pairs and employ a set covering technique to find the reduced link set.

The appropriate pattern classification criterion when using this method is

$$D_{m',t} = \min_{j \epsilon J} \{ \sum_{\ell \epsilon L'} \mid v_{\ell,t} - s_{\ell j} \mid + \sum_{\ell \epsilon \bar{L}} \mid \tilde{v}_{\ell,t} - s_{\ell j} \mid \} \qquad (17)$$

where L' is the subset of links used to derive volume estimates for the remaining links and $\tilde{v}_{\ell,t}$ denotes the transfer function estimate for link ℓ obtained from the model of Eq. (14).

The procedure described above results in the subset $L' = \{72, \quad 113, \quad 172, \quad 421, \quad 611, \quad 612, \quad 614 \quad 732, \quad 742\}$.

3. EVALUATION

In the previous sections, four techniques have been described for reducing the dimensionality of the network flow vector V_t. Recall that in deriving the link subset L', only one day's data was used. The remaining data (7 days) was used to evaluate the results.

To set the basis for comparison, Table 1 illustrates the pattern number (j = 1,2,3,4) selection using the field volume data for all 36 network links. Each row represents the 96 daily 15-minute time intervals. Each procedure was evaluated by comparing the pattern number identified using the volume counts from all 36 links (see Eqs. (1) and (2)) with the pattern selected via the reduction technique. Tables 2 through 5 contain the results. The last column in each row of each table indicates the number of times a pattern mismatch occurred in a given day. A mismatch is denoted by a letter (A=1, B=2, C=3, D=4) indicating the pattern number identified using the full link set (shown in Table 1). Certain entries in the tables have been underlined to indicate that during the corresponding time intervals, operator intervention took place and the volume counts may not be accurate.

An examination of the tables reveals that the mismatch percentage for the estimation techniques (linear model and transfer function) with 9 predictor links is 8-9%. By contrast, the zero-one programming solution with 9 links results in 11% mismatch. This small difference and the fact that the transfer functions did not outperform the linear models is surprising. However, one should note that the set covering technique does not yield a unique solution so that the link set used to produce the results in Table 5 may be non-optimal. Moreover, many of the links exhibited very similar traffic volume patterns so that linear models may

```
W OCT 14  4444444444444444444444444444444443111111111111111111112222222222222222222222222222222222222222223333333333334444444444444
T NOV  3  4444444444444444444444444444444443111111111111111111112222222222222222222222222222222222222222223333333333333333444344444
W NOV  4  5444444444444444444444444444444443311122222224231111111222222222222222222222222222222222222222112233333333333344444344444
T NOV  5  4444444444444444444444444444444444441111111111111111111112222222222422222222222222222222222222233333333333333333333334444
F NOV  6  4444444444444444444444444444444444331111111111111111111111122222222222222222222222222222222222233333333333333333333333444
S NOV  7  4444444444444444444444444444444444444444444444433333333332222222222222222222222222222222222233433333333333333333333333444
S NOV  8  4444444444444444444444444444444444444444444444444333333332222222222222222222222222222223222311133333333333344444444444444
M NOV  9  444444444444444444444444444444444444444444444311114224111222222222222222222222222222222222222333333333444444444444444444444
```

Table 1.   Pattern selection: full link set (36 links)

```
W OCT 14  44444444444444444444444444444444431111111111111111111122222222222222222222222222222222222222222233333333333333334444444444444444444  15
T NOV  3  4444444444444444444444444444444D31111DDDDDDDDDDD2222222C2222222222222222C33333D3D3444344444444444444444  8
W NOV  4  5444444444444444444444444BBB2222B4211111A2222222222222222A222222C33333344444D44444444444444C44444444  5
T NOV  5  4444444444444444444444444444C1111111111112A22222222222222222C3333333333333333C1333CC  6
F NOV  6  444444444444444444444444D31111111111111A22222222222222222223C133333333333333C1313CC  10
S NOV  7  444444444444444444444444444444444D3133332C222222C22222222223122BB13B33333333CC34C4  23
S NOV  8  4444444444444444444444444444444444D341113133331B1BBBBBBBBBB3B3333DCCC44444444444444  +8
M NOV  9  4444444444444444444444444444444444D31111AAA22222222222222222B233333DC3C44444444444444  ──
                                                                                             75
                           TOTAL NUMBER OF MIS-MATCHED TIMING PLANS
```

Table 2.   Pattern selection: zero-one programming (9 links)

Table 3. Pattern selection: linear programming (15 links)

	Pattern	Mis-matched
W OCT 14	44444444444444444443111111111111111112222222222222222223333333333444444444	2
T NOV 3	44444444444444444444444D31111111111A22222221111A22222221222AA222A222222223333333344444444	13
W NOV 4	444444444444444444444443CC2222CB42111C2222A222224222222A2A2222222C3333333330D44444D44444444	7
T NOV 5	444444444444444444444444C111111111C2222222222222222C333333333333C44444C44444444	7
F NOV 6	44444444444444444444444D31111111BB2A2222222222222222323333333333333C333CD44444	6
S NOV 7	4444444444444444444444444D3333333222222222222222BB33B33333333333334CD4444	11
S NOV 8	444444444444444444444444443D43333333332333333B3B3BBBBB33B33333333CC4444444444444	+10
M NOV 9	44444444444444444444444443D3111114C24111AAA2A22222A22AA12222222223333333333C44444444444	56

TOTAL NUMBER OF MIS-MATCHED TIMING PLANS   56

Table 4. Pattern selection: linear model and set covering (9 links)

	Pattern	Mis-matched
W OCT 14	44444444444444444444444C3111111111B22222A22A222A22222222222222333333333444444444	6
T NOV 3	444444444444444444444443311111111BBBBB2222222222222222222222B33333333444344444444	6
W NOV 4	444444444444444444444D3CC2CCC34C3B1122222221222222C3333333334444444444444	9
T NOV 5	4444444444444444444444C111111111BCCCCCCCCC4222222222222222CB3333333C44C44444444444	17
F NOV 6	444444444444444444444444331111111111111111B22222222222222222B33333333D3333C333CC	5
S NOV 7	44444444444444444444444444444444443333333C22222222222222223CB33333333D333D333D333C3444	5
S NOV 8	44444444444444444444444444444444440D433333333CC33333333333333C444444444444	5
M NOV 9	44444444444444444444444444331114CC411B2222222222222222222333333333C444444444	5

TOTAL NUMBER OF MIS-MATCHED TIMING PLANS   60

have been sufficient. Figure 2 illustrates the traffic volume time series on link 111. The predicted volume for link 441 using the transfer function model with link 111 as predictor is shown in Fig. 3. For this particular link pair, it is apparent that a simple linear model would suffice.

A further comparison is provided in Tables 6 and 7 which illustrate the results of the transfer function with 11 links and the linear model with 12 links. Note the marked reduction in the mismatch percentage in both cases, 4.9% and 5.8%, respectively. Moreover, in this case the transfer function technique produces better results than the linear model with one less link in the predictor set.

As an additional test of performance, the linear model-set covering technique was used in a test case with 7 predefined patterns (4 daytime, 3 nighttime plans). The 12 links which were generated were used in the pattern matching algorithm with a resulting mismatch of 5.8%.

It is noteworthy that if one discounts the data corresponding to time periods during which operator intervention occurred, the resulting mismatch frequency is considerably reduced.

4. CONCLUSIONS AND APPLICATION

It is noted that in this paper we do not address the prediction problems arising from the fact that plan selection is based on data which lags the time interval for implementation. On the basis of the foregoing results, however, it appears that any one of several techniques can be employed to reduce the dimensionality of the flow vector with a reliability in excess of 0.90. In particular, the techniques which are based on link volume estimation in conjunction with set covering produce excellent results as demonstrated by a pattern mismatch percentage of less than 5% with approximately one third of the original number of links. Furthermore, it is suspected that the set covering technique can be improved upon by essentially formulating the problem as a p-median problem in which one attempts to find the subset of $p = L'$ links so as to cover the entire set of L links with the minimum total distance (square error).

Table 5. Pattern selection: transfer function model and set covering (9 links)

Day		Count
W OCT 14	...B2222A2222...B3333...3D4444D4444	3
T NOV 3	...C11BB2222...B3333...3D4444D4444	5
W NOV 4	...C23C3111...B3333...3C4444D4444	7
T NOV 5	...CCCCCC2A...B3333...3C44C4444	16
F NOV 6	...B2A2222...B2333...334 3334	3
S NOV 7	...CC2222A2...2122B3333...3DDDDDDDDD4D444	20
S NOV 8	...3C3333...3DD4444444	5
M NOV 9	4CC4111122A2222...B3333...34 4444444444	4

TOTAL NUMBER OF MIS-MATCHED TIMING PLANS 63

Table 6. Pattern selection: transfer function (11 links)

Day		Count
W OCT 14	...B1A2222...B3333...34 4444444444	1
T NOV 3	...B1111222...B3333...34 3444444444	3
W NOV 4	D3CAAA2234C3111...B3333...3C44444444	8
T NOV 5	...C222C2222...B3333...3C444444444	5
F NOV 6	...3A111111...2222...2B33333334 33334	2
S NOV 7	4C4...333333...322B3333...34 4430444	4
S NOV 8	4C4...4444DD433333...3C23333...3D33333 4304444	+6
M NOV 9	4C4111AA2222...B3333...3C4444444444	5

TOTAL NUMBER OF MIS-MATCHED TIMING PLANS 34

Table 7. Pattern selection: linear model (12 links)

Day		Count
W OCT 14	...C3111111111B2222A222A222...2223333...3334 4444444444	5
T NOV 3	4C433B1B1111...BB1112222...2223333...34 44444444444	4
W NOV 4	...4D31A2A22B4C3111222...212A22...A2222...2233333...3C44444444	8
T NOV 5	...4C11111111BCCCCCC2C224222...CB3333...3C4C4444444444	12
F NOV 6	...11111111122A2222...2233333...33334 433344	1
S NOV 7	...44443C3333...32233333...3DC4444444444	5
S NOV 8	...33333333333333D44304444	3
M NOV 9	4C24111112A222A22...2B3333...34 4444444444	+6

TOTAL NUMBER OF MIS-MATCHED TIMING PLANS 44

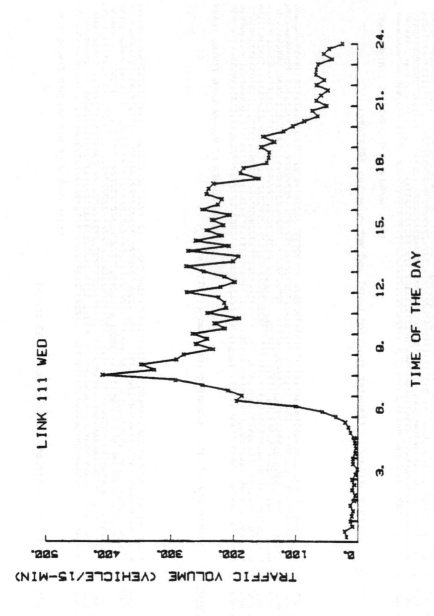

Figure 2. Traffic volume for link 111.

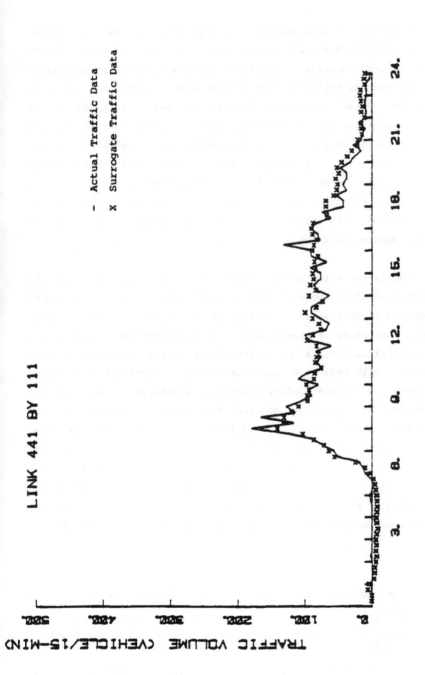

Figure 3.   Predicted traffic volume for link 441 using transfer function from link 111

4.1 Potential Applications of Dimensionality Reduction

Computerized traffic control systems require extensive detectorization for gathering traffic flow information both for performance monitoring and traffic control purposes. The level of instrumentation depends on the particular control system and ranges approximately from one in every four links for First Generation Systems to every link for Third Generation and Adaptive Control Systems. The control concept of Adaptive Systems (distributed) obviously does not permit much flexibility in the level of instrumentation. Centralized systems such as First and Second Generation Control, however, can be operated with fewer detectors if, in the former case, the reliability of the pattern matching algorithm can be maintained, and in the latter case, surrogate traffic data can be generated for undetected flows.

The dimensionality reduction schemes discussed in this paper are directly or indirectly related to both of the aforementioned cases. Additional work is required in this area to verify the results on larger networks. In addition, sensitivity analyses should be conducted to determine the effect of seasonal variations in traffic flows on the resulting control effectiveness. With regard to implementation, the primary difficulty associated with the dimensionality reduction techniques is the need for extensive traffic data which is not available prior to the installation of the system. Consequently, it would not be possible to use these techniques to reduce the level of instrumentation unless the detection subsystem is installed in stages. On the other hand, these techniques may prove very useful during detector downtime to provide information for pattern matching or signal timing plan optimization. As this relates to Second Generation Control, the prediction aspects have to be addressed.

REFERENCES

Box, G.E.P., G.M. Jenkins (1976). Time Series Analysis: Forecasting and Control (Holden-Day, San Francisco), pp. 370-402.
Chandler, M.J.H. (1979). Progress of Urban Traffic Control in London. In: Proceedings of Engineering Foundation Conference on Research Directions in Computer Control of Urban Traffic Systems, Pacific Grove, California.
Chin, S.M., A. Eiger (1984). Generating Surrogate Link Traffic Volumes Using Transfer Functions. Presented at the Annual Meetings of the Transportation Research Board, Washington, D.C.

FHWA, U.S. Department of Transportation (1976). Traffic Control Systems Handbook. Stock No. 050-001-001144.

FHWA, U.S. Department of Transportation (1979). Computerized Signal Systems: Instructor's Guide.

Larson, R.C., A.R. Odoni (1981). Urban Operations Rsearch (Prentice-Hall, New York).

Ninth International Symposium on
Transportation and Traffic Theory
© 1984 VNU Science Press, pp. 197–216

OPTIMAL SIGNAL CONTROLS ON CONGESTED NETWORKS

CAROLINE S. FISK

Department of Civil Engineering, University of Illinois at Urbana-Champaign, Urbana, IL, U.S.A.

ABSTRACT

A formulation of the optimal signal control problem is studied in which network flows are assumed to satisfy the user optimal route choice criterion. The formulation consists of expressing the route choice conditions in the form of a single nonlinear constraint. It is then shown that an approximate penalty formulation can be easily obtained and an algorithm for solving the problem is presented.

1.   INTRODUCTION

The optimal signal control problem consists of determining signal timings
at signalized intersections of a street network so that some network performance function such as total travel time is minimized. In the most
general problem the signal variables include green splits, cycle times,
and offsets. The performance function is obtained by specifying the
travel time function for each link as a function of these variables and
the link flows which, in early formulations of this problem, were fixed
at observed values obtained from traffic count data. However this
approach does not recognize the impact that changing signal controls
will have on route choice because of changes in travel times. In fact
once the new timings are implemented, the network link flows will change
and the timings will no longer be optimal.

The desired optimal solution has to be derived from a model which incorporates user route choice behavior either in the form of constraints, or
implicitly in the objective function by recognizing the dependence of
the link flows on the signal variables. In the next section a general
formulation of this problem is given from which several alternative formulations are derived and discussed in relation to previous work. One
formulation can be approximated in a natural way by a penalty function
and computational results for a simple example are given by way of
illustration. An algorithm for solving this problem is then proposed
and applied to a test problem.

2.   GENERAL FORMULATION OF THE OPTIMAL SIGNAL CONTROL PROBLEM

The street network is represented by a directed graph $G(N,L)$ where N is
the set of nodes and L is the set of directed links with N and L elements
respectively. A network performance function $P(f, s)$ is specified as a
function of the network link flow vector $f \epsilon R^L$ and the vector of signal
variables $s \epsilon R^S$ with S components.

The objective of the traffic operations analyst is to determine s so that

P is minimized, the link flows being constrained to satisfy conditions representing route choice behavior assumed here to be user optimal. Each user seeks to minimize his/her own travel time resulting in the conditions that travel times on used paths between a given origin/destination (O/D) pair are equal and travel times on unused paths are at least as high.

As shown by Aashtiani (1979) these conditions can be formulated in a general manner as a nonlinear complementarity problem (NCP). Let $h \epsilon R^K$ be a path flow vector for the K network paths, and $u \epsilon R^J$ be the O/D travel time vector, where J is the number of O/D pairs. The NCP expressing the user optimality conditions for a given signal vector s is

$$F(y, s) \cdot y = 0 \tag{1}$$

$$F(y, s) \geq 0 \tag{?}$$

$$y \geq 0 \tag{3}$$

where $F = \begin{bmatrix} F^1 \\ F^2 \end{bmatrix} \qquad y = \begin{bmatrix} y^1 \\ y^2 \end{bmatrix}$

$$F_r^1 = C_r - u_j \qquad \text{all } r \epsilon P_j, \ j \epsilon J \tag{4}$$

$$F_j^2 = \sum_{r \epsilon P_j} h_r - d_j \qquad \text{all } j \epsilon J \tag{5}$$

$$y_r^1 = h_r \qquad \text{all } r \epsilon P \tag{6}$$

$$y_j^2 = u_j \qquad \text{all } j \epsilon J \tag{7}$$

J is the set of O/D pairs

P_j is the set of paths between $j \epsilon J$

$$P = \bigcup_{j \epsilon J} P_j$$

$C_r(h, s)$ = travel time on path r

d_j = given number of trips between $j \epsilon J$

If A is the link path incidence matrix for the network, i.e.

$$a_{\ell r} = \begin{cases} 1 \text{ if } \ell \text{ lies on path } r \\ 0 \text{ otherwise,} \end{cases}$$

then h induces the link flow pattern

$$f = Ah \tag{8}$$

and C can be obtained from the link travel time vector c as

$$C = A^t c \tag{9}$$

The problem then is to solve

$$PO: \quad \min_{\substack{s \in S \\ y \in R^{K+J}}} \quad P(f, s) \tag{10}$$

s.t.

$$F(y, s) \cdot y = 0 \tag{11}$$

$$F(y, s) \geq 0 \tag{12}$$

$$y \geq 0 \tag{13}$$

where S designates the set of feasible signal variables. In this paper we consider only the green times as variables; the inclusion of offsets introduces integer variables (Gartner et al. 1975) and will be investigated in a future paper. The cycle times could also be included as variables; normally all intersections will have the same cycle time in a given period so that only one additional variable would be needed. For simplicitly in this presentation it is considered fixed.

The performance function P is taken to be the total travel time on the network and has the form

$$P(f, g) = \sum_{\ell} f_{\ell} c_{\ell}(f, g_{\ell})$$

where g_{ℓ} is the green time on link ℓ. Each component c_{ℓ} of c is a specified function of flows which interact on link ℓ and the green time g_{ℓ} (if the intersection is signalized). To model travel times realistically, each turning movement at the intersection should be incorporated into a separate link, which can be accomplished by placing the network nodes mid-block. Let I be the set of intersections in the network, then the green times for links approaching a given signalized intersection $i \epsilon I$ will satisfy some constraints, the form of which depends on the phase structure. For example at a four-way intersection with phases R-G-Y on each approach (with the yellow time included as part of the 'green' time), the following conditions will hold:

$$g_1 = g_2 = g_3 \qquad\qquad g_4 = g_5 = g_6$$
$$g_7 = g_8 = g_9 \qquad\qquad g_{10} = g_{11} = g_{12}$$
$$g_2 = g_8 \qquad\qquad\quad g_5 = g_{11}$$
$$g_2 + g_5 = T_i$$

g_ℓ is the green time on link ℓ with the turning movement as indicated in Fig. 1 and T_i is the cycle time.

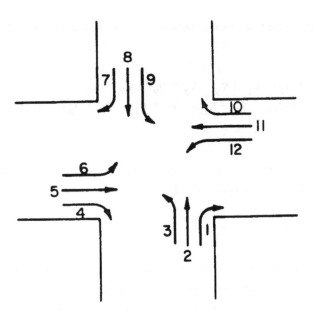

Fig. 1. Links approaching a four-way intersection.

These conditions are analogous to flow conservation constraints for flows on the network in Fig. 2.

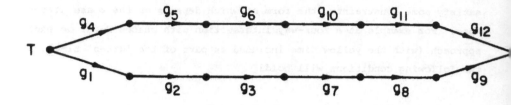

<u>Fig. 2.</u> Signal network for simple R-G-Y phases.

Let b_1 and b_2 designate respectively the flow on the lower and upper path, then

$$b_1 + b_2 = T_i$$

If the intersection has a left turn arrow phase, then the constraints are:

$$g_1 = g_2, \quad g_4 = g_5, \quad g_3 = g_9, \quad g_7 = g_8$$
$$g_{10} = g_{11}, \quad g_6 = g_{12}, \quad g_5 = g_{11}, \quad g_8 = g_2,$$
$$g_1 + g_3 + g_4 + g_6 = T_i$$

This has the network flow representation shown in Fig. 3. If b_k is the flow on path k $(k = 1, \ldots, 4)$ then

$$b_1 + b_2 + b_3 + b_4 = T_i$$

Since there will be fewer paths than links in these intersection networks, it is convenient to transform the optimization problem into the path flow variables b_k. Let I_s designate the set of signalized intersections in the street network, L_i is the set of approaching links for $i \in I_s$, and M_i is the arc chain incidence matrix for the network representing the signal constraints at i. PO can now be written more specifically as

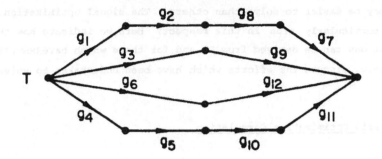

Fig. 3. Signal network for intersection with a left turn arrow phase.

P1: min $\sum\limits_{i \in I} \sum\limits_{\ell \in L_i} f_\ell c_\ell (f, g_\ell)$
$b \in B$
$y \in R^{K+J}$

s.t.

$F(y, g) \cdot y = 0$ $\qquad\qquad\qquad\qquad\qquad\qquad$ (14)

$F(y, g) \geq 0$ $\qquad\qquad\qquad\qquad\qquad\qquad\quad$ (15)

$y \geq 0$ $\qquad\qquad\qquad\qquad\qquad\qquad\qquad\quad$ (16)

If K_i is the set of paths for the signal network at i, B contains vectors b such that

$$\sum\limits_{k \in K_i} b_k = T_i \qquad\qquad i \in I_s \qquad\qquad\qquad (17)$$

and $b_k \geq 0$ $\qquad\qquad\qquad k \in K_i, \ i \in I_s$ $\qquad\qquad$ (18)

Also $g_\ell = \sum\limits_{k} m_{i\ell k} b_k$ $\qquad\qquad$ all $\ell \in L, \ i \in I_s$ $\qquad$ (19)

and we define G to be the set of g obtained from Eg. (19) for all $b \in B$.

3. ALTERNATIVE FORMULATIONS

Any given problem may often have several alternative mathematical formulations. This is important for solution purposes because some formula-

tions may be easier to solve than others. The signal optimization problem is particularly rich in this respect. Here we indicate how these
formulations can be derived from P1 and for those which have been studied
previously, discuss the efforts which have been undertaken to solve them.

3.1 Hybrid Optimization Formulation

This is the formulation given in Tan et al. (1979) and is the first
attempt at including the route choice conditions in the form of constraints. Conditions (14)-(16) imply the relationship $F_r y_r = 0$ for
all $r = 1,\ldots, K+J$ which can be used to eliminate the vector u from
the constraints. We have

$$\sum_{r \in P_j} F_r y_r = \sum_{r \in P_j} (c_r - u_j) h_r = 0$$

$$\text{so that } u_j = \frac{1}{d_j} \sum_{r \in P_j} c_r h_r$$

Substituting into inequality (15)

$$c_r \geq \frac{1}{d_j} \sum_{q \in P_j} c_q h_q \qquad \text{all } r \in P_j,\ j \in J \qquad (20)$$

It is straightforward to show that inequaltiy (20) also implies $F_r y_r = 0$,
all $r \in P_j$, $j \in J$, and consequently P1 is equivalent to

P2:   min $P(f, g)$
       $g \in G$
       $h \in H$
       s.t. $c_r(h, g) \geq \frac{1}{d_j} \sum_{q \in P_j} c_q(h, g) h_q \qquad$ all $r \in P_j$, $j \in J$

together with the definitional constraints (8) and (19). H is the set of
feasible path flow variables, i.e. $h \in H$ if $F_j^2 = 0$ for all $j \in J$. In Tan et
al. (1979), an augmented Lagrangian approach for solving P2 is investigated but found to be impractical. They consequently proposed a heuristic (iterative optimization assignment) which involves alternating
between the signal optimization problem (with flows held fixed) and the
user optimal problem (with signal variables fixed). But as discussed

further below, this will only converge to the solution for a restricted
class of problems.

3.2 The infinitely constrained formulation

The NCP (14)-(16) is equivalent to the variational inequality (VI)

$$F(y, g) \cdot (v - y) \geq 0 \qquad \forall v \epsilon R_+^{K+J}$$

(see Karamardian (1971)). As shown in Fisk and Boyce (1983), this simplifies to

$$c(f, g) \cdot (e - f) \geq 0 \qquad \forall e \epsilon F \qquad (21)$$

where F is the set of feasible link flow variables; i.e. $F = AH$. This
formulation of the user optimal travel choice conditions was first derived in a different manner by Smith (1979).

P1 is then equivalent to:

P2:   min   P(f, g)
      $f \epsilon F$
      $g \epsilon G$
      s.t. $c(f, g) \cdot (e - f) \geq 0$ $\forall e \epsilon F$

This is an infinitely constrained problem, generally known as the Fritz
John problem (John 1949), and has been investigated by Marcotte (1981)
in connection with the continuous network design problem. One convergent
approach and five heuristic algorithms for solving this problem are suggested. The former consists in relaxing the constraints, and at each
step solving a finitely constrained problem. If the solution at this
step does not satisfy all of the constraints, then a constraint is added
and the step is repeated. However, as stated by Marcotte, this method
would not be practical on moderately sized networks due to the large number of constraints which would have to be considered simultaneously.
Also the optimization problem to be solved in each step would require a
considerable amount of computation time.

3.3 Implicit substitution formulation

The NCP conditions (14)-(16) imply a relationship between the signal controls and the user optimal network flows corresponding to these controls. In the case where there is a unique flow vector for a given g, then the correspondence is one-to-one and can be represented by the mapping

$$f = T(g) \tag{22}$$

where $T: R^{L_g} \to F \subset R^L$

and L_g is the number of links which contain signals.

Equation (22) can replace f in the objective function resulting in the problem

P3: $\min_{g \in G} P(T(g), g)$

The functional form of T(g) is not known explicitly, and is not differentiable at all points $g \in G$ (see Marcotte (1981)). In connection with a similar formulation of the network design problem, Abdulaal and LeBlanc (1979), have studied solution methods of Powell (1964) and Hooke and Jeeves (1961), which do not require derivatives of the objective function. However in each iteration of these algorithms a large number of user optimal assignment problems require solution which would be impractical in the case of moderately sized networks.

3.4 Maxmin formulation

If the cost functions c(f, g) satisfy the monotonicity condition

$$[c(f^1, g) - c(f^2, g)] \cdot [f^1 - f^2] \geq 0 \qquad \forall f^1, f^2 \epsilon F \atop f^1 \neq f^2 \tag{23}$$

for any $g \epsilon G$, then (21) is equivalent to the maxmin problem

$$\max_{f \in F} \min_{e \in F} c(f, g) \cdot (e - f) \tag{24}$$

(see Fisk and Nguyen (1982)). Let

$$w(f, g) = \min_{e \in F} c(f, g) \cdot (e - f)$$

then f^* is a solution to (24) if and only if $w(f^*, g) = 0$ (see Zukhovitshii et al. 1973). Also if f is not a solution, $w(f, g) < 0$. P2 is then equivalent to

P4: $\min P(f, g)$
    $f \in F$
    $g \in G$
    s.t. $w(f, g) = 0$

This formulation first appeared in Fisk (1983) and has not been investigated for solution algorithms.

Alternatively (24) is equivalent to

$$\min_{f \in F} \max_{e \in F} c(e, g) \cdot (f - e) \tag{25}$$

(see Zukhovitshii et al. 1973). Let

$$W(f, g) = \max_{e \in F} c(e, g) \cdot (f - e),$$

then f^* is a solution to (25) if and only if $W(f^*, g) = 0$, and if f is not a solution, $W(f, g) > 0$. P4 is then equivalent to

P5: $\min P(f, g)$
    $f \in F$
    $g \in G$
    s.t. $W(f, g) = 0$

3.5 Penalty formulations of the signal optimization problem

P4 and P5 can be approximated in a natural way by an unconstrained problem using properties of $w(f, g)$ and $W(f, g)$. These are:

   i)   Both $w(f, g)$ and $W(f, g)$ are continuous (assuming $c(f, g)$ to
        be continuous),

   ii) $w(f, g) \le 0$ and $W(f, g) \ge 0$ for all $f \epsilon F$, $g \epsilon G$,

   iii) $w(f, g) = 0$ and $W(f, g) = 0$, if and only if f is the user
        optimal solution corresponding to g.

These are the properties required for a penalty function (Luenberger
1973) so that an approximate solution to P4 can be found by solving

P6:   max $\{-P(f, g) + \mu w(f, g)\}$
       $f \epsilon F$
       $g \epsilon G$

where μ is a positive constant. In the limit $\mu \to \infty$ the solution to P6
will approach that of P4.

Similarly the penalty formulation of P5 is

P7:   min $\{P(f, g) + \mu W(f, g)\}$
       $f \epsilon F$
       $g \epsilon G$

4. SIMPLE APPLICATION OF PENALTY APPROACH

To obtain a better understanding of the penalty approach, it is applied
to the simple network shown in Fig. 4.

The set J of O/D pairs is $\{(a, b), (c, d)\}$ and there is a signal at the
intersection of links 1 and 3. The link travel time functions are

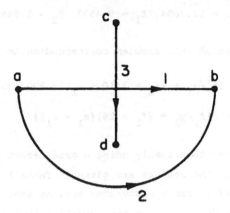

Fig. 4. Test network.

$$c_1 = \frac{f_1}{g_1} + 2 \qquad c_2 = 2f_2 \qquad c_3 = \frac{2f_3}{g_3}$$

where g_ℓ is the green time on link ℓ. In this problem

$$g_1 + g_3 = 20$$
$$d_1 = d_2 = 10$$

Formulation P3 can easily be obtained using the user optimal conditions between O/D pair 1 to obtain $T(g)$. Solving the equations

$$\frac{f_1}{g_1} + 2 = 2f_2$$
$$f_1 + f_2 = 10$$

gives

$$f_1 = 18g_1/(1 + 2g_1)$$
$$P = 10c_1 + 10c_3 = \frac{200 + 4g_1}{1 + 2g_1} + \frac{200}{20 - g_1}$$

where g_3 has been replaced by $20 - g_1$.

Solving for the optimality conditions $\frac{\partial P}{\partial g_1} = 0$ leads to the solution

$$g_1 = 7.7306, \quad g_3 = 12.2694, \quad f_1 = 8.4533, \quad f_2 = 1.5467$$

The maxmin formulation of this problem corresponding to P6 is

$$\max_{\{f_1, \ g_1\}} \quad \min_{e_1} \{-[f_1^2/g_1 + 2f_1 + 2(10 - f_1) + 200/(20 - g_1)]$$
$$+ \mu[f_1/g_1 + 2f_1 - 18][e_1 - f_1]\}$$

This problem was solved numerically using a grid search procedure for different values of μ; the results are given in Table 1. As μ increases the value of the penalty term w approaches zero as expected. For $\mu = 1$ the solution corresponds closely to the 'exact' solution given above.

μ	f_1	g_1	P	w	c_1	c_2
0	8.923	7.737	46.766	-8.9167	3.1533	2.1540
.005	8.90	7.737	46.767	-8.4578	3.1503	2.20
.01	8.877	7.736	46.77	-8.0025	3.1475	2.2460
.05	8.699	7.732	46.873	-4.5501	3.1251	2.6020
.07	8.616	7.732	46.967	-2.9840	3.1143	2.7680
.1	8.496	7.731	47.154	-0.7727	3.0990	3.0080
1	8.453	7.727	47.236	0.6767×10^{-4}	3.0940	3.0940

Table 1.   Solution of P6 with different values of μ.

5.   POSSIBLE ALGORITHMS FOR LARGE SCALE PROBLEMS

In this section two algorithms are described which would be suitable for solving large problems of the type P6 and P7; convergence can be guaranteed if the link travel time functions satisfy certain convexity requirements.

5.1 General description

These methods originate from game theory; the following description is
taken from Ermol'ev (1966). They have been proposed to solve the problem

$$\min_{x \in X} \max_{u \in \mathcal{U}} \phi(x, u)$$

where ϕ is continuous, convex in x, concave in u, and X and $\mathcal{U}$ are closed,
convex, bounded sets. The following iterative schemes can be used:

a). $x^{k+1} = \lambda_k x(u^k) + (1 - \lambda_k)x^k$

$u^{k+1} = \lambda_k u(x^k) + (1 - \lambda_k)u^k$

b). $x^{k+1} = \lambda_k x(u^k) + (1 - \lambda_k)x^k$

$u^{k+1} = u(x^{k+1})$

$x(u^k)$ and $u(x^k)$ are such that

$$\phi(x(u^k), u^k) = \min_{x \in X} \phi(x, u^k) \tag{26}$$

$$\phi(x^k, u(x^k)) = \max_{u \in \mathcal{U}} \phi(x^k, u) \tag{27}$$

and λ_k is chosen so that the following conditions are satisfied:

$$0 < \lambda_k \leq 1, \quad \lim_{k \to \infty} \lambda_k = 0, \quad \sum_{k=1}^{\infty} \lambda_k = \infty$$

Convergence of a) is guaranteed if $x(u^k)$ and $u(x^k)$ are unique; for b)
only $x(u^k)$ is required to be unique.

5.2 Application to signal optimization

The above algorithms can be applied to formulations P6 and P7 of the
signal optimization problems with

$$x = (f, g), \quad u = e.$$

For P6

$$\phi(f, g, e) = c(f, g) \cdot f - \mu c(f, g) \cdot (e - f)$$

Solving (27) amounts to finding the shortest O/D paths on the network with travel time vector $c(f^k, g^k)$. (26) is the minimization problem

$$\min_{\substack{f \in F \\ g \in G}} \{c(f, g) \cdot f - \mu c(f, g) \cdot (e^k - f)\}$$

which could be solved using the Frank-Wolfe algorithm, since by assumption, the objective function is convex on $F \times G$.

For P7

$$\phi(f, g, e) = c(f, g) \cdot f + \mu c(e, g) \cdot (f - e)$$

Conditions (26) and (27) are respectively equivalent to solving

$$\min_{\substack{f \in F \\ g \in G}} c(f, g) \cdot f + \mu c(e^k, g) \cdot (f - e^k) \tag{28}$$

and
$$\min_{e \in F} c(e, g^k) \cdot (e - f^k) \tag{29}$$

Both (28) and (29) could be solved using the Frank-Wolfe method.

Roughly speaking, each iteration of the above algorithms would be similar in terms of computation time, to solving two standard traffic assignment problems, i.e. traffic assignment with separable travel time functions, or in terms of total computation time, would take approximately twice as long as solving the asymmetric traffic assignment problem using relaxation schemes which require solution of a standard problem in each iteration.

5.3 Results for the test network

As an initial step in exploring properties of these methods, they are applied to the network problem in section 4. Formulation P7 is solved using method b) with μ = 1 and 10 and $\lambda_k = \frac{1}{k}$. Because

$$\lim_{g_\ell \to 0} c_\ell(f_\ell, g_\ell) = \infty$$

for ℓ = 1, 3, it was necessary to include a nonzero lower bound constraint for g; this would also be desirable in practice to ensure that there is sufficient green time for pedestrians to cross. The results of these calculations are given in Table 2.

μ	f_1	g_1	P	W	c_1	c_2
1	8.846	7.739	46.942	.082	3.143	2.308
10	8.453	7.417	46.46	$.5 \times 10^{-3}$	3.139	3.095

Table 2. Solution of P7 using algorithm b).

6. ITERATIVE OPTIMIZATION AND ASSIGNMENT ALGORITHM

In this section it is shown that this heuristic solution approach will not converge to the solution of the signal optimization problem. Similar results are presented in a different manner in Dickerson (1981). The steps involved can be summarized as follows using formulation P3:

Step 0. Set k = 0. Start with an initial approximation $g^0 \epsilon G$

Step 1. Obtain $f^k = T(g^k)$; i.e. solve the traffic assignment problem with fixed signal values g^k.

Step 2. If solution has converged, stop. Otherwise set k = k + 1 and solve

$$\min_{g \in G} P(f^{k-1}, g) = P(f^{k-1}, g^k)$$

Return to step 1.

Now let (f^*, g^*) be the solution to P3 and set $(f^k, g^k) = (f^*, g^*)$. For the above algorithm to converge to (f^*, g^*), it is necessary that $(f^{k+1}, g^{k+1}) = (f^*, g^*)$. This is not normally the case because g^{k+1} is obtained from

$$\min_{g \in G} P(f^*, g)$$

which can result in a value $P(f^*, g^{k+1})$ smaller than $P(f^*, g^*)$ because g is not constrained to satisfy the relationship $f^* = T(g)$.

The procedure may converge to a point which is the solution of the problem in which the optimality conditions for

$$\min_{g \in G} P(f, g)$$

and the user optimal conditions $f = T(g)$ are satisfied simultaneously. The above algorithm corresponds to the block Gauss-Seidel method for solving this problem and will converge to the solution if the mapping which takes a point (f^k, g^k) to (f^{k+1}, g^{k+1}), is a contraction mapping. This problem is not the same as the constrained optimization problems presented in the previous sections. A further discussion of this latter point can be found in Fisk (1983).

7. DISCUSSION

This paper has presented all known and some new formulations of the signal optimization problem which includes route choice behavior as a component of the model. Some approximate formulations are shown to have the form of minmax problems arising from game theory, and two solution algorithms are presented which would be feasible for large problems. These are applied to a test problem and found to perform satisfactorily. To ascertain their suitabilty for real world problems it would be necessary to examine the convexity properties of the function ϕ using realistic travel time functions.

Acknowledgements

This work was supported in part by a grant from the National Science Foundation (grant no. CEE 81-19772).

REFERENCES

Aashtiani, H. (1979). The multimodal traffic assignment problem. Ph.D. Dissertation, M.I.T., Sloan School of Management.
Abdulaal, M., L. LeBlanc (1979). Continuous equilibrium network design models. Transpn. Res. 13B, pp. 19-32.
Dickerson, T. J. (1981). A note on traffic assignment and signal timings in a signal controlled road network. Transpn. Res. 15B, pp. 267-271.
Ermol'ev, Y. M. (1966). Methods of solution of nonlinear extremal problems. Kibernetika, 2, pp. 1-17.
Fisk, C., S. Nguyen (1982). Solution algorithms for network equilibrium models with asymetric user costs. Transpn. Sci. 16, pp. 361-381.
Fisk, C., D. Boyce (1983). A general variational inequality formulation of the network equilibrium-travel choice problem. To appear in Transpn. Sci.
Fisk, C. (1983). Game theory and transportation systems modelling. To appear in Transpn. Res.
Gartner, N. H., J. D. C. Little, H. Gabbey (1975). Optimization of traffic signal settings by mixed integer linear programming. Transpn. Sci. 9, pp. 321-363.
Hooke, R., T. A. Jeeves (1961). Direct search solution of numerical and statistical problems. J. Ass. Comp. Mach. 8, pp. 212-229.
John, F. (1949). Extremum problems with inequalities as subsidiary conditions. Studies and Essays, Courant Anniversary Volume, (Interscience, New York), pp. 187-204.
Karamardian, S. (1971). Generalized complementarity problem. JOTA 8, pp. 161-168.
Luenberger, D. G. (1973). Introduction to Linear and Nonlinear Programming. (Addison-Wesley, Massachusettes).
Marcotte, P. (1981). Design optimal d'un reseau de transport en présence d'effects de congestion. Ph.D. Dissertation, CRT, Université de Montreal.
Powell, M. J. (1964). An efficient method for finding the minimum of a function of several variables without using derivatives. Brit. Computer J. 9, pp. 155-162.
Smith, M. J. (1979). Existence, uniqueness and stability of traffic equilibria. Transpn. Res. 13B, pp. 295-304.

Tan, H., S. Gershwin, M. Athans (1979). Hybrid optimization in urban traffic networks. Final Report, US DOT-TSC-RSPA-79-7.
Zukhovitshii, S., R. Poliak, M. Primak (1973). Concave multiperson games: numerical methods. Matekon, pp. 11-30.

Ninth International Symposium on
Transportation and Traffic Theory
© 1984 VNU Science Press, pp. 217–231

STABILITY AND SENSITIVITY ANALYSIS FOR THE GENERAL NETWORK EQUILIBRIUM–TRAVEL CHOICE MODEL

STELLA DAFERMOS[1] and ANNA NAGURNEY[2]

[1] *Division of Applied Mathematics, Brown University, Providence, RI, U.S.A.*
[2] *School of Management, University of Massachusetts, Amherst, MA, U.S.A.*

Abstract

We consider the combined asymmetric network equilibrium problem where the travel demand function may depend upon the travel cost and we introduce a new link formulation for this problem which has the desirable feature that uses directly the travel demand function (and not its inverse). Assuming that the travel cost and demand functions are monotone we first show that the equilibrium travel pattern depends continuously upon the travel cost and demand functions. We then focus on the delicate question of predicting the direction of the change of the travel pattern and the incurred travel costs and travel demands from changes in the travel cost and demand functions and attempt to elucidate certain counter intuitive phenomena such as "Braess' paradox". Our analysis depends crucially on the fact that the governing equilibrium conditions can be formulated as a variational inequality.

1.   Introduction

In this paper we consider a combined asymmetric traffic equilibrium problem
in which the travel demands associated with every mode and every origin-destination (O/D) pair depend in a prescribed way on the travel cost associated
with all O/D pairs and all modes; the travel cost depends in a prescribed
way on the travel pattern and one has to determine the equilibrium travel
pattern with the property that, once established, no user may decrease his
travel cost by making a unilateral decision to change his travel choices.
Variations of this general model have been studied before by Beckmann et. al
(1956), Evans (1976), Florian (1977), Aashtiani and Magnanti (1981), Dafermos (1982), Fisk and Boyce (1982), and Fernandez and Friesz (1983), among
others.   In particular, Aashtiani and Magnanti (1981) considered the most
general case where the travel cost may depend upon the entire link pattern
while the travel demands associated with each O/D pair and mode may depend
upon travel costs associated with every O/D pair and every mode. For this
model, they gave a nonlinear complementarity path formulation of the equilibrium conditions, they established existence and uniqueness of the equilibrium using complementarity theory and they proposed an algorithm for computing the equilibrium. An alternative path formulation of the same problem in the form of a variational inequality was given by Dafermos (1982).
Unfortunately, it requires that the travel demand functions be invertible,
a constraint that may induce theoretical and/or computational difficulties.
Subsequently, Fisk and Boyce (1982), using the well known equivalence between complementarity and variational inequality problems, derived from
the aforementioned results of Aashtiani and Magnanti an alternative path
formulation of the problem in the form of a variational inequality that does
not involve the inverse of the travel demand function.

One of the disadvantages of a "path" formulation is that it may lead to a
costly computation considering the storage requirement.

In the paper mentioned above, Dafermos (1982) gives also a variational inequality link formulation of the equilibrium conditions and she establishes
the existence and uniqueness of the equilibrium using the theory of variational inequalities. She also derives convergent algorithms for the computation of the equilibrium pattern. However, the link formulation in
Dafermos (1982) involves the inverse of the travel demand function. Fisk
and Boyce (1982) rederive this link formulation, attempt to remove

pendence on the inverse of the travel demand function but succeed only in special case.

this paper, we introduce a <u>new</u> link formulation of the symmetric integra- d network equilibrium problem which has the desirable feature that uses rectly the travel demand function (and not its inverse) and for which we e able to establish as powerful existence, uniqueness, stability and sen- tivity and convergence results as those that can be obtained for the link rmulation of the problem when the inverse of the travel demand function available. The crucial step in our analysis is the realization that the sic variables should not be the pair of flow and demand, as it has been ssumed in earlier works, but rather the triple of flow, demand and O/D avel cost.

Section 2 of this paper we outline the general combined asymmetric net- rk equilibrium model and, in Section 3, we write down the path formula- on of the equilibrium conditions in the form of a variational inequality.

en, in Section 4 we establish (Theorem 1) a variational inequality formu- ation of the combined network equilibrium problem in terms of link flows. e discuss for this problem the issues of existence and uniqueness of the uilibrium travel pattern. We show that under appropriate monotonicity ssumptions on the travel cost functions and the travel demand functions, e equilibrium travel pattern is, indeed, unique.

e then proceed to study the stability, as well as the sensitivity of the uilibrium with respect to changes in the travel cost functions and the avel demand functions. Our analysis here depends crucially on the vari- tional inequality formulation of the combined network equilibrium prob- em.

t is important to know whether the equilibrium travel pattern depends con- inuously upon the travel cost and travel demand functions; in other words, hether small changes in the travel demands or in the travel cost functions duce small changes in the equilibrium travel pattern. In Section 5 we vestigate these questions for the combined network equilibrium model with avel cost functions and travel demand functions each of which satisfy a notonicity condition and establish that the equilibrium travel pattern pends continuously on the cost functions as well as the demand functions Theorem 2).

e then proceed in Section 6 to the delicate question of whether one is

algorithms are presented which would be feasible for large problems. These are applied to a test problem and found to perform satisfactorily. To ascertain their suitabilty for real world problems it would be necessary to examine the convexity properties of the function ϕ using realistic travel time functions.

Acknowledgements

This work was supported in part by a grant from the National Science Foundation (grant no. CEE 81-19772).

REFERENCES

Aashtiani, H. (1979). The multimodal traffic assignment problem. Ph.D. Dissertation, M.I.T., Sloan School of Management.

Abdulaal, M., L. LeBlanc (1979). Continuous equilibrium network design models. Transpn. Res. 13B, pp. 19-32.

Dickerson, T. J. (1981). A note on traffic assignment and signal timings in a signal controlled road network. Transpn. Res. 15B, pp. 267-271.

Ermol'ev, Y. M. (1966). Methods of solution of nonlinear extremal problems. Kibernetika, 2, pp. 1-17.

Fisk, C., S. Nguyen (1982). Solution algorithms for network equilibrium models with asymetric user costs. Transpn. Sci. 16, pp. 361-381.

Fisk, C., D. Boyce (1983). A general variational inequality formulation of the network equilibrium-travel choice problem. To appear in Transpn. Sci.

Fisk, C. (1983). Game theory and transportation systems modelling. To appear in Transpn. Res.

Gartner, N. H., J. D. C. Little, H. Gabbey (1975). Optimization of traffic signal settings by mixed integer linear programming. Transpn. Sci. 9, pp. 321-363.

Hooke, R., T. A. Jeeves (1961). Direct search solution of numerical and statistical problems. J. Ass. Comp. Mach. 8, pp. 212-229.

John, F. (1949). Extremum problems with inequalities as subsidiary conditions. Studies and Essays, Courant Anniversary Volume, (Interscience, New York), pp. 187-204.

Karamardian, S. (1971). Generalized complementarity problem. JOTA 8, pp. 161-168.

Luenberger, D. G. (1973). Introduction to Linear and Nonlinear Programming. (Addison-Wesley, Massachusettes).

Marcotte, P. (1981). Design optimal d'un reseau de transport en présence d'effects de congestion. Ph.D. Dissertation, CRT, Université de Montreal.

Powell, M. J. (1964). An efficient method for finding the minimum of a function of several variables without using derivatives. Brit. Computer J. 9, pp. 155-162.

Smith, M. J. (1979). Existence, uniqueness and stability of traffic equilibria. Transpn. Res. 13B, pp. 295-304.

consider a transportation network G. Links are denoted by a, b, etc.,
ths by p, q, etc., and origin-destination pairs by w, ω, etc. Let L
the set of links with K elements, P the set of paths with Q elements,
the set of O/D pairs with Z elements. Let P_w be the set of paths
ining an O/D pair. Flows in the network are generated by n different
des of transportation, denoted by i,j and contained in the set J.

flow on path p generated by mode i will be denoted by F_p^i and the
ow on link a generated by mode i will be denoted by f_a^i. The travel
st of a user of mode i on path p will be denoted by c_p^i and the tra-
l cost of a user of mode i on link a will be denoted by c_a^i. The tra-
l demand of potential users of mode i generated between an O/D pair
will be denoted by d_w^i and the travel cost associated with traveling by
de i between the O/D pair w will be denoted by v_w^i. We arrange the
th flows and travel costs into n-tuples of vectors $F^1,\ldots,F^n$, $C^1,\ldots,C^n$
R^Q and we incorporate these n-tuples into column vectors F and C
R^{nQ}. Furthermore, we arrange the link flows and travel costs into n-
ples of vectors $f^1,\ldots,f^n$, $c^1,\ldots,c^n$ in R^K and we incorporate the
ove n-tuples into column vectors f and c in R^{nK}. We also arrange
e O/D travel demands and travel costs into n-tuples of vectors $d^1,\ldots,d^n$,
$,\ldots,v^n$ in R^Z and we incorporate the above n-tuples into colum vec-
rs in R^{nZ}. We consider the most general situation where the travel cost
a user of any mode on any link may, in general, depend upon flows by
ery mode on every link of the network, that is,

$$c = c(f) \tag{1}$$

ere c(f) is a known smooth function.

milarly, for the demand functions we consider the most general situation
ere the demand associated with any mode and any O/D pair may depend upon
e travel costs associated with all O/D pairs and all modes in the network;
at is

$$d = g(v) \tag{2}$$

ere g(v) is a known smooth function.

e travel demand vector d induces a <u>path flow</u> vector F with components
fined on every path p. Thus

$$d_w^i = \sum_{p \in P_w} F_p^i \tag{3}$$

, in matrix form,

$$d = BF \tag{4}$$

where B is the $(nZ \times nQ)$ matrix whose $(i,p)-(j,w)$ entry is 1 if $i=j$ and p connects w and 0 otherwise. The flow vector F induces a <u>link flow</u> vector f with components defined on every link a by

$$f_a^i = \sum_{p \in P} \delta_{ap} F_p^i \tag{5}$$

or, in matrix form,

$$f = AF \tag{6}$$

where A is the $nK \times nQ$ link-path incidence matrix whose $(i,a)-(j,p)$ entry is 1 if $i=j$ and a is contained in path p and 0 otherwise.

A user of mode i traveling on path p incurs travel cost

$$c_p^i = \sum_{a \in L} \delta_{ap} c_a^i \tag{7}$$

where δ_{ap} is 1 if a is contained in path p and 0 otherwise.

In view of (1),

$$c_p^i = c_p^i(f) \overset{\text{def}}{=} \sum_{a \in L} \delta_{ap} c_a^i(f). \tag{8}$$

By virtue of (8) and (6)

$$c_p^i = c_p^i(F) \overset{\text{def}}{=} \sum_{a \in L} \delta_{ap} c_a^i(AF) \tag{9}$$

3. Path formulation of the combined network equilibrium problem

We will say that a <u>travel path pattern</u> (F,v) in R^{nQ+nZ} is feasible if it is contained in the set $\mathscr{H}$, where

$$\mathscr{H} = \{(F,v) \mid F \geq 0, \ v \geq 0\}. \tag{10}$$

The set $\mathscr{H}$ of feasible travelpath patterns is a closed convex subset of R^{nQ+nZ}.

A travel path pattern $(\hat{F},\hat{v}) \in \mathscr{H}$ is in <u>equilibrium</u> if, once established, no user has any incentive to alter his travel choices. This state is characterized by the following equilibrium conditions which must hold for every mode $i \in J$, every O/D pair $w \in W$ and every path $p \in P_w$:

$$c_p^i(\hat{F}) - \hat{v}_w^i \begin{cases} = 0, & \text{if } \hat{F}_p^i > 0 \\[2mm] \geq 0, & \text{if } \hat{F}_p^i = 0 \end{cases} \tag{11}$$

$$\sum_{p \in P_w} \hat{F}_p^i = g_w^i(\hat{v}) \tag{12}$$

(see Aashtiani and Magnanti (1981))

llowing similar arguments as in Dafermos (1982), one can easily verify
nat the equilibrium conditions (11),(12) can be expressed as the <u>varia-
ional inequality</u>

$$\sum_{i,w} \sum_{p \in P_w} [C_p^i(\hat{F}) - \hat{v}_w^i][F_p^i - \hat{F}_p^i]$$

$$+ \sum_{i,w} \sum_{p \in P_w} [\hat{F}_p^i - g_w^i(\hat{v})] [v_w^i - \hat{v}_w^i] \geq 0, \ (F,v) \in \tag{13}$$

r, in vector form,

$$\Phi(\hat{x})^T(X-\hat{x}) \geq 0, \ X \in \mathscr{H} \tag{14}$$

here $X \stackrel{def}{=} (F,v)$ is a vector in R^{nQ+nZ} and $\Phi(X) = (C(F) - v, \sum_{p \in P} F_p - g(v))$
s a function from R^{nQ+nZ} to R^{nQ+nZ}.

n alternative, indirect, way for deriving (13) (see Fisk and Boyce (1982))
s to note that the equilibrium conditoins (11) can be formulated as a non-
inear complementarity problem (Aashtiani and Magnanti (1981)) and then to
se the well known equivalence between complementarity problems and varia-
ional inequalities.

Link formulation of the combined network equilibrium problem.

e now proceed to establish our link formulation of the problem. We will
ay that a <u>travel link pattern</u> (f,v,d) is feasible if it is contained in
 set κ defined by

$$\kappa = \{(f,v,d) | \text{there is } (F,v) \mathscr{H} \text{ such that } f = AF, \ d = BF\} \tag{15}$$

ne set κ of feasible travel patterns is a closed convex subset of
nK+2nZ
.

ne new element that distinguishes our approach from earlier ones is the
ealization that the feasible set κ should be described in terms of
f,v,d) rather than (f,d) alone.

e can then easily establish the following.

heorem 1. A travel link pattern $(\hat{f},\hat{v},\hat{d})$ in κ is an equilibrium travel
oad pattern if and only if

$$\sum_{i,a} c_a^i(\hat{f})(f_a^i - \hat{f}_a^i) - \sum_{i,w} \hat{v}_w^i(d_w^i - \hat{d}_w^i)$$

$$+ \sum_{i,w} [\hat{d}_w - g_w(\hat{v})] [v_w^i - \hat{v}_w^i] \geq 0, \ (f,d,v) \in \kappa \tag{16}$$

r, in vector form,

$$\varphi(\hat{x})^T(x - \hat{x}) \geq 0, \ x \in \kappa, \tag{17}$$

here $x \stackrel{def}{=} (f,v,d)$ is a vector in R^{nK+2nZ} and $\varphi(x) \stackrel{def}{=} (c(f),-v,d-g(v))$

is a function from R^{nK+2nZ} to R^{nK+2nZ}.

The equivalence of the path variational inequality formulation (13) (or (14))
with the link variational inequality formulation (16) (or (17)) follows from
the observation that, the lefthand sides of the two variational inequalities
(13),(16) coincide if and only if $(f,d,v),(\hat{f},\hat{d},\hat{v})$ in κ and $(F,v),(\hat{F},\hat{v})$
in $\mathcal{H}$ are related by $f = AF$, $\hat{f} = \hat{A}\hat{F}$, $d = BF$, $\hat{d} = \hat{B}\hat{F}$.

We recall that the function $\varphi(x)$ is called <u>strictly monotone</u> if

$$[\varphi(x^1)-\varphi(x^2)]^T[x^1-x^2] > 0, x^1, x^2 \in \kappa,\ x^1 \neq x^2 \tag{18}$$

and <u>strongly monotone</u> if

$$[\varphi(x^1)-\varphi(x^2)]^T[x^1-x^2] \geq \mu[x^1-x^2]^T[x^1-x^2],\ x^1,x^2 \in \kappa \tag{19}$$

where μ positive.

By the standard theory of variational inequalities the strict monotonicity
condition (18) implies uniqueness but not necessarily existence of an equilibrium pattern. The stronger, strong monotonicity assumption (18) implies
both existence and uniqueness of equilibrium. A short calculation yields

$$[\varphi(x^1)-\varphi(x^2)]^T[x^1-x^2] = [c(f^1)-c(f^2)]^T[f^1-f^2]$$
$$- [g(v^1)-g(v^2)]^T[v^1-v^2]. \tag{20}$$

It follows then that a necessary condition for $\varphi(x)$ to be strictly monotone is that both $c(f)$ and $-g(v)$ be strictly monotone, viz.

$$[c(f^1)-c(f^2)]^T[f^1-f^2] > 0,\ f^1,f^2 \geq 0,\ f^1 \neq f^2$$
$$-[g(v^1)-g(v^2)]^T[v^1-v^2] > 0,\ v^1,v^2 \geq 0,\ v^1 \neq v^2. \tag{21}$$

Similarly, a necessary condition for $\varphi(x)$ to be strongly monotone is that
both $c(f)$ and $-g(v)$ be strongly monotone, i.e.,

$$[c(f^1)-c(f^2)]^T[f^1-f^2] \geq \alpha[f^1-f^2]^T[f^1-f^2]$$
$$-[g(v^1)-g(v^2)]^T[v^1-v^2] \geq \beta[v^1-v^2]^T[v^1-v^2]. \tag{22}$$

where α and β positive.

Monotonicity conditions such as (21) and (22) have been discussed extensive-
ly in the transportation literature (see e.g., Dafermos (1980), (1982a),
(1982b), Bertsekas and Gafni (1982)). Such assumptions are reasonable even
though somewhat restrictive. In particular, (22) implies that $g(v)$ is
invertible. However, computing the inverse is generally a numerically
difficult task.

By virtue of (20), (21) or (22) implies that $\varphi(x)$ is monotone, that is,

$$[\varphi(x^1)-\varphi(x^2)]^T[x^1-x^2] \geq 0, \quad x^1,x^2 \in \kappa. \tag{23}$$

However, strict or strong monotonicity of $\varphi(x)$ is a rare occurrence and holds only in the exceptional situation where the null space of the matrix B is contained in the null space of the matrix A. Indeed, whenever the null space of B is contained in the null space of A an inequality

$$(d^1-d^2)^T(d^1-d^2) \leq \eta(f^1-f^2)^T(f^1-f^2), \tag{24}$$

with η positive, holds for every feasible pair $(f^1,d^1,v^1),(f^2,d^2,v^2)$ in which case (21) implies (18) and (22) implies (19).

On the contrary, if the null space of B contains elements that do not belong to the null space of A it is easy to construct feasible pairs $x^1 = (f^1,d^1,v^1)$, $x^2 = (f^2,d^2,v^2)$ such that $f^1 = f^2$, $v^1 = v^2$ but $d^1 \neq d^2$. For such a pair the right-hand side of (20) vanishes, even though $x^1 \neq x^2$, and so strict or strong monotonicity of $\varphi(x)$ fails.

Even though $\varphi(x)$ generally fails to be strictly or strongly monotone, due to the special structure of the problem it turns out that results on existence and uniqueness of the equilibrium may be obtained under appropriate monotonicity assumptions on the travel cost function $c(f)$ and the travel demand function $g(v)$. As an illustration, let us show that (21) guarantees uniqueness of the equilibrium travel pattern. To this end, assume $\hat{x}^1 = (f^1,d^1,v^1)$ and $\hat{x}^2 = (f^2,d^2,v^2)$ are any two equilibrium patterns, that is any two solutions of the variational inequality (17). Let us write (17) twice, first with $\hat{x} = \hat{x}^1$, $x = \hat{x}^2$ and then with $\hat{x} = \hat{x}^2$, $x = \hat{x}^1$. Adding up the resulting inequalities we obtain

$$[\varphi(\hat{x}^1)-\varphi(\hat{x}^2)]^T[\hat{x}^1-\hat{x}^2] \leq 0 \tag{25}$$

Combining (25), (20) and (21) we deduce $f^1 = f^2$, $v^1 = v^2$. At the same time, at equilibrium, $d^1 = g(v^1)$, $d^2 = g(v^2)$. Therefore $d^1 = d^2$ and the two equilibria $\hat{x}^1,\hat{x}^2$ must coincide.

Along the same lines we show in a forthcoming publication that the strong monotonicity assumptions (22) implies existence as well as uniqueness of the equilibrium travel pattern. Furthermore, under the same assumptions we construct a "projection" algorithm which yields a sequence of iterations that converges at least linearly fast to the equilibrium travel pattern. Here and in the following section we perform, under the strong monotonicity assumptions (22), stability and sensitivity analysis.

5. Stability

In this section we study how changes in the transportation cost functions and the travel demand functions affect the equilibrium travel pattern. On a fixed network we change the travel cost functions from $c(\cdot)$ to $c^*(\cdot)$, and the travel demand functions from $g(\cdot)$ to $g^*(\cdot)$ and we are to compare the corresponding equilibrium travel patterns $x = (f,d,v)$ and $x^* = (f^*,d^*,v^*)$.

For convenience, we use $|\cdot|$ to denote the Euclidean norm in R^{nK}, $||\cdot||$ to denote the norm in R^{nZ} and $|||\cdot|||$ to denote the norm in R^{nK+2nZ}. In particular, if $x = (f,d,v) \in R^{nK+2nZ}$,

$$|f|^2 = f^T f, \quad ||d||^2 = d^T d, \quad ||v||^2 = v^T v, \quad |||x|||^2 = |f|^2 + ||d||^2 + ||v||^2 \tag{26}$$

We impose upon $c(\cdot)$ and $-g(\cdot)$ the strong monotonicity conditions (22) and show that a small change in the travel cost and demand functions induces a small change in the equilibrium travel pattern.

Theorem 2. Under the above assumption and given that x and x^* lie in an a priori fixed compact convex set $\mathscr{O}$ of R^{nK+2nZ}

$$|||x^*-x|||^2 = |f^*-f|^2 + ||d^*-d||^2 + ||v^*-v||^2 \leq \gamma\{|c^*(f^*)-c(f^*)|^2 + \\ + ||g^*(v^*)-g(v^*)||^2\} \tag{27}$$

where γ is a positive constant.

Proof: x and x^* must satisfy the variational inequality (17). Let us write (17) twice, first with $\hat{x} = x^*, x = x$ and then with $\hat{x} = x, x = x^*$. Adding up the resulting inequalities we obtain

$$[\varphi(x) - \varphi^*(x^*)]^T [x-x^*] \leq 0 \tag{28}$$

or

$$[\varphi^*(x^*) - \varphi(x^*)]^T [x-x^*] \geq [\varphi(x) - \varphi(x^*)]^T [x-x^*] \tag{29}$$

where

$$\varphi(x) = (c(f), -v, d-g(v)), \quad \varphi^*(x^*) = (c^*(f^*), -v^*, d^*-g^*(v^*)) \tag{30}$$

Using now (30), (20) and the strong monotonicity conditions (22), (29) yields

$$[\varphi(x) - \varphi(x^*)]^T [x-x^*] \geq \alpha |f-f^*|^2 + \beta ||v-v^*||^2 \tag{31}$$

By virtue of (30) and Schwarz's inequality we obtain

$$[\varphi^*(x^*) - \varphi(x^*)]^T [x-x^*] = [c^*(f^*) - c(f^*)]^T [f-f^*]$$

$$-[g*(v*)-g(v*)]^T[v-v*]$$
$$\leq |c*(f*)-c(f*)| \, |f-f*| + ||g*(v*)-g(v*)|| \, ||v-v*|| \tag{32}$$

Combining (29), (31) and (32) we deduce

$$\alpha|f-f*|^2 + \beta||v-v*||^2 \leq \frac{1}{\alpha}|c*(f*)-c(f*)|^2 + \frac{1}{\beta}||g*(v*)-g(v*)||^2 \tag{33}$$

Note that at equilibrium $d = g(v)$ and $d* = g*(v*)$. Hence

$$||d-d*|| = ||g(v)-g*(v*)|| \leq ||g(v)-g(v*)|| + ||g(v*)-g*(v*)|| \tag{34}$$
$$\leq \delta||v-v*|| + ||g(v*)-g*(v*)|| \tag{35}$$

where δ is a Lipschitz constant.

Observe that (35) implies also that

$$||d-d*||^2 \leq 2\delta^2||v-v*||^2 + 2||g(v*)-g*(v*)||^2 \tag{36}$$

Combining (33) and (36) and after a short calculation we deduce

$$|f-f*|^2 + ||d-d*||^2 + ||v-v*||^2 \leq \left\{\frac{1}{\alpha^2} + \frac{2\delta^2+1}{\alpha\beta}\right\}|c*(f*)-c(f*)|^2$$
$$+\left\{\frac{1}{\alpha\beta} + \frac{2\delta^2+1}{\beta^2} + 2\right\}||g(v*)-g*(v*)||^2$$
$$\leq \gamma\{|c*(f*)-c(f*)|^2 + ||g*(v*)-g(v*)||\}^2. \tag{37}$$

This completes the proof.

6. Sensitivity Analysis

We now proceed to study how changes in the travel cost functions affect the direction of the change in the equilibrium travel pattern and the incurred travel costs, and how changes in the travel demand functions affect the direction of the change in the travel costs and the incurred travel demands.

We assume that the travel costs $c(f)$ and the travel demands $g(v)$ satisfy the strong monotonicity conditions (22).

Theorem 3. Consider the network with two travel cost functions $c(\cdot)$, $c*(\cdot)$, which induce corresponding path travel cost functions $C(\cdot), C*(\cdot)$; and demand functions $g(\cdot)$, $g*(\cdot)$. Let (f,F,v) and $(f*,F*,v*)$ be the equilibrium load, flow, and travel cost patterns associated with $(c(\cdot)$, $g(\cdot))$ and $(c*(\cdot),g*(\cdot))$.

Then

$$\sum_{i,a} [c*_a^i(f*)-c_a^i(f)][f*_a^i-f_a^i] - \sum_{i,w} [g*_w^i(v*)-g_w^i(v)][v*_w^i-v_w^i] \leq 0 \tag{38}$$

$$\sum_{i,p} [C*_p^i(f^*)-C_p^i(f)] [F*_p^i-F_p^i] - \sum_{i,w} [g*_w^i(v^*)-g_w^i(v)] [v*_w^i-v_w^i] \leq 0 \qquad (39)$$

and

$$\sum_{i,a} [c*_a^i(f^*)-c_a^i(f^*)] [f*_a^i-f_a^i] - \sum_{i,w} [g*_w^i(v^*)-g_w^i(v^*)] [v*_w^i-v_w^i] \leq 0. \qquad (40)$$

$$\sum_{i,p} [C*_p^i(f^*)-C_p^i(f^*)] [F*_p^i-F_p^i] - \sum_{i,w} [g*_w^i(v^*)-g_w^i(v^*)] [v*_w^i-v_w^i] \leq 0. \qquad (41)$$

<u>Proof</u>: Inequalities (38) and (40) have already been established in the course of proving Theorem 2. Inequalities (39) and (41) follow from (38) and (40) upon using (5) and (7). The proof is complete.

Observe that inequalities (38), (39), (40) and (41) show that, under the monotonicity conditions (22), as regards changes in the incur-red travel costs brought about by "improving" the travel cost situation, the following counterintuitive phenomenon may prevail: An improvement in the travel cost situation for a mode may result in an increase in some of the incurred travel cost and a decrease in some of the flows. This is us-ually called "Braess' paradox". Nevertheless, it follows from Theorem 3 that Braess' paradox can never occur under the following circumstances:

<u>Corollary 1.</u> Assume that the travel cost functions and the travel demand functions satisfy the monotonicity conditions (22) and that one link for one mode in the network is improved while the rest remain unchanged, i.e., $c*_a^i(f') < c_a^i(f')$ for some $a \in L$, $i \in J$ and $c*_b^j(f') = c_b^j(f')$ for all $b \neq a$ and $j \neq i$ and $f' \in \kappa$. Assume also that $\frac{\partial c_b^j(f')}{\partial f_a^i} = 0$ for all $b \neq a$ and $j \neq i$. If we fix the travel demand functions for all O/D pairs and all modes, that is $g*_w^j(v') = g_w^j(v')$, for all $w \in W$, $j \in J$ and $v' \in \kappa$, then the load on link a for mode i cannot decrease and the incurred travel cost cannot increase, i.e.,

$$f*_a^i \geq f_a^i \quad \underline{and} \quad c*_a^i(f^*) \leq c_a^i(f).$$

<u>Proof</u>: Since $c*_b^j(f')=c_b^j(f')$ for all $b \neq a$, $j \neq i$ and $f' \in \kappa$ and $g*_w^j(v')=g_w^j(v')$ for all $w \in W$, $j \in J$ and $v' \in \kappa$, inequality (40) yields

$$\sum_{j,b} [c*_b^j(f^*)-c_b^j(f^*)] [f*_b^j-f_b^j]=[c*_a^i(f^*)-c_a^i(f^*)] [f*_a^i-f_a^i] \leq 0 \qquad (42)$$

which implies, since $c*_a^i(f^*) < c_a^i(f^*)$, that $f*_a^i \geq f_a^i$.

From inequality (38) it follows that

$$[c*_a^i(f^*)-c_a^i(f)] [f_a^i-f*_a^i] \geq \sum_{\substack{j,b \\ j \neq i \\ b \neq a}} [c_b^j(f^*)-c_b^j(f)] [f*_b^j-f_b^j] \qquad (43)$$

$$- \sum_{j,w} [q_w^j(v^*) - q_w^j(v)] [v^{*j}_w - v^j_w]. \tag{43}$$

The last term on the right-hand side of (43) is nonnegative by the monotonicity condition (22). Applying the mean value theorem to the first term on the right-hand side of (43) and using the assumption $\dfrac{\partial c_b^j(f')}{\partial f_a^i} = 0$ for all $b \neq a$, $j \neq i$, $f' \in \kappa$, we obtain

$$\sum_{\substack{j,b \\ j \neq i \\ b \neq a}} [c_b^j(f^*) - c_b^j(f)] [f^{*j}_b - f^j_b] = \sum_{\substack{j,b \\ j \neq i \\ b \neq a}} \sum_{\substack{k,c \\ k \neq i \\ c \neq a}} \int_0^1 \frac{\partial c_b^j(f')}{\partial f_c^k} \bigg| dt (f^{*k}_c - f^k_c)(f^{*j}_b - f^j_b) \quad f' = [(1-t)f + tf^*] \tag{44}$$

On account of the monotonicity condition (22), the matrix $\left[\dfrac{\partial c}{\partial f}\right]$ is positive definite and hence the right-hand side of (43) is nonnegative.

Thus (43) implies

$$[c^{*i}_a(f^*) - c^i_a(f)] [f^i_a - f^{*i}_a] \geq 0. \tag{45}$$

Since we showed that $f^{*i}_a \geq f^i_a$, we conclude that

$$c^{*i}_a(f^*) \leq c^i_a(f). \tag{46}$$

Finally, inequalities (38) and (40) suggest that, as regards changes in the incurred demands brought about by changes in the demand mechanism, the following counterintuitive phenomena may prevail: An increase in the demand of a mode for an O/D pair may result in a decrease in some of the travel costs. However, it follows from Theorem 3 that the above-mentioned counterintuitive phenomena can never occur under the following circumstances:

<u>Corollary 2</u>. Here we assume that the travel demand for origin/destination pair w and mode i is increased, while all other travel demand functions remain fixed, that is, $g^{*i}_w(v') > g^i_w(v')$ for some $w \in W$, $i \in J$ and $v' \in \kappa$ and $g^{*j}_\omega(v') = g^j_\omega(v')$ for all $\omega \neq w$, $j \in J$ and $v' \in \kappa$. Assume also that $\dfrac{\partial g^j_\omega(v')}{\partial v^i_w} = 0$, for all $\omega \neq w$, $j \neq i$ and $v' \in \kappa$. If we fix the travel cost functions for all links and all modes, that is, $c^{*j}_a(f') = c^j_a(f')$, for all $a \in L$, $j \in J$ and $f' \in \kappa$, then the travel cost for mode i traveling between O/D pair w cannot decrease and the incurred travel demand cannot decrease, i.e., $v^{*i}_w \geq v^i_w$ and $g^{*i}_w(v^*) \geq g^i_w(v)$.

<u>Proof</u>: Since $c^{*j}_a(f') = c^j_a(f')$ for all $a \in L$, $j \in J$ and $f' \in \kappa$, and $g^{*j}_\omega(v') = g^j_\omega(v')$ for all $\omega \neq w$, $i \neq j$ and $v' \in \kappa$, inequality (40) yields

$$-\sum_{j,w} [g^{*j}_\omega(v^*)-g^j_\omega(v^*)][v^{*j}_\omega-v^j_\omega] = -[g^{*i}_w(v^*)-g^i_w(v^*)][v^{*i}_w-v^i_w] \le 0 \qquad (47)$$

which implies, since $g^{*i}_w(v^*) > g^i_w(v^*)$, that $v^{*i}_w \ge v^i_w$. From inequality (38) we get

$$-[g^i_w(v)-g^{*i}_w(v^*)][v^{*i}_w-v^i_w] \ge -\sum_{\substack{j,\omega \\ j\neq i \\ \omega\neq w}} [g^j_\omega(v^*)-g^j_\omega(v)][v^{*j}_\omega-v^j_\omega]$$

$$+ \sum_{j,a} [c^j_a(f^*)-c^j_a(f)][f^{*j}_a-f^j_a]. \qquad (48)$$

The last term on the right-hand side in (48) is nonnegative under the assumption of monotonicity. Applying the mean value theorem to the first term on the right-hand side and using the assumption $\dfrac{\partial g^j_\omega(v')}{\partial v^i_w} = 0$ for all $\omega \neq w$, $i \neq j$, $v' \in \kappa$, we obtain

$$-\sum_{\substack{j,\omega \\ j\neq i \\ \omega\neq w}} [g^j_\omega(v^*)-g^j_\omega(v)][v^{*j}_\omega-v^j_\omega] = -\sum_{\substack{j,\omega \\ j\neq i \\ \omega\neq w}} \sum_{\substack{k,u \\ k\neq i \\ u\neq w}} \int_0^1 \left.\frac{\partial g^j_\omega(v')}{\partial v^k}\right| dt \, (v^{*k}_u-v^k_u)(v^{*k}_\omega-v^k_\omega)$$
$$v'=[(1-t)v+tv^*] \qquad (49)$$

On account of monotonicity condition (22), the matrix $-\left[\dfrac{\partial g}{\partial v}\right]$ is positive definite and hence the right-hand side of (48) is nonnegative.

Therefore, (48) implies

$$-[g^i_w(v)-g^{*i}_w(v^*)][v^{*i}_w-v^i_w] \ge 0 \qquad (50)$$

and since, as shown above, $v^{*i}_w \ge v^i_w$, (50) implies that

$$g^{*i}_w(v^*) \ge g^i_w(v). \qquad (51)$$

References

Aashtiani, M., T. L. Magnanti (1981). Equilibrium on a congested transporta tion network, SIAM J. Algebraic and Discrete Methods 2, pp. 213-216.
Beckmann, M. J., C.B. McGuire, C.B. Winsten (1956). *Studies in the Economic of Transportation*, Yale University Press, New Haven, Conn.
Bertsekas, D.P., E. Gafni (1982). Projection mehtods for variational inequalities and application to the traffic assignment problem, Mathematical Programming Study 17, pp. 139-159.
Dafermos, S. (1980). Traffic equilibrium and variational inequalities, Transportation Science 14, pp. 42-54.
Dafermos, S. (1982a). The general multimodal network equilibrium problem with elastic demand, Networks 12, pp. 57-72.
Dafermos, S. (1982b). Relaxation algorithms for the general asymmetric traffic equilibrium problem, Transportation Scinece 16, pp. 231-240.
Dafermos, S. (1983). An iterative scheme for variational inequalities, Mathematical Programming 26, pp. 40-47.

fermos, S., A. Nagurney (1983). Sensitivity analysis for the general ymmetric network equilibrium problem, Mathematical Programming (in press).
ans, S. (1976). Derivation and analysis of some models for combining trip stribution and assignment, Transportation Research 10, pp. 37-57.
sk, C. (1979). More paradoxes in the equilibrium assignment problem, ansportation Research 13B, pp. 305-309.
sk, C., D. Boyce (1983). Alternative variational inequality formulation f the network equilibrium-travel choice problem, Transportation Science 17, . 454-463.
orian, M. (1977). A traffic equilibrium model of travel by car and public ansit modes, Transportation Science 11, pp. 166-179.
ll, M. A. (1978). Properties of the equilibrium state in transportation tworks, Transportation Science 12, pp. 208-216.
einberg, R., W. Zangwill (1983). The prevalence of Braess' paradox, ansportation Science 17, pp. 301-318.

Ninth International Symposium on
Transportation and Traffic Theory
© 1984 VNU Science Press, pp. 233–252

BOUNDING THE SOLUTION OF THE CONTINUOUS EQUILIBRIUM NETWORK DESIGN PROBLEM

PATRICK T. HARKER[1] and TERRY L. FRIESZ[2]

[1] *Department of Geography, University of California at Santa Barbara, Santa Barbara, CA 93106, U.S.A.*
[2] *Department of Civil Engineering, University of Pennsylvania, Philadelphia, PA 19104, U.S.A.*

ABSTRACT

The solution of the optimal equilibrium network design problem
is perhaps the most computationally intensive problem encountered
in transportation network analysis. This paper illustrates through
the use of game-theoretic concepts how one can bound the solution
of the continuous version of this problem by using two computationally
efficient heuristics. This bounding procedure can then be used to
assess whether or not it is necessary to solve for the exact solution
to this problem. A numerical example is presented which illustrates
this procedure.

1. INTRODUCTION

The solution of the optimal equilibrium network design problem, either with continuous or discrete improvement variables, is
perhaps the most computationally intensive problem encountered
in transportation network analysis. The seminal paper on the discrete formulation of this problem is due to LeBlanc (1975), while
the seminal paper on the continuous formulation is due to Abdulaal
and LeBlanc (1979). These papers and the work by Tan et al (1979)
and Marcotte (1981) suggest that only problems of relatively modest
size may be solved exactly in practice due, in both formulations,
to the necessity of computing the Wardropian equilibrium virtually
each time a change in the network design is considered.

Numerous heuristics and approximation methods for this problem have been proposed over the last several years; in fact the
adaptation of the Hooke and Jeeves algorithm proposed by Abdulaal
and LeBlanc (1979), which is considered the standard technique for
solving the continuous equilibrium design problem, is itself a
heuristic since it ignores nonnegativity constraints on improvement variables. The purpose of this paper is to explore, through
the use of game theory concepts, the relationship between the continuous equilibrium network design model and two heuristics: the
so-called iterative optimization-equilibrium (or optimization-
assignment) algorithm (Allsop 1974; Tan et al 1979; Friesz and
Harker 1983) and a heuristic which is based upon the work by Marcotte
(1981). The next section states the problem and reviews the exact
algorithms which researchers have attempted to use on the continuous
equilibrium network design problem. Section 3 analyzes the
iterative optimization-equilibrium heuristic. Section 4 presents a
heuristic first proposed by Marcotte (1981), and Section 5 shows
through the use of a game-theoretic interpretation of these heuristics
that they can be used to bound the solution of the exact problem.

Section 6 presents numerical experiments which illustrate the points made in the previous sections, and conclusions are drawn in Section 7.

2. STATEMENT OF THE PROBLEM

The following notation will facilitate our exposition:

$G(N,A)$ = the network, where N and A are the set of nodes and arcs respectively

W = the set of origin-destination (O-D) pairs on the network

w = a specific O-D pair $w \in W$

P = the complete set of available paths in the network

P_w = the set of paths between O-D pair $w \in W$

h_p = the flow on path p

$\delta_{ap} = \begin{cases} 1 \text{ if arc } a \in A \text{ is on path } p \in P \\ 0 \text{ otherwise} \end{cases}$

f_a = the flow on arc $a \in A$; note that $f_a = \sum_{p \in P} \delta_{ap} h_p$

$$\forall a \in A \qquad (1)$$

$f = (f_a | a \in A)$

y_a = the improvement made to arc $a \in A$

$y = (y_a | a \in A)$

$c_a(f_a, y_a)$ = the unit cost of transportation on arc $a \in A$, assumed to be twice continuously differentiable in both f_a and y_a

c_p = the unit cost of transportation on path $p \in P$; note that

$$c_p = \sum_{a \in A} \delta_{ap} c_a(f_a, y_a) \qquad \forall p \in P \qquad (2)$$

β_a = the unit cost of improvement on arc $a \in A$

T_w = the fixed demand for transportation between O-D pair $w \in W$; note that

$$T_w = \sum_{p \in P_w} h_p \qquad \forall w \in W \qquad (3)$$

    B = the total amount of dollars allocated for network
       improvements.

Note that we have taken the simplest case of separable cost functions, constant unit improvement costs and fixed demands in order
to facilitate our arguments. One could, of course, relax any of
the above assumptions.

    Given the above notation, the continuous equilibrium network
design problem can be written as:

$$\begin{aligned} \underset{y}{\text{minimize}} \quad & \sum_{a \in A} c_a(f_a(y), y_a) f_a \\ \text{subject to} \quad & \sum_{a \in A} \beta_a y_a \leq B \qquad\qquad\qquad (P) \\ & y_a \geq 0 \quad \forall a \in A \end{aligned}$$

where $f(y) = (f_a(y) | a \in A)$ is the solution to the Wardropian User
Equilibrium problem:

$$\begin{aligned} \underset{y}{\text{minimize}} \quad & \sum_{a \in A} \int_0^{f_a} c_a(s, y_a) \, ds \\ \text{subject to:} \quad & \sum_{p \in P_w} h_p = T_w \quad \forall w \in W \\ & f_a = \sum_{p \in P} \delta_{ap} h_p \quad \forall a \in A \qquad (U) \\ & f_a \geq 0 \quad \forall a \in A \\ & h_p \geq 0 \quad \forall p \in P. \end{aligned}$$

That is, the planner (P) attempts to minimize the total costs
incurred in the transportation system, taking into account the
behavior of the users (U) of the transportation system via
Beckmann et al's (1956) formulation of Wardrop's User Equilibrium
model.

    The algorithms by Marcotte (1983), Tan et al (1979) and
Abdulaal and LeBlanc (1979) have been proposed for the exact solution
of the two-level mathematical program described by (P) and (U). These

algorithms, however, are very computationally intensive and, thus, it is doubtful that they can be used on large-scale problems.

How can we characterize this problem in game-theoretic terms? The users (U) take the improvements (the planner's (P) strategy set) as given when deciding upon their strategy (the flow vector). However, the planner, through f(y), anticipates the reactions of the users when choosing his strategy set. Thus, this problem is a leader-follower or Stackelberg (see, for example pp. 70-72 of Moulin 1982) game in which the leader is the planner and the users are the followers.

3. CHARACTERIZATION OF THE ITERATIVE OPTIMIZATION-EQUILIBRIUM HEURISTIC

The iterative optimization-equilibrium heuristic (Allsop 1974; Friesz and Harker 1983) is an attempt to solve the (P)-(U) two-level math program by iteratively solving (P), assuming f is fixed, and then solving (U), assuming y is fixed, until convergence is reached. Figure 1 illustrates this procedure. Friesz and Harker (1983) show that this algorithm leads not to a solution of the Stackelberg game defined in the previous section, but rather, is an exact algorithm for a Cournot-Nash model of the planner-users interaction. In this section, we briefly illustrate this result.

A Cournot-Nash game is one in which each player attempts to maximize his utility or payoff noncooperatively and assumes that his action or strategy will have no effect on the actions or strategies of the other players. Notice that each player in the Cournot-Nash game has a myopic view of the other player's actions (see, for example, p. 24 of Friedman 1977). An equilibrium to this game is defined as a point at which no player may increase his utility by unilaterally altering his strategy.

The users are already defined to be myopic in the sense that they take the planner's strategy y as given when choosing their

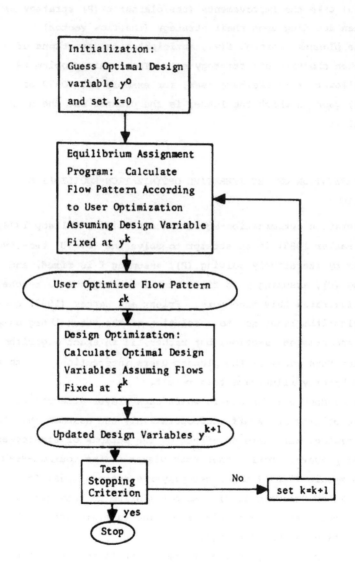

Fig. 1.   Iterative Optimization-Equilibrium Heuristic

strategies f.   To place the planner into the mold of a Cournot-
Nash player, let us no longer treat flows as a function of improvement, but rather, treat the flows as fixed. Problem (P) thus
becomes

$$\underset{y}{\text{minimize}} \ \sum_{a \in A} c_a(f_a, y_a) f_a \qquad\qquad\qquad (\text{P}')$$

$$\text{subject to} \ \sum_{a \in A} \beta_a \, y_a \leq B$$

$$y_a \geq 0 \qquad\qquad \forall \, a \in A$$

     Given a Cournot-Nash game, Harker (1983a, 1983b) has shown
that this game can be stated mathematically as a variational inequality
problem (see also Gabay and Moulin 1980) and the diagonalization/
relaxation algorithm (Dafermos 1983; Pang and Chan 1982) can be
applied to calculate the exact solution. Friesz and Harker (1983)
show how this approach can be applied to a variant of (P')-(U).
The result of applying the diagonalization/relaxation algorithm
is that at each iteration, the problem separates into two subproblems.
The first subproblem is precisely (P'), where f is held fixed at
its value at the previous iteration, and the second subproblem is
precisely (U) where y is held fixed at its value at the previous
iteration. This algorithm continues in this fashion until convergence
is reached. Thus, the diagonalization/relaxation algorithm, which
can be proven to be an exact algorithm for the Cournot-Nash problem
(Harker 1983b), follows the same steps as the iterative optimization-
equilibrium heuristic. One can conclude that this heuristic does
not solve (P)-(U), but in fact gives a solution to (P')-(U), the
Cournot-Nash game. Therefore, we will characterize a solution to
the iterative optimization-equilibrium heuristic as a Cournot-Nash
equilibrium.

4.   A SYSTEM OPTIMALITY-BASED HEURISTIC

Marcotte (1981), in his study of heuristic and exact solution algorithms for (P) - (U), suggested the use of a heuristic which precisely solves for the planner's optimality conditions and only approximately solves for the User Equilibrium conditions. Although Marcotte (1981) has shown theoretically and empirically that this heuristic may often perform poorly in terms of representing the exact model (P) - (U), this heuristic can be used to find a lower bound for the solution to the exact problem. Thus, let us briefly describe this heuristic model and its solution in this section, and in the next section we will prove that this model and (P') - (U) can be used to bound the solution to (P) - (U).

Let us assume that the objective function in (P) is an adequate approximation of Beckmann's (1956) User Equilibrium objective function in (U). In this case, let us consider the following problem:

$$\underset{f,y}{\text{minimize}} \quad \sum_{a \in A} c_a(f_a, y_a) f_a$$

subject to:
$$\sum_{p \in P_w} h_p = T_w \qquad \forall w \in W$$

$$f_a = \sum_{p \in P} \delta_{ap} h_p \qquad \forall a \in A$$

$$f_a \geq 0 \qquad \forall a \in A \qquad \text{(SO)}$$

$$h_p \geq 0 \qquad \forall p \in P$$

$$\sum_{a \in A} \beta_a y_a \leq B$$

$$y_a \geq 0 \qquad \forall a \in A.$$

How can we characterize (SO) in game-theoretic terms? In (SO),
the planner has both strategy vectors, f and y, under his control.
That is, the planner does not have to be concerned in any way with
the behavior of the users. Thus, (SO) represents a trivial game
in which there is only one player, the planner.

To solve this problem, let us apply the Frank-Wolfe feasible
direction algorithm to (SO) (see, for example, LeBlanc, Morlok and
Pierskalla 1975). In each iteration of this algorithm, the following
linear program (LP) would result

$$
\begin{array}{ll}
\underset{f,y}{\text{minimize}} & \sum_{a\in A} \overline{MC}_a\, f_a + \sum_{a\in A} \overline{C}_a^{\,y}\, y_a
\end{array}
$$

$$
\begin{array}{lll}
\text{subject to:} & \sum_{p\in P_w} h_p = T_w & \forall w\in W \\[2mm]
& f_a = \sum_{p\in P} \delta_{ap} h_p & \forall a\in A \\[2mm]
& f_a \ge 0 & \forall p\in P \\[2mm]
& h_p \ge 0 & \forall a\in A \\[2mm]
& \sum_{a\in A} \beta_a y_a \le B & \\[2mm]
& y_a \ge 0 &
\end{array} \tag{4}
$$

where

$$\overline{MC}_a = c_a(\overline{f}_a,\overline{y}_a) + \overline{f}_a\,(\partial c_a(\overline{f}_a,\overline{y}_a)/\partial f_a),$$

$$\overline{C}_a^{\,y} = f_a(\partial c_a(\overline{f}_a,\overline{y}_a)/\partial y_a),$$

and

(——) denotes variables held at the previous iteration's
values.

As the reader can easily see, (4) separates into two disjoint LP's

$$\underset{f}{\text{minimize}} \quad \sum_{a\in A} \overline{MC}_a\, f_a$$

subject to: $\quad \sum_{p \in P_w} h_p = T_w \qquad \forall w \in W$

$\qquad\qquad f_a = \sum_{p \in P} \delta_{ap} h_p \qquad \forall a \in A \qquad\qquad (5)$

$\qquad\qquad f_a \geq 0 \qquad\qquad\qquad \forall a \in A$

$\qquad\qquad h_p \geq 0 \qquad\qquad\qquad \forall p \in P$

and

$\qquad \underset{y}{\text{minimize}} \quad \sum_{a \in A} \bar{c}_a^y \, y_a$

subject to: $\quad \sum_{a \in A} \beta_a \, y_a \leq B \qquad\qquad\qquad (6)$

$\qquad\qquad\qquad y_a \geq 0 \qquad \forall a \in A.$

From LeBlanc, Morlok and Pierskalla (1975), it is well-known that
the solution to (5) is an all-or-nothing assignment in which all
the flow between each O-D pair is assigned to the shortest path,
based on $\overline{MC}_a$'s, between that O-D pair. Problem (6) is a version
of the well-known knapsack problem without the integrality constraints usually associated with the decision variables (see, for
example pp. 116-117 of Shapiro 1979). The solution to (6) is to
set

$$y_a^* = \begin{cases} -\bar{c}_{a^*}^y/\beta_a & \text{if } -\bar{c}_{a^*}^y/\beta_a \geq 0 \\ 0 & \text{otherwise} \end{cases} \qquad (7)$$

for a^* such that

$$a^* = \operatorname{argmax}\left\{ \frac{-\bar{c}_a^y}{\beta_a} \right\}, \qquad (8)$$

and set $y_a = 0$ for all $a \in A$, $a \neq a^*$. Therefore, both subproblems
resulting from the application of the Frank-Wolfe algorithm to (SO)
are very easy to solve.

The second step of the Frank-Wolfe algorithm is a unidimensional
line search along the direction of descent formed from the LP
solution and the previous iteration's solution. If the objective

function of (SO) were convex, then unidimensional search methods such as a binary search or a Fibonacci search can be used. However, the objective function in (SO) is typically nonconvex in the improvement variables y. In this case, a more complicated line search procedure such as the Armijo rule described in Bertsekas (1982) must be used. Avriel (1976) and Luenberger (1973) discuss the convergence of this type of feasible direction algorithm for nonconvex problems in some detail. Thus, the Frank-Wolfe algorithm for (SO) would consist of the following steps:

Step 0   Set $y^o = 0$ and find an initial feasible flow vector f^o. Set k=0. Let $x^o = (f^o, y^o)$

Step 1   Linearize the objective function in (SO) about x^k to create the two subproblems (5) and (6). Solve (5) by an all-or-nothing assignment and solve (6) by the rule stated in (7)-(8). Call the resulting solution $\overline{x} = (\overline{f}, \overline{y})$

Step 2   Perform a unidimensional line search along the direction $d^k = \overline{x} - x^k$. This line search must be performed with a method such as the Armijo rule to insure convergence. Call the solution to this line search x^{k+1}.

Step 3   Check the convergence criteria. If convergence is reached, stop. Else, set k = k+1, and go to Step 1.

Therefore, the solution to (SO) is very easy to compute. As we shall now see, the solutions to (P') - (U) and (SO) can be used to bound the solution to the exact model (P') - (U).

5. BOUNDING THE EXACT MODEL'S SOLUTION

Consider the following optimization problem:

$$\text{minimize}_{x,z} \quad F(x,z) \tag{9}$$

$$\text{subject to:} \quad z \in \Omega \subseteq R^m$$

$$x \in \Lambda_1 \cap \Lambda_2 \subseteq R^n$$

where $F: R^m \times R^n \to R$. Also, consider two variants of problem (9):

$$\text{minimize}_{x,z} \quad F(x,z) \tag{10}$$

$$\text{subject to:} \quad z \in \Omega \subseteq R^m$$

$$x \in \Lambda_1, \subseteq R^n$$

and

$$\text{minimize}_{x,z} \quad F(x,z) \tag{11}$$

$$\text{subject to:} \quad z \in \Omega \subseteq R^m$$

$$x = x^* \in \Lambda_1 \cap \Lambda_2 \subseteq \Lambda_1. \tag{12}$$

Since it is true that

$$x^* \in \Lambda_1 \cap \Lambda_2 \subseteq \Lambda_1,$$

then it must be the case that the solution to (11) has at its solution
an objective function value which is at least as great as the solution
to (9) which is in turn at least as great as the solution to (10).
Thus, as the feasible set for the vector of variables x becomes
larger, the value of the objective function at the optimal solution
cannot exceed its value for a more restrictive feasible set.

Given this fact, let us consider the three models (P) - (U),
(P') - (U) and (SO). The exact model (P) could be rewritten as

$$\text{minimize}_{f,y} \quad \sum_{a \in A} c_a(f_a, y_a) f_a \tag{13}$$

$$\text{subject to:} \quad y \in \Omega$$

$$f \in \Lambda_1 \cap \Lambda_2$$

where the set Ω consists of the nonnegativity and budget constraints on the improvement vector y, Λ_1 is the set of nonnegativity arc flows which obey constraints (1) and (3), and Λ_2 is the
set of arc flows which are User Equilibrium flow patterns. Thus, we
have simply replaced the function $f = f(y)$ by a set of constraints
(see Marcotte 1981 or Tan et al 1979). Therefore, the exact design
model takes the same form as problem (9). The heuristic model (SO)
relaxes the requirement that the flow pattern be a User Equilibrium
and hence the flow pattern is only required to be an element of Λ_1,
or $f \in \Lambda_1$. Thus, the heuristic (SO) corresponds to problem (10).
Finally, the planner's problem in (P') - (U) assumes that the
planner take the arc flows as fixed. Since the arc flows must constitute a User Equilibrium flow pattern (be a solution to the second
player's optimal strategy problem (U) in the Nash equilibrium framework we are considering), the constraints on the arc flows in (P')
must take the form

$$f = f^* \in \Lambda_1 \cap \Lambda_2, \tag{14}$$

where f^* is the flow level which the planner takes as fixed. Thus,
the heuristic (P') - (U) corresponds directly to problem (11). Given
the discussion in the previous paragraph, it must be the case that:

Total System Cost $\leq$ Total System Cost $\leq$ Total System Cost
in (SO)   in (P') - (U)

Therefore, the heuristics (SO) and (P') - (U) bound the solution to
the exact model (P) - (U).

The above result can be intuited from game-theoretic concepts.
Since the planner has all variables under his control in (SO), the
solution to (SO) must achieve the lowest total system costs. That
is, (SO) must achieve the lowest value of the objective function in
(P). At the other extreme, the iterative optimization-equilibrium
heuristic which is based upon (P') - (U) assumes that the planner is
totally myopic in his view of the change in arc flows with respect
to a change in investment levels. Therefore, the solution to
(P') - (U) must achieve the highest total system costs. In between
is the exact problem formulation (P) - (U) in which the planner can
foresee the changes in arc flows, but cannot fully control these flows.
Thus, the ordering of these models which is given above is not only

mathematically but also intuitively correct.

The purpose behind the development of (SO) and (P') - (U) should now be clear. Harker (1983b) has shown that an efficient algorithm exists for the calculation of the solution to a Cournot-Nash game and thus problem (P') - (U). It has also been demonstrated that an efficient algorithm exists for the solution of (SO). Thus, if we have a network design problem, apply the solution procedures to both (SO) and (P') - (U), and find that the value of the total system cost and the improvement variables do not significantly differ between the two solutions, then it is not necessary to go to the expense of solving the difficult problem (P) - (U). Therefore, the result of this paper can be used to calculate bounds of the continuous design problem which given the complexity of problem (P') - (U), are extremely useful results. It may be that for most realistic network design problems the two bounds are so close that it is not necessary to expend the time and money to exactly solve (P) - (U).

6. NUMERICAL EXAMPLES

To illustrate our discussion in the previous sections, let us consider the network depicted in Fig. 2 where there is flow from node 1 to node 6 $(T_{(1,6)})$ and from 6 to 1 $(T_{(6,1)})$. Table 1 contains the arc cost function information (Λ_a, X_a, K_a) and the per unit investment costs for each arc (β_a). The total budget B for improvements is assumed to be equal to 100.000.

Table 2 contains the results of solving both the trivial game (SO) and the Cournot-Nash game (P')-(U) for various levels of O-D flows, and Fig. 3 plots these results. At low flow levels, the network is not very congested (the nonlinear portion of the arc cost function does not contribute significantly to the cost on an arc) and the trivial game solution (the lower bound) is very close to the Cournot-Nash solution (the upper bound). In fact, for the case of total flow equal to 7.5, these two solutions are exactly equal. As the flow level increases, the network becomes congested and the bounds begin to diverge. Table 3 also illustrates this result. At low flow levels, the arc improvements are identical, and as the flow level increases, so does the divergence between the two solutions.

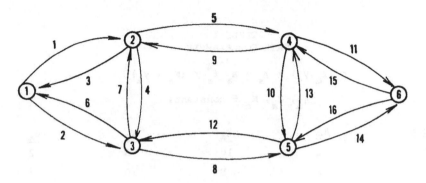

Fig. 2 Network Example

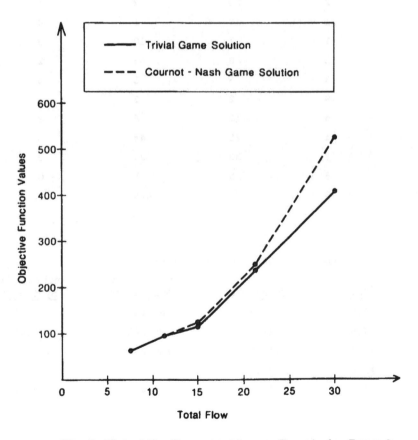

Fig. 3 Plot of the Upper and Lower Bounds for Example

TABLE 1
DATA FOR EXAMPLE

$$c_a(f_a, y_a) = \Lambda_a + X_a \, f_a^4 \, / \, (K_a + y_a)^4$$

$$\Lambda_a, \; X_a, \; K_a \equiv \text{constants}$$

Arc a	Λ_a	X_a	K_a	β_a
1	1	10	3	2
2	2	5	10	3
3	3	3	9	5
4	4	20	4	4
5	5	50	3	9
6	2	20	2	1
7	1	10	1	4
8	1	1	10	3
9	2	8	45	2
10	3	3	3	5
11	9	2	2	6
12	4	10	6	8
13	4	25	44	5
14	2	33	20	3
15	5	5	1	6
16	6	1	4.5	1

TABLE 2
FLOW VS. SYSTEM COST FOR EXAMPLE

$T_{(1,6)}$	$T_{(6,1)}$	Total Flow	Trivial Game System Cost	Cournot-Nash Game System Cost
2.5	5.0	7.5	63.282	63.282
3.75	7.5	11.25	99.143	99.685
5.0	10.0	15.0	140.206	147.594
7.5	15.0	22.5	260.061	273.506
10.0	20.0	30.0	411.440	526.245

ARC IMPROVEMENT SOLUTIONS

Arc a	Total Flow=7.5		Total Flow=11.25		Total Flow=15.0		Total Flow=22.5		Total Flow=30.0	
	y_a^{so}	y_a^{cn}	y_a^{so}	y_a^{cn}	y_a^{so}	y_a^{cn}	y_a^{so}	y_a^{cn}	y_a^{so}	y_a^{cn}
1										
2					0.214		0.911	0.226	3.701	
3	4.237	4.237	4.169	4.237	3.648	4.237	6.298	3.640	9.902	4.244
4										
5										
6			0.773		6.073		7.505	6.129	10.398	7.483
7										
8									0.044	
9									0.044	
10										
11										
12										
13										
14									0.212	
15	13.136	13.136	13.064	13.136	12.507	13.136	8.391	12.500	1.212	11.476
16							7.925		20.951	3.438

* y_a^{so} = trivial game (SO) arc improvements

y_a^{cn} = Cournot-Nash game (P')-(U) arc improvements

7. CONCLUSION

The equilibrium network design model is an important tool
in transportation network planning, but is perhaps the most computationally intensive problem encountered in this field. This
paper has shown that two relatively efficient heuristics can be
used to bound the solution to the optimal design problem. These
bounds are very good for uncongested or slightly congested networks,
and begin to diverge as the flow level on the network increases.

The point to be made in this paper is simply that for many
realistic problems, these two heuristics may be used to assess
whether or not it is necessary to expend the time and resources to solve
for an exact solution of this problem. In many realistic networks,
it is only a few arcs which are highly congested and in this case,
these two bounds should perform rather well. Also, this paper
has shown that the solution to the iterative optimization-equilibrium
heuristic is always greater than or equal to the solution to the
true model. This fact along with the fact that the solution to (SO)
is at least as low as the solution to the exact model is very
valuable in assessing the value of these two heuristics even when
the network is heavily congested.

In conclusion, it is doubtful that in the near future a solution technique will be developed for (P) - (U) which is efficient
enough to allow one to compute an optimal network investment
strategy for problems of realistic size. Thus, the planner-analyst
must rely on heuristic methods of solving (P) - (U). Using two
efficient heuristics, one can bound the exact solution. Thus, the
first task in any optimal network investment study should be to
apply the strategy presented in this paper, and then assess whether
or not additional resources need to be used to solve exactly the
problem (P) - (U).

8. REFERENCES

Abdulaal, M.S., L.J. LeBlanc (1979). Continuous equilibrium network design models, Transportation Research 13B, pp. 19-32.

Allsop, R.E. (1974). Some possibilities for using traffic control to influence trip distribution and route choice. In: Transportation and Traffic Theory: Proceedings of the Sixth International Symposium on Transportation and Traffic Theory, edited by D.J. Buckley (American Elsevier, New York), pp. 345-375.

Avriel, M. (1976). Nonlinear Programming: Analysis and Methods (Prentice-Hall, Englewood Cliffs, New Jersey).

Beckmann, M.J., C.B. McGuire, C.B. Winston (1956). Studies in the Economics of Transportation (Yale University Press, New Haven, Conn).

Bertsekas, D.P. (1982). Constrained Optimization and Lagrange Multiplier Methods (Academic Press, New York).

Dafermos, S. (1983). An iterative scheme for variational inequality. Mathematical Programming 26, pp. 40-47.

Friedman, J.W. (1977). Oligopoly and the Theory of Games (North-Holland, Amsterdam).

Friesz, T.L., P.T. Harker (1983). Properties of the iterative optimization-equilibrium algorithm, Report No. CE-UTND-1982-10-1, Department of Civil Engineering, University of Pennsylvania, Philadelphia, Pennsylvania.

Gabay, D., H. Moulin (1980). On the uniqueness and stability of Nash-equilibria in noncooperative games. In: Applied Stochastic Control in Econometrics and Management Science, edited by P. Kleindorfer and T.S. Tapiero (North-Holland, Amsterdam), pp. 271-293.

Harker, P.T. (1983a). A variational inequality formulation of the Nash equilibrium problem: towards an efficient and easily implementable computational procedure. Presented at the 1983 Joint National Meeting of TIMS/ORSA, Chicago, Illinois, April 1983.

Harker, P.T. (1983b). A variational inequality approach for the determination of oligopolistic market equilibrium, Mathematical Programming (forthcoming).

LeBlanc, L.J. (1975). An algorithm for the discrete network design problem, Transportation Science 9, pp. 183-199.

LeBlanc, L.J., E.K. Morlok, W.P. Pierskalla (1975). An efficient approach to solving the road network equilibrium traffic assignment problem. Transportation Research 9, pp. 309-318.

Luenberger, D.G. (1973). Introduction to Linear and Nonlinear Programming (Addison-Wesley, Reading, Mass.).

Marcotte, P. (1981). An analysis of heuristics for the network design problem. Presented at the Eighth International Symposium on Traffic Flow and Theory, Toronto, Canada.

Marcotte, P. (1983). Network optimization with continuous control parameters, Transportation Science 17, pp. 181-197.

Moulin, H. (1982). Game Theory for the Social Sciences (New York University Press, New York).

Pang, J.S., D. Chan (1982). Iterative methods for variational and complementarity problems, Mathematical Programming 24, pp. 284-313.

Shapiro, J.F. (1979). Mathematical Programming: Structures and Algorithms (John Wiley, New York).

Tan, H.N., S.B. Gershwin, M. Athans (1979). Hybrid Optimization in Urban Traffic Networks, LIDS Report DOT-TSC-RSPA-79-7, Laboratory for Information and Decision Systems, M.I.T., Cambridge, Mass.

Ninth International Symposium on
Transportation and Traffic Theory
© 1984 VNU Science Press, pp. 253–271

EQUILIBRIUM FLOWS IN A NETWORK WITH CONGESTED LINKS

IWAO OKUTANI

Department of Civil Engineering, Shinshu University, 500 Wakasato, Nagano 380, Japan

ABSTRACT

The existing traffic assignment technique which employs a link cost function, $f(X)$, such that $f(X) \to \infty$ as the flow approaches the capacity only yields an approximate solution to the problem which is produced by adding explicit capacity restraints to the mathematical programming problem, P-1, for obtaining the equilibrium flow pattern. P-1 is not necessarily effective if the network involves some link(s) where the flow is in the congested flow reqime since the maximum solution of P-1 can offer the equilibrium flow pattern if all paths between each origin-destination pair are utilized. For this reason it is proposed to deal with the dual problem, D-1, of P-1 whose maximum solution is proved to offer the equilibrium flow pattern. If the congested flow regime is taken into account, the equilibrium flow pattern is not unique since the objective function of D-1 is not concave and further we can produce many different D-1's depending on the combinations of links in the congested flow regime.

1. INTRODUCTION

In most existing techniques for obtaining the equilibrium
traffic flow pattern according to the Wardrop's first princi-
ple(1952), it is assumed either that the link cost, $f(X)$, is
an increasing function of the flow, X, on it over the range
$[0,\infty)$ or that $f(X)$ tends to infinity as X approaches the
link capacity(Davidson 1966, Daganzo 1977, Assad 1978) so
that $f(X)$ acts as a penalty function to account for capacity
restraint. However, under the former assumption it may occur
that the flow on the link involved in the utilized path exceeds its capacity in a great degree when alternative paths
are too costly. Also the solution under the latter assumption offers merely an approximate solution to the mathematical program, P-2, which is constructed by adding explicit
capacity restraints to the traffic equilibrium mathematical
program, P-1, enunciated by Beckmann *et al*(1956).   In the
real life if a path involves some link(s) where the flow attains the capacity is still more attractive than alternative
paths insofar as cost is concerned, driver may continue to
choose that path until the traffic situation on the link(s)
under consideration gets into the congested flow regime and
costs on alternative paths become comparable to the cost on
that path. Note that the congested flow regime is defined as
the flow regime where the traffic density prevails which is
greater than the optimum density corresponding to the maximum
flow. Saishu(1976), Okutani and Matsuzaki(1976, 1977, 1978),
Okutani(1980), Okutani and Komachiya(1981) have taken into
account such phenomena in the analyses of the traffic equilibrium in a surface road network. Newell(1977) has conducted the analyses on the queue configuration at freeway
bottleneck and/or the ramps and route choice between the
freeway and parallel streets. In the present paper some
properties of the solution to P-2 are exhibited first. On
the basis of a simple example, the difference between the
solution derived from the existing techniques and the
solution derived by taking the congested flow regime into
account is demonstrated.   It is also shown that
P-1 is not effective for solving the same example since

the objective function takes maximum at the equilibrium despite the fact that P-1 is a minimization problem.   This is
followed by a presentation of the dual problem of P-1 and
its effectiveness in solving the traffic equilibrium problem.
Finally, a solution algorithm utilizing the properties of
the dual problem and a numerical example are presented.

2. SOME PROPERTIES OF THE SOLUTION TO P-2

We consider a road network with m directed links numbered 1
to m and assume there exist N origin-destination pairs numbered 1 to N.   The mathematical program P-1 for obtaining
the equilibrium flow pattern is described as follows:

P-1:

$$\phi(x) = \sum_{j \in L} \int_0^{X_j} f_j(X)\,dX \to \min. \tag{1}$$

subject to

$$\sum_{k \in P_i} x_k^i - s^i = 0 \qquad i \in I, \tag{2}$$

$$x \geqq 0, \tag{3}$$

where
  L = set of directed links in the network,
  $|L|$ = m,
  I = set of O-D pairs,
  $|I|$ = N,
  P_i = set of independent paths of O-D pair i $i \in I$,
  s^i = vehicle trips of O-D pair i $i \in I$,
  x_k^i = vehicle trips of O-D pair i $i \in I$,
  x = vector of which entries are x_k^i's,
  $X_j = \sum\limits_{i \in I} \sum\limits_{k \in P_i} r_{kj}^i x_k^i$ = flow on link j $j \in L$,
  $r_{kj}^i = \begin{cases} 1 & \text{if link j is on path k of O-D pair i,} \\ 0 & \text{otherwise,} \end{cases}$
  $f_j(X)$ = travel cost(travel time) of link j when the
         flow is X $j \in L$.

The link cost function $f_j(X_j)$ can be depicted as shown in

in Fig.1 if we consider the congested
flow regime as well as the free flow
regime. We suppose that this two valu-
ed function of X_j, $f_j(X_j)$, can be approximated by two convex functions as
illustrated in Fig.2 throughout this
study,i.e.,

$$f_j(X_j) = \begin{cases} f_j^I(X_j) \text{ if } X_j \text{ is in the} \\ \text{free flow regime,} \\ f_j^{II}(X_j) \text{ if } X_j \text{ is in the} \\ \text{congested flow regime.} \end{cases}$$

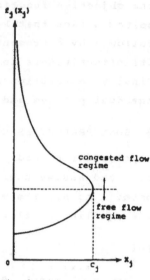

Note that $f_j(X_j)$ is taken such that it
intersects the ordinate since otherwise
the integral calculus of eq.(1) cannot
be executed.

Fig. 1. Link cost function.

We add the capacity restraints
to P-1 to produce P-2:

$$X_j - C_j \leq 0 \qquad j\epsilon L, \qquad (4)$$

where C_j is the capacity of link
j. Suppose P-2 is feasible and
let $\bar{x}_k^i$, $\bar{x}_j$ denote the path and
link flow respectively which
minimize $\phi(x)$ of P-2. Since all
of the constraints are linear
and therefore satisfy the reverse constraint qualification,
there exist vectors u of dimension m and v of dimension N such
that $(\bar{x}, u, v)$ solves the Kuhn-
Tucker stationary point problem;

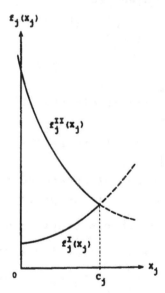

Fig. 2. Link cost function approximated
by two convex functions.

$$\sum_{j\epsilon L} r_{kj}^i f_j(\bar{x}_j) + \sum_{j\epsilon L} r_{kj}^i u_j + v_j \geq 0 \quad k\epsilon P_i, \ i\epsilon I, \qquad (5)$$

$$\bar{x}_k^i [\sum_{j\epsilon L} r_{kj}^i f_j(\bar{x}_j) + \sum_{j\epsilon L} r_{kj}^i u_j + v_j] = 0 \quad k\epsilon P_i, \ i\epsilon I, \qquad (6)$$

$$u_j(\bar{x}_j - C_j) = 0 \quad j\epsilon L, \tag{7}$$

$$u \geq 0, \tag{8}$$

and eqs.(2) to (4). From eqs.(5) and (6) one can readily deduce

$$\sum_{j\epsilon L} r^i_{kj} f_j(\bar{x}_j) + U^i_k = -v^i \quad \text{if } \bar{x}^i_k > 0, \tag{9}$$

$$\sum_{j\epsilon L} r^i_{kj} f_j(\bar{x}_j) + U^i_k \geq -v^i \quad \text{if } \bar{x}^i_k = 0, \tag{10}$$

where

$$U^i_k = \sum_{j\epsilon L} r^i_{kj} u_j. \tag{11}$$

It follows from eqs.(7) and (8) that

$$u_j = 0 \quad \text{if } \bar{x}_j < C_j, \tag{12}$$

$$u_j \geq 0 \quad \text{if } \bar{x}_j = C_j. \tag{13}$$

We group the links into two subsets; one, L_s, for links where the capacity is reached, the other for the remaining links. It can be seen from eqs.(11) through (13) that U^i_k equals zero if path k of O-D pair i does not involve any link $j\epsilon L_s$ and U^i_k is greater than or equal to zero if it does. Therefore one can conclude that

(1) the costs of utilized paths which do not involve any link $j\epsilon L_s$ are equal(let the cost be denoted by t^i for O-D pair i),

(2) the costs of utilized paths which involve link(s)ϵL_s are less than or equal to t^i,

(3) the costs of nonutilized paths which do not involve any link $j\epsilon L_s$ are greater than or equal to t^i,

(4) it cannot be known whether the costs of nonutilized paths which involve link(s)ϵL_s are greater than or equal to t^i or not.

As the above properties are quite different from those of the equilibrium flow pattern according to Wardrop's first principle, it is not appropriate to employ such a link cost function as tends to infinity as the flow approaches the capacity because such a solution technique merely yields the approximate solution to P-2. However this does not imply that the existing method employing the increasing link cost function over the range of the flow $[0, \infty)$ is sufficient since the link flow may exceed the capacity in a great degree under some circumstances as mentioned in the preceding section. In the subsequent section we will make comparisons between solutions obtained through the existing techniques and the new method considering the congested flow regime.

3. COMPARISONS OF THE FLOW PATTERNS IN A SIMPLE NETWORK OBTAINED BY THE EXISTING AND THE NEW TECHNIQUES

We obtain the equilibrium flow patterns in a simple network through the existing techniques and the new technique to clarify the drawbacks of the former and the necessity of the latter. The subject network is as shown in Fig.3. For simplicity the link cost function is assumed to be a linear function of the flow both in the free flow regime and the congested flow regime;

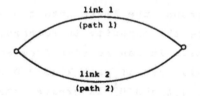

$$f_j(X_j) = a_j X_j + b_j.$$

Fig. 3. The network with two parallel links.

a_j, b_j are listed in Table 1 along with the capacity C_j. Let traffic demand between the nodes be 1425. It is noted that the superscript i indicating O-D pair is omitted for brevity.

Table 1. Link parameters.

Link #	Regime	a_j	b_j	C_j
1	f.f.	0.04	35	1500
	c.f.	-0.05	170	
2	f.f.	0.01	20	1000
	c.f.	-0.08	110	

Note: f.f. = free flow
c.f. = congested flow

The equilibrium flow pattern which conforms to the solution to P-1 without considering the congested flow regime is

Solution 1: $x_1 = X_1 = 0$ $(t_1 = 35)$,

$x_2 = X_2 = 1425$ $(t_2 = 34.25)$,

where t_k $(k = 1, 2)$ is the cost of path k. Inspecting the above solution we notice that the flow on link 2 exceeds its capacity more than 40 percent. In order for the flow on link 2 to be within the capacity limitation we make a version of the cost function as follows;

$$\tilde{f}_j(X_j) = \begin{cases} f_j(X_j) & \text{if } 0 \le X_j \le 0.95C_j, \\ \dfrac{\beta_j}{C_j - X_j} & \text{if } 0.95C_j \le X_j < C_j, \end{cases} \quad (14)$$

where β_j's are given as

$$\beta_1 = 6900, \quad \beta_2 = 1475,$$

which are computed from the continuity condition .at $X_j = 0.95C_j$ (see Fig.4). The resulting flow pattern derived by putting $f_j(X_j)$ for $\tilde{f}_j(X_j)$ represented as eq.(14) is

Solution 2: $x_1 = X_1 = 453$

$(t_1 = 53)$,

$x_2 = X_2 = 972$

$(t_2 = 53)$.

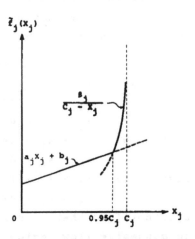

Fig. 4. Modified cost function.

It should be noted that the cost of link 2 is different from the original cost(= 0.01×972 + 20) by about 70 percent. It is noted that the portion of the link cost function $\tilde{f}_j(X_j)$ over the range $[0.95C_j, C_j)$ plays a role of penalty function which is often employed in solving a constrained non-linear programming problem and hence it is natural that the Solution 2 is an approximation of the solution(Solution 3) to P-2;

Solution 3: $x_1 = X_1 = 425$ $(t_1 = 52)$,

$x_2 = X_2 = 1000$ $(t_2 = 30)$.

In the computations performed to this point the congested flow regime is not taken into account. We introduce next the link cost function in the congested flow regime as well as that in the free flow regime. As path 2 is still more attractive for the drivers than path 1 even when the capacity is reached on link 2, it may be natural that we suppose for the flow on link 2 to get into the congested flow regime. The resulting flow pattern is as

Solution 4: $x_1 = X_1 = 975$ $(t_1 = 74)$,

$\quad\quad\quad\quad\quad x_2 = X_2 = 450$ $(t_2 = 74)$.

We notice that the above solution is rational in the sense that the flow does not exceed the capacity on any link and the given cost functions are employed as they are.

It should be mentioned that there exist three more euilibrium flow patterns, Solution 5, Solution 6 and Solution 7, i.e.

Solution 5: $x_1 = X_1 = 1425$ $(t_1 = 92)$,

$\quad\quad\quad\quad\quad x_2 = X_2 = 0$ $(t_2 = 110)$,

Solution 6: $x_1 = X_1 = 1425$ $(t_1 = 98.75)$,

$\quad\quad\quad\quad\quad x_2 = X_2 = 0$ $(t_2 = 110)$,

Solution 7: $x_1 = X_1 = 1338$ $(t_1 = 103)$,

$\quad\quad\quad\quad\quad x_2 = X_2 = 87$ $(t_2 = 103)$.

It is notified that Solution 5 is derived when link 2 is in the congested flow regime while Solution 6 and 7 are derived when both links are in the congested flow regime. Solution 5 and 6 are not realistic since it appears not to occur in practice that vehicles keep stopping on link 2. Solution 7 can be realized if link 2 becomes more congested for some reasons starting from the flow pattern corresponding to Solution 4 and some trips divert to path 1 temporarily causing the entering flow to exceed the capacity on link 1. Concluding, only two solutions, Solution 4 and 7, yield the equilibrium flow patterns which are plausible to occur. Further, these patterns are derived under the condition that one or two links are in the congested flow regime. The discussions to this point imply that it is required for us to develop some traffic assignment techniques taking the congested flow regime into account.

4. INTRODUCTION OF THE DUAL PROBLEM OF P-1 FOR SEEKING THE
   EQUILIBRIUM

In the preceding section we present that there exist two
equilibrium flow patterns making sense from the practical
viewpoint. However, these patterns cannot be derived from
applications of P-1.
This is because both So-
lution 4 and 7 are maxi-
mum solutions of P-1 de-
spite the fact that P-1
is a minimization program.
The graphical illustra-
tion is given in Fig.5
where x_1 is eliminated by
utilizing the origin-
destination demand condi-
tion. Note that this
graph is obtained for the
case where only link 2

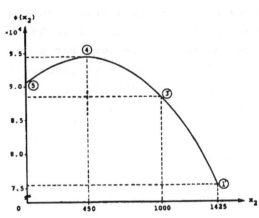

Fig. 5. Curve of the objective function.

is in the congested flow regime. The numbers 4, 5 correspond to the solution numbers presented in the preceding
section and further the numbers 1', 3' correspond to the
solution to P-1 and P-2, respectively.

The reason why $\phi(x)$ takes maximum at the equilibrium flow is
owing to the property 1 as expressed below:

Property 1: The maximum solution of P-1 yields the equilibrium flow pattern if all paths connecting each
of the O-D pairs are utilized.

Proof: Putting the minus sign on the objective function and
developing the Kuhn-Tucker's optimality conditions, we obtain

$$-\sum_{j \in L} r_{kj}^{i} f_j(\bar{x}_j) = -v^i \quad \text{if } \bar{x}_k^{i} > 0 \qquad i \in I, \quad k \in P_i, \qquad (15)$$

$$-\sum_{j \in L} r_{kj}^{i} f_j(\bar{x}_j) \geq -v^i \quad \text{if } \bar{x}_k^{i} = 0 \qquad i \in I, \quad k \in P_i. \qquad (16)$$

where $\bar{x}_k^{i}$ is the maximum solution, $\bar{x}_j$ the corresponding link
flow and v^i the Lagrange multiplier. If all paths between
each of the O-D pairs are utilized, eq.(16) vanishes. Elim-
inating the minus sign from both sides of eq.(15) and noting

that the left hand side of the resulting equation represents
the cost of path k, we complete the proof.

The property indicates that P-1 is not always effective for
obtaining the equilibrium since generally it is difficult
for us to foresee if all paths are utilized or not. Consequently,we need to look for other mahtematical tool replacing P-1. To this end we propose to employ the dual problem,
D-1, of P-1 in this study;

D-1:

$$\psi(x,\lambda) = \phi(x) + \sum_{i \in I} \lambda^i s^i - \sum_{i \in I} \sum_{k \in P_i} x_k^i \sum_{j \in L} r_{kj}^i f_j(X_j) \to max. \quad (17)$$

subject to

$$\sum_{j \in L} r_{kj}^i f_j(X_j) \geq \lambda^i \qquad i \in I, \; k \in P_i, \qquad\qquad (18)$$

where

$$\lambda = (\lambda^1, \lambda^2, \cdots, \lambda^N) = \text{Lagrange mutiplier.}$$

The properties that follow imply that D-1 is worthy to be
considered when we obtain the equilibrium flow pattern in
the network with congested links.

Property 2: The maximum solution of D-1 yields the equilib-
          rium flow pattern.

Proof: Let $(\bar{x},\bar{\lambda})$ denote the maximum solution to D-1 and $\bar{x}_j$(
$j \in L$) denote the corresponding link flow. Since the con-
straints given by eq.(18) fulfil the reverse constraint qualification because $f_j(X_j)$ is assumed convex both in the free
flow regime and the congested flow regime as mentioned ear-
lier, $(\bar{x},\bar{\lambda})$ fulfils the Kuhn-Tucker's optimality condition
along with some Lagrange multiplier v of which element is v_k^i

$(i \in I, \; k \in P_i)$;

$$\sum_{j \in L} r_{kj}^i \frac{df_j(\bar{x}_j)}{dx_j} (\bar{x}_j - v_j) = 0 \qquad i \in I, \; k \in P_i, \qquad (19)$$

$$\sum_{k \in P_i} v_k^i = s^i \qquad i \in I, \qquad\qquad (20)$$

$$v \geq 0, \qquad\qquad\qquad\qquad\qquad\qquad\qquad (21)$$

$$\sum_{j \in L} r^i_{kj} f_j(\bar{x}_j) \geq \bar{\lambda}^i \quad i \in I, \ k \in P_i, \tag{22}$$

$$v^i_k [\bar{\lambda}^i - \sum_{j \in L} r^i_{kj} f_j(\bar{x}_j)] = 0 \quad i \in I, \ k \in P_i, \tag{23}$$

where

$$V_j = \sum_{i \in I} \sum_{k \in P_i} r^i_{kj} v^i_k \quad j \in L. \tag{24}$$

If we consider the specific O-D pair between the endpoints of link j and the path consisting of link j only, eq.(19) is written as

$$\frac{df_j(\bar{x}_j)}{dx_j} (\bar{x}_j - v_j) = 0.$$

Since $f_j(x_j)$ is either an increasing function or a decreasing function, the derivative of it with respect to x_j never becomes zero. Hence

$$\bar{x}_j = v_j \tag{25}$$

Substituting eq.(25) into eqs.(22) and (23) and taking into account eqs.(20) through (24) we see that v^i_k yields the equilibriumm path flow of path k of O-D pair i, V_j the equilibriumn link flow and $\bar{\lambda}^i$ the cost of the utilized path of O-D pair i. Therefore from eq.(25) $\bar{x}_j$ proves to offer the equilibrium flow of link j (j∈L). This completes the proof.

Property 3: D-1 does not have any minimum solution.
Proof: Suppose $(\tilde{x}, \tilde{\lambda})$ yields the minimum solution of D-1. Then there exists some open ball $B_\delta(\tilde{x}, \tilde{\lambda})$ around $(\tilde{x}, \tilde{\lambda})$ with radius δ > 0 such that

$$(x,\lambda) \in B_\delta(\tilde{x}, \tilde{\lambda}) \cap \Gamma \rightarrow \psi(x,\lambda) \geq \psi(\tilde{x}, \tilde{\lambda}), \tag{26}$$

where Γ is the constraint set. If we take a given $\overset{*}{\lambda}$ such that $\overset{*}{\lambda} < \tilde{\lambda}$ and $\| \overset{*}{\lambda} - \tilde{\lambda} \| < \delta$, $(\tilde{x}, \overset{*}{\lambda})$ is apparently a point of Γ. Further, from the relationship

$$\| (\tilde{x}, \overset{*}{\lambda}) - (\tilde{x}, \tilde{\lambda}) \| = \| \overset{*}{\lambda} - \tilde{\lambda} \| < \delta,$$

$(\tilde{x}, \overset{*}{\lambda})$ proves to be an interior point of $B_\delta(\tilde{x}, \tilde{\lambda})$. However,

$$\psi(\tilde{x},\overset{*}{\lambda}) - \psi(\tilde{x},\bar{\lambda}) = S(\overset{*}{\lambda} - \bar{\lambda}) < 0,$$

where $S = (S^1, S^2, \cdots, S^N)$. This contradicts eq.(6).
Hence D-1 does not have any minimum solution. The property
is proved.

Property 2 and property 3 ensure that all that is necessary
for obtaining the equilibrium flow pattern is to maximize
$\psi(x,\lambda)$ of D-2 whereas it is reminded that we need to pay
attention to the maximum solution as well as the minimum
solution when handling P-1. However, we cannot prove the
existence of the maximum solution since the concavity of
$\psi(x,\lambda)$ cannot be known. Further it seems impossible for us
to foretell the existence of the equilibrium flow pattern,
$\hat{X}$, in which the flow is within the capacity limitation on
any link in the network. Osanaga(1970) and Iri(1971) presented the necessary and sufficient condition for the constraint domain of P-2 to be feasible. However this makes a
necessary condition only for the existence of $\hat{X}$.

We assume the existence of the solution in this study. It
may be interesting to know that there may exist plural equilibrium flow patterns. This is because $\psi(x,\lambda)$ may have plural local maxima and because different programs of D-1(total $\sum_{r=1}^{m} {}_m C_r$) can be produced depending on which links are
in the congested flow regime.

Since we employ the link cost function in the congested flow
regime such that it has an intercept on the ordinate as depicted in Fig.2, there is a possibility that we obtain the
meaningless solution such as Solution 5 presented in the
preceding section when solving D-1. The subsequent property ensures that such a solution is precluded from the solutions.

Property 4: The maximum solution of D-1 does not yield the
              flow pattern in which the flow equals zero in
              the congested flow regime on some link(s) in the
              network.

Proof: Suppose $(\bar{x},\bar{\lambda})$ is the maximum solution of D-1. Let us assume there exists some link ℓ where the flow, $\bar{x}_\ell$, takes the value of zero in the congested flow regime. We can consider an open ball $B_\delta(\bar{x},\bar{\lambda})$ around $(\bar{x},\bar{\lambda})$ with radius $\delta > 0$ such that

$$(\bar{x},\bar{\lambda}) \epsilon B_\delta(\bar{x},\bar{\lambda}) \cap \Gamma \rightarrow \psi(x,\lambda) \leq \psi(\bar{x},\bar{\lambda}). \tag{27}$$

Let the O-D pair between the endpoints of link ℓ be O-D pair e and the path consisting of link ℓ only be path q. We put $\bar{x}_q^e$ for $\bar{x}_q^e + \eta$ ($|\eta| < \delta$, $\eta < 0$) and let the resulting path flow vector be denoted by $\overset{*}{x}$ and the corresponding link flow by $\overset{*}{x}_j$ ($j\epsilon L$). First $(\overset{*}{x},\bar{\lambda})$ proves to be an interior point of $B_\delta(\bar{x},\bar{\lambda})$ from

$$\| (\bar{x},\bar{\lambda}) - (\overset{*}{x},\bar{\lambda}) \| = |\eta| < \delta.$$

Since $(\bar{x},\bar{\lambda})\epsilon\Gamma$

$$\bar{\lambda}^e < \sum_{j\epsilon L} r_{qj}^i f_j(\bar{x}_j) = f_\ell^\Pi(\bar{x}_\ell) = f_\ell^\Pi(0). \tag{28}$$

It holds for $\overset{*}{x}$

$$\sum_{j\epsilon L} r_{qj}^i f_j(\overset{*}{x}_j) = f_\ell^{\Pi}(\overset{*}{x}) = f_\ell^\Pi(\eta). \tag{29}$$

Taking into account that $f_\ell^\Pi(X)$ is a decreasing function of X and η is negative, one can easily deduce from eqs. (28) and (29) that

$$\bar{\lambda}^e \leq \sum_{j\epsilon L} r_{qj}^i f_j(\overset{*}{x}_j).$$

As long as we take paths independently to each other, the change of $\bar{x}_q^e$ by $\eta < 0$ has no effect on other paths of O-D pair e. Concerning other O-D pairs that change increases the cost of paths involving link ℓ. Thus all the constraints represented by eq. (18) are fulfilled. Therefore $(\overset{*}{x},\bar{\lambda})\epsilon\Gamma$. However

$$\psi(\bar{x},\bar{\lambda}) - \psi(\overset{*}{x},\bar{\lambda}) = - \int_\eta^0 X \frac{df_\ell^\Pi(X)}{dX} dX < 0. \tag{30}$$

Equation (30) contradicts eq. (27). Consequently the flow does not become zero in the congested flow regime on any link for the solution of D-1. This completes the proof.

Properties 1 through 4 encourage us to employ D-1 for obtaining the equilibrium flow pattern in the network involving

some links in the congested flow regime. Further, it should
be pointed out that D-1 does not have any O-D demand constraint and it may make another attractive feature of D-1
since it permits us to assign traffic by any increment as
long as eq.(6) is satisfied.

5. ALGORITHM

The algorithm we develop to compute the equilibrium flow
pattern is a version of the conventional incremental assignment technique to take D-1 into account. It is briefly stated as follows;

Step 1. Determine the initial path flow pattern x and the
shortest path cost λ. Let D^i($i \epsilon I$) denote the remaining traffic demand to be assigned. Put $\Delta x^i = \alpha D^i$ (
$0 < \alpha < 1$, $i \epsilon I$) and go to Step 2.

Step 2. Put $i = 1$ and go to Step 3.

Step 3. If $D^i = 0$, go to Step 9; otherwise go to Step 4.

Step 4. Find the shortest path, p^i, and its cost, t^i. If it
does not involve any link in the congested flow regime, go to Step 5; otherwise go to Step 6.

Step 5. If $\Delta x^i > 0$, assign Δx^i onto path p^i; otherwise assign $-\Delta x^i$ onto path p^i, find the second shortest
path, q^i, assign $2\Delta x^i$ onto q^i and go to Step 8.

Step 6. Let t^{*i} denote the cost of p^i when $|\Delta x^i|$ is assigned
onto it. If $t^i \geq t^{*i}$, go to Step 7; otherwise go to
Step 5.

Step 7. If $\Delta x^i > 0$, assign $-\Delta x^i$ onto path p^i and $2\Delta x^i$ onto
the second shortest path; otherwise assign Δx^i onto
path p^i and go to Step 8.

Step 8. Let x, λ denote the resulting path flow and the cost
of the shortest path. If $\psi(x,\lambda) \geq \psi(x,\lambda)$, put $x = x$,
$\lambda = \lambda$, $D^i = D^i - \Delta x^i$ and go to Step 9; otherwise go
to Step 10.

Step 9. Let X_j($j \epsilon L$) denote the link flow corresponding to x.
If $|X_j - C_j| < \gamma$ (where $\gamma > 0$ is a tolerance), replace
the link cost function $f_j^I(X_j)$($f_j^{II}(X_j)$) by $f_j^{II}(X_j)$($f_j^I(X_j)$
) when the link j is currently in the free flow(con-
gested flow) regime and go to Step 10.

Step 10. If all the traffic demands are assigned to the network, go to Step 11; otherwise put i = i + 1 modulo
N + 1 and go to Step 3.

Step 11. Compute the maximum and minimum utilized path costs
, λ^i_{max}, λ^i_{min} (i∈I). If $\lambda^i_{max} - \lambda^i_{min} < \varepsilon$ (where ε is a
tolerance, i∈I), terminate; otherwise go to Step 12.

Step 12. Increase or decrease each utilized path flow by
100ξ percent(0 < ξ < 1) so that the cost of it is
decreased. Put D^i = (total decreased path flow of
O-D pair i) - (total increased path flow of O-D
pair i), $\Delta x^i = D^i$(i∈I), i = 1 and go to Step 3.

It is noted that we judge that there exists no equilibrium
flow pattern in which any capacity restraint is not violated if the solution does not converge when the above computation is repeated by the predetermined number.

6. NUMERICAL EXAMPLE

We compute the equilibrium link flow pattern in the network
as shown in Fig.6. The number by the link denotes its number
and the number in circle denotes the node number. The
following three O-D pairs are considered in this example;

> O-D pair 1: node 1 → node 3,
> O-D pair 2: node 6 → node 2,
> O-D pair 3: node 4 → node 6,

of which traffic demands
are given as S^1 = 1200,
S^2 = 1000, S^3 = 800. We
employ the piecewise lin-
ear link cost function as
depicted in Fig.7. Let

$$f_j(X_j) = a_j X_j + b_j$$

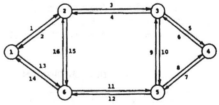

Fig. 6. The network for the numerical example.

and suppose a_j, b_j take the value of a^I_j, a^{II}_j, a^{III}_j, b^I_j, b^{II}_j,
b^{III}_j, each corresponding to the line segment I, II, III in
Fig.7 respectively. Those parameters are listed in Table 2
along with the capacity C_j and D_j(see Fig.7). We derive sev-
en equilibrium flow patterns by starting from twenty-eight

initial conditions(see Table 3).
It may be interesting to know
that the costs of the utilized
paths of O-D pair 1, 2, 3 are
10.1, 4.2, 9.9, respectively for
the solution 1 while those values are 13.3, 4.7, 12.6 for the
solution 7; and values for other
solutions lie between the above
two sets of the path cost. Thus
it is concluded that the equilibrium flow pattern corresponding
to the solution 1 is most prefer-
able for users among seven equi-
ribrium flow patterns we obtain.
However, it is noticed that
there can be more solutions to be

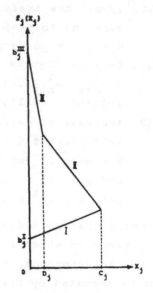

Fig. 7. Piecewise linear cost function.

Table 2. Link parameters for the numerical example.

Link #	a_j^I	b_j^I	a_j^{II}	b_j^{II}	a_j^{III}	b_j^{III}	C_j	D_j
1,2	0.25	2.0	-7.0	9.25	-15.0	409.25	1000	50
3,4	2.00	6.0	-9.0	12.60	-20.0	562.60	600	50
5,6	0.40	1.8	-8.0	10.20	-15.0	360.20	1000	50
7	0.25	2.0	-10.0	7.13	-18.0	407.13	500	50
8	0.25	2.0	-9.0	11.10	-16.0	361.10	1200	50
9,10	0.40	1.6	-8.0	14.20	-20.0	614.20	1500	50
11,12	0.95	5.5	-5.0	18.00	-12.0	368.00	2100	50
13,14	0.40	1.8	-7.0	9.94	-14.0	359.94	1100	50
15,16	1.50	1.5	-8.0	9.10	-17.0	459.10	800	50

Table 3. Obtained link flow patterns

Link # Efp #	1	2	3	4	5	6	7	8	9	10	11	12	13	14	15	16
1	905	0	523	118	0	460	340	0	677	342	677	682	677	382	118	618
2	700	0	247	82	0	462	338	0	954	380	954	718	845	344	191	657
3	874	0	521	221	0	715	85	0	680	494	680	579	680	353	221	648
4	913	0	230	50	0	485	315	0	970	435	970	750	744	456	276	544
5	728	0	221	221	0	715	85	0	980	494	980	579	817	344	384	657
6	962	0	547	123	0	467	333	0	653	344	653	678	653	415	123	585
7	921	0	221	141	0	716	84	0	978	574	978	659	726	447	393	553

Note: 1) Efp = equilibrium flow pattern
2) The underlined value represents the flow is of congested
flow regime.

derived by starting from other initial conditions than those
we employed in the above. However it is not our purpose to
derive all solutions but to present that there are substantially different equilibrium flow patterns even in the small
network if the congested flow regime is taken into account.

7. CONCLUSION

This paper described some features of the equilibrium traffic assignment which takes the capacity restraints and the
congested flow regime into account. Major findings are;
(1) The existing traffic assignment technique which employs
the link cost function $f_j(X_j)$ such that $f_j(X_j) \rightarrow \infty$ as X_j
approaches the capacity yields the flow pattern different
from the equilibrium flow pattern.
(2) We can find the equilibrium flow pattern (if it exists)
not violating any capacity restraint if we take the congested flow regime into account whereas we cannot find it
taking into account the free flow regime only.
(3) The mathematical programming problem for obtaining the
equilibrium flow pattern enunciated by Beckmann *et al* is not
always effective if the congested flow regime is considered
while its dual problem is effective in that the maximum solution always yields the equilibrium, the objective function
does not have any minimum and the derived flow does not become zero in the congested flow regime.
(4) Generally, the equilibrium flow pattern is not unique
since we can produce different mathematical progamming problems for obtaining it depending on the combinations of the
links in the congested flow regime and further each problem
may have some local optimum solutions.

Since it appears that the proposed algorithm is not efficient enough to apply to large networks, more effective algorithms are currently being sought. One thing that can be
stated at present is that the Frank-Wolf decomposition technique seems not valid for our problem because the objective
function of D-1(P-1) is not proved to be concave(convex).
One of the most important issues which require further investigations is how to find the equilibrium flow pattern

which is most probable to occur in the real world networks.
It may be dissolved by combining some appropriate initial
condition and simulation like algorithms.

The link cost function employed in this paper is a theoretical type of function deduced from conventional speed-flow
relationships while Robertson(1979) and Akcelik(1981) suggest that there may not be such a simple cost flow function
if we take into account a time-dependent queueing process on
a congested link. Extensive analysis of real traffic data
is required for determining the link cost function in the
congested flow regime prior to applications of the proposed
method to actual networks.

Also the new model developed by Dafermos(1971) would make an
effective tool to improve the proposed method since in congested cases cost on a link may be influenced by flows on
downstream links.

Finally, it should be mentioned that we have not examined
whether equilibrium flow patterns are stable in the present
study. A stability analysis deserves future research as it
should serve to delimit the equilibrium flow patterns that
can be realized in practice.

ACKNOWLEDGMENT

The author wishes to thank Mr. Komachiya for his assistance
in obtaining the results of the numerical example.

REFERENCES

Akcelik, R.(1981). Traffic signals - capacity and timing
analysis, Australian Road Research Board, Research Report,
No. 123.
Assad,A.A.(1978). Multicommodity network flows-a survey,Net-
works 8, pp.37-91.
Beckmann,M.,McGuire,C.B. and Winsten,C.(1956). Studies in
the economics and transportation (Yale University Press, New
Haven), pp.59-79.
Dafermos, S.C.(1971). An extended traffic assignment model
with applications to two-way traffic, Transportation Science,
7, pp.211-223.
Daganzo,C.F.(1977). On the traffic assignment problem with
flow dependent costs -I, Transportation research 11, pp.433-
437.
Daganzo,C.F.(1977). On the traffic assignment problem with

flow dependent costs-$\amalg$, Transportation research <u>11</u>, pp.439-441.

Davidson,K.B.(1966). A flow-travel time relationship for use in transportation planning, Proc. Aus. Road Res. Board <u>3</u>, pp. 183-194.

Iri,M.(1971). On an extension of the maximumflow-minimum cut theorem to multicommodity flows, Journal of Operations Research Society of Japan, Vol.<u>13</u> No.3, pp.129-135.

Newell,G.F.(1977). The effect of queues on the traffic assignment to freeways, Proc. of the 7th international symposium on transportation and traffic theory, edited by T.Sasaki and T.Yamamoto(The institute of SSR, Kyoto), pp.315-340.

Okutani,I., Matsuzaki,T.(1976). Equilibrium traffic assignment to freeways, Proc. of the 7th international symp. of the annual meeting of JSCE, $\amalg$-157.

Okutani,I., Matsuzaki,T.(1977). Traffic equilibrium considering the congested flow regime, Proc. of the annual meeting of JSCE, $\amalg$- 32.

Okutani,I., Matsuzaki,T.(1978). On the convergency of the traffic assignment algorithm considering the congested flow regime, Proc. of the annual meeting of Chubu branch of JSCE, $\amalg$- 14.

Okutani,I.(1980). An aspect on the traffic equilibrium considering the congested flow regime, Proc. of the annual meeting of Chubu branch of JSCE, $\amalg$-1.

Okutani,I., Komachiya,A.(1981). Studies on the non-uniqueness of the traffic equilibrium solution considering the congested flow regime, Proc. of the annual meeting of Chubu branch of JSCE, $\amalg$-11.

Osanaga,K.(1970). A theorem of multicommodity flows, Proc. of Japan Electronics and Communication Society, Vol.<u>53-A</u>, No. 7, pp.350-356.

Robertson, D.I.(1979). Traffic models and optimum strategies of control, Inst. Transp. Studies and U.S. Dept. Transp. 'Proc. Int. Symp. on Traffic Control Systems', Vol. 1. pp. 262-288.

Saishu, K.(1976). Traffic assignment based on the equal travel time principle, Traffic engineering, Vol.<u>11</u>, No.5, pp.23-30.

Ninth International Symposium on
Transportation and Traffic Theory
© 1984 VNU Science Press, pp. 273–297

A DESCENT ALGORITHM FOR SOLVING A VARIETY OF MONOTONE EQUILIBRIUM PROBLEMS

M. J. SMITH

Department of Mathematics, University of York, York, U.K.

ABSTRACT

A particular equilibrium problem, problem (E), is derived as a slight variation of a fixed demand equilibrium problem. An algorithm, algorithm (D), is stated which solves problem (E) when that problem is smooth and monotone. The paper then proceeds to show that three elastic demand problems may be cast in the form of a problem (E); so that algorithm (D) essentially solves these problems too. The last of the elastic demand problems considered has queues on saturated links and signal controls to resolve conflicts at junctions. Computational results for a spectrum of simple fixed demand problems are given.

1.   INTRODUCTION

1.1   Context

Smith (1983a, 1984) gives two descent algorithms for solving a certain fixed
demand equilibrium problem (E) with a smooth monotone network cost-flow
function. The algorithm in Smith (1984) is here called algorithm (D) and
is likely to be faster than the algorithm in Smith (1983a) from which it was
developed. The algorithm should be interleaved with a cheapest route
algorithm as discussed in Smith (1983b).

This paper shows that, under natural conditions, three elastic demand
equilibrium problems may be written in the form of problem (E). Thus,
provided the cost-flow function in the ensuing problem (E) is smooth and
monotone, algorithm (D) essentially solves these problems too.

Others have cast certain elastic problems into fixed demand form. In this
paper we look especially at the properties of the ensuing fixed demand
problem (E); we find conditions on the elastic problem which ensure that
the cost-flow function of the ensuing problem (E) is smooth and monotone.
The conditions make sure that algorithm (D) works.

Strict monotonicity is not needed for a proof of convergence of algorithm (D).
Thus the algorithm is likely to converge for a wider range of problems than
either the cutting plane algorithm of Sang Nguyen and Dupuis (1981) or the
contraction mapping algorithms of Dafermos (1980, 1982a, 1982b). (There
are as yet, as far as I am aware, no computational results which compare
these algorithms.)

The problem (E) which ensues from the last problem considered in this paper
is never strictly monotone; so, for this problem, an algorithm which requires
only monotonicity for convergence is certainly needed.

The first three equilibrium problems need little introduction. But the last
involves queueing delays and an adaptive signal-setting policy $\bar{P}_0$; both of
these factors are unusual elements in equilibrium problems and so we give a
brief discussion of this problem and the related bounded flow equilibrium
problem.

1.2 Equilibrium with queues and signal controls

Payne and Thompson (1975) extended the standard independent link, uncapacitated, model of traffic assignment. They consider a network of capacitated links and allow queues, which impose delays to traffic, on saturated links. They prove that any feasible assignment problem of this form has, under weak natural conditions, an equilibrium pattern of flows and queues.

Smith (1979c) gives a condition (18) which, together with continuity, ensures the existence of an equilibrium solution to an assignment problem on a network with more general capacity constraints, without queues. This condition is strengthened in Smith (1981a) and leads to the control policy P_o stated in the conclusion of Smith (1981b). Policy P_o has the property that any feasible assignment problem, satisfying weak continuity conditions, has a solutions consistent with P_o. Queues do not arise in this model.

It is natural to combine these two extensions of the standard theory.

The queueing version of P_o is $\tilde{P}_o$ (defined in Smith (1981c)). Smith (1983c) shows that any 'continuous' feasible assignment problem on a capacitated network with queueing has an equilibrium solution consistent with $\tilde{P}_o$. In this model interacting capacity constraints are allowed at junctions. (Policies P_o and $\tilde{P}_o$ were both first given, for a simple junction, in Smith (1979b). That paper is a simple introduction to some of the ideas behind the two policies P_o and $\tilde{P}_o$, although in that paper P_o and $\tilde{P}_o$ are not distinguished.)

The final equilibrium problem in this paper has a capacitated network with queues, and employs $\tilde{P}_o$ to resolve junction conflicts. As far as I am aware, this is the first elastic model to allow queues and an adaptive control policy. It is shown that, as it is formulated in the paper and under reasonable conditions, the problem gives rise to a smooth monotone problem (E) and so algorithm (D) may again be employed.

1.3 The bounded flow equilibrium problem

Daganzo (1977) has considered the solution of the independent link traffic assignment problem in which there is a limit to the flow possible on each link, and the link cost tends to $+\infty$ as this limit is approached. He gives a truncated version of the Frank-Wolfe algorithm for solving this "bounded flow equilibrium problem".

Hearn and Ribera (1980) and Polak (1982) have considered the above bounded flow equilibrium problem when the link cost does not tend to + ∞ as the bound on the link flow is approached. They suggest penalty Lagrangian methods for the solution of this problem, in which the bounds will often be active constraints at the solution. (The bounds cannot be active at the solution of Daganzo's bounded flow problem.)

The work in this paper shows how the Lagrangian, or multiplier, approach may be extended to cover a wide variety of junction interactions; and also shows how to modify this approach so that algorithm (D) works. An important element of the modification is the determination of appropriate bounds on the Lagrange multipliers.

1.4   Suggestion for further work

It is obviously useful to have a single algorithm which converges under weak conditions and solves a variety of equilibrium problems. Algorithm (D) does converge under quite weak conditions, and does solve a variety of problems. Algorithm (D) needs now to be tested against the cutting plane and contraction algorithms, to see how fast it is.

Algorithm (D) is a descent algorithm relative to a certain (quadratic) penalty function. The gap function G introduced by Hearn (1981) is the limit of a sequence of similar (but non-quadratic) penalty functions, each with its own descent algorithm similar to algorithm (D). Thus it would be natural to modify algorithm (D) to obtain a descent algorithm relative to Hearn's gap function G. I do not see how to do this; the difficulty resides in the non-differentiability of G. (Hearn defines the gap function of a minimisation problem but this definition may clearly be applied to define the gap function of an equilibrium problem, whether this arises from a minimisation problem or not.)

It would also be natural to seek to combine the ideas behind algorithm D and the idea behind the cutting plane algorithm of Sang Nguyen and Dupuis (1981).

2. A FIXED DEMAND PROBLEM AND PROBLEM (E)

Suppose that we are given a network with K origin-destination pairs and N_k routes joining the kth O-D pair. Suppose that the demand for travel between the kth O-D pair is ρ_k.

A flow ρ_k along the jth route joining O-D pair k may be represented by a vector

$$(0,0,\ldots,0,\rho_k,0,\ldots 0,0)$$

in $R_+^{N_k}$, where the ρ_k occurs in the jth place. Call this vector P_j^k. Any flow along the N_k routes meeting the demand ρ_k may then be represented by a vector of the form

$$\sum_{j=1}^{N_k} \mu_j P_j^k ,$$

where $\mu_j \geqslant 0$ and $\Sigma \mu_j = 1$. Let $\mathscr{G}_k$ denote the set of all vectors of this form. Then $\mathscr{G}_k$ is a simplex in $R_+^{N_k}$ with N_k vertices

$$P_1^k, \ P_2^k, \ldots, P_{N_k}^k .$$

$\mathscr{G}_k$ is the convex hull of the P_j^k:

$$\mathscr{G}_k = \text{conv}(P_1^k, P_2^k, \ldots, P_{N_k}^k).$$

Any flow along the $M = \sum_{k=1}^{K} N_k$ routes, meeting all demands ρ_k, may be represented by a vector

$$(F^1, F^2, \ldots, F^K)$$

where each $F^k \in \mathscr{G}_k$. The set of all such vectors is denoted by

$$\mathscr{G} = \mathscr{G}_1 \times \mathscr{G}_2 \times \ldots \times \mathscr{G}_K ,$$

and this is a subset of R_+^M where $M = \sum_{k=1}^{K} N_k$. Here, of course, the kth co-ordinate F^k of F is itself a vector in $R_+^{N_k}$.

We shall call any vector F in R_+^M a <u>route-flow</u>.

Let $\mathscr{S}$ be the set of route-flows in R_+^M which are within the capacity limitations of the network, and let $C : \mathscr{S} \to R_+^M$ be the network cost-flow function; thus $C_r(F)$ is the cost of travel along route r when the route-flow is F.

Define $U : \mathscr{S} \to \mathbb{R}^M$ by putting

   $U(F) = -C(F)$ for all F in $\mathscr{S}$.

2.1   Definition of equilibrium

The route-flow F is an **equilibrium** if and only if

   $F \in \mathscr{D} \cap \mathscr{S}$ and

   $U(F)$ is normal, at F, to $\mathscr{D}$;   (1)

or, in other words,

   $F \in \mathscr{D} \cap \mathscr{S}$ and

   $U(F) \cdot (G-F) \leqslant 0$ for all $G \in \mathscr{D}$.   (2)

This definition of equilibrium follows Wardrop (1952) and Smith (1979c).

It follows fairly easily that if $\mathscr{D} \subset \mathscr{S}$, and $U(\cdot)$ (or $C(\cdot)$) is continuous, there is an equilibrium F in $\mathscr{D}$. (See Smith (1979c).) This is a standard result in the theory of variational inequalities, since $\mathscr{D}$ is convex, closed and bounded.

2.2   Problem (E)

Suppose given, for each $k = 1,2,\ldots,K$, a set

   $$P_1^k, \ P_2^k, \ldots, P_{N_k}^k$$

of points in $\mathbb{R}_+^{M_k}$;   let

   $\mathscr{D}_k = \mathrm{conv}(P_1^k, P_2^k, \ldots, P_{N_k}^k)$

and let   (3)

   $\mathscr{D} = \mathscr{D}_1 \times \mathscr{D}_2 \times \ldots \times \mathscr{D}_K.$

This is a subset of $\mathbb{R}_+^M$ where $M = \sum_k M_k$.

Suppose also given a subset $\mathscr{S}$ of $\mathbb{R}_+^M$ and a "cost-flow" function $C : \mathscr{S} \to \mathbb{R}^M$.

Finally suppose that

$$\mathcal{D} \subset \mathcal{S}.$$

Problem (E) may now be stated in terms of $U = -C$ as follows:

> Under the above conditions, find $F \in \mathcal{D}$ such that
> $U(F)$ is normal, at F, to $\mathcal{D}$. $\hspace{3cm}$ (4)

Thus problem (E) is very close to the problem of finding an F satisfying (1). The main differences are that, in (E), $\mathcal{D} \subset \mathcal{S}$, $\mathcal{D}_k$ may not be a simplex and M_k may not be equal to N_k.

If $\mathcal{D} \subset \mathcal{S}$ the problem of finding a fixed demand equilibrium F becomes a problem (E).

2.3 The existence of a solution of (E)

The problem (E) has a solution if $U(\cdot)$ (or $C(\cdot)$) is continuous. (This is the standard result mentioned earlier.)

2.4 Optimisation formulation of (E)

Suppose henceforth that $U(\cdot)$ is defined throughout $\mathcal{S}$.

In order to verify (2) it is sufficient to check that

$$U(F) \cdot (G-F) \leq 0$$

for only some of the G in $\mathcal{D}$.

Given $F \in R_+^M$, let $F = (F^1, F^2, \ldots, F^K)$ where each $F^k \in R_+^{M_k}$ and

$$G_i^k = (F^1, F^2, \ldots, F^{k-1}, P_i^k, F^{k+1}, \ldots, F^K).$$

Then it is easy to check that (2) is satisfied if and only if

$$U(F) \cdot (G_i^k - F) \leq 0 \hspace{3cm} (5)$$

for all i, k with $1 \leq i \leq N_k$ and $1 \leq k \leq K$.

Let

$$x_+ = \max\{0,x\}$$

and $x_+^2 = (x_+)^2$ for all real x. Then all the conditions (5) are satisfied if and only if

$$V(F) = \sum_{i,k} [U(F) \cdot (G_i^k - F)]_+^2 = 0.$$

Also $V(F) \geq 0$ for all F in $\mathcal{D}$, so we now have a minimisation problem. (Hearn (1982) introduced the gap function G as an objective function. The problem with G is its non-differentiability; our V is differentiable.)

2.5 A descent direction, when C(·) is monotone.

Let, for $F \in \mathcal{D}$,

$$\Delta(F) = \sum_{k=1}^{K} \sum_{i=1}^{N_k} [U(F) \cdot (G_i^k - F)]_+ (G_i^k - F).$$

Then $\Delta(F)$ is a descent direction for V at F if $F \in \mathcal{D}$ and C(·) is smooth and monotone (unless $V(F) = 0$ already). The following lemma gives a precise statement of this result.

Lemma. Let $\mathcal{D} \subset \mathcal{S}$ and let $C : \mathcal{S} \rightarrow \mathbb{R}^M$ be monotone and continuously differentiable. Then, if $F \in \mathcal{D}$

$$\text{grad } V(F) \cdot \Delta(F) \leq -2 \ W(F)$$

where:

$$W(F) = \sum_{i,k,\ell} [(U(F) \cdot (G_i^k - F)]_+^2 [U(F) \cdot (G_\ell^k - F)]_+.$$

Proof. This is given in Smith (1984).

The important thing about $\Delta(F)$ is that it does not involve the derivative of C(·). Thus it is reasonably easy to calculate, provided the routes in the network are known. The same comment applies to V(F) and W(F).

(The need to know the routes does mean that it is natural, when methods based on the lemma are used, to generate new routes, having at least partially equilibrated among the previously generated routes. See Leventhal, Nemhauser and Trotter (1973), Smith (1983b) or Hearn and Lawphongpanich (1982).)

In problem (E), $\mathcal{D} \subset \mathcal{S}$ and so U(F), V(F) and Δ(F) are defined at each route-flow in $\mathcal{D}$.

2.6   An application of a method of Lyapunov

Let C($\cdot$) = -U($\cdot$) be monotone and continuously differentiable (and $\mathcal{D} \subset \mathcal{S}$).
Then for any $F_o \in \mathcal{D}$ there is a function F($\cdot$) defined for all t $\geqslant$ 0,
differentiable at each t > 0 and continuous at t = 0, such that F(0) = F_o
and

$$\dot{F}(t) = \Delta(F(t)).$$

For this function F($\cdot$), Smith (1983a) has shown that, whatever F_o is,

$$V(F(t)) \leqslant V(F_o)e^{-2t}$$

for all t $\geqslant$ 0 and so V(F(t)) $\rightarrow$ 0 as t $\rightarrow \infty$.

Hence (since $\mathcal{D}$ is compact) F(t) tends to the set

$$\mathcal{E} = \{F; \quad F \in \mathcal{D} \text{ and } V(F) = 0\}$$

of equilibria, as t $\rightarrow \infty$.

2.7   V as a Lyapunov function

The above proof works because V is a <u>Lyapunov function</u> for the dynamical system

$$\dot{F}(t) = \Delta(F(t))$$

Any function V is called a Lyapunov function for the dynamical system

$$\dot{F}(t) = \Delta(F(t))$$

if V is continuously differentiable and the following conditions are satisfied:
   (i)   V(F) $\geqslant$ 0 for all F in $\mathcal{D}$.
   (ii)   V(F) = 0 if and only if F is an equilibrium.
   (iii) grad V(F). Δ(F) < 0 if F is not an equilibrium.
Property (iii) ensures that, if $\dot{F}(t) = \Delta(F(t))$,

$$\frac{d}{dt}(V(F(t))) = \{grad\ V(F(t))\} \cdot \dot{F}(t)$$
$$= \{grad\ V(F(t))\} \cdot \Delta(F(t))$$
$$< 0$$

if $F(t)$ is not an equilibrium. Thus V declines along any solution of $\dot{F} = \Delta(F)$ and it follows from the continuity of grad V that $V(F(t)) \to O(t \to \infty)$.

Remark. In our case, if $\mathcal{G} \not\subset \mathcal{S}$ a solution of $\dot{F} = \Delta(F)$ may leave $\mathcal{S}$ and, at the point where $F(t)$ is first outside $\mathcal{S}$ (assuming that $\mathcal{S}$ is open in $\mathbb{R}_+^M$), $\Delta(F(t))$ is undefined. Thus, if $\mathcal{G} \not\subset \mathcal{S}$ it may well be impossible to find an $F(\cdot)$, defined for all $t \geqslant 0$, and satisfying $\dot{F} = \Delta(F)$. If $\mathcal{G} \subset \mathcal{S}$ (and $C(\cdot)$ is continuously differentiable) then there is always a solution of $\dot{F} = \Delta(F)$, defined for all $t \geqslant 0$, starting at any $F(0) \in \mathcal{G}$.

Further remark. The definition of a Lyapunov function given here is appropriate when solution trajectories are necessarily bounded (as they are here).

3.   ALGORITHM (D)

Algorithm (D) is a discrete version of a solution of the differential equation $\dot{F} = \Delta(F)$.

Algorithm (D) either terminates at an equilibrium, or yields a sequence $\{F_n\}$ of route-flows converging to the set of equilibria, provided

    (i) $C(\cdot)$ is monotone and

    (ii) $C(\cdot)$ is continuously differentiable.

A proof of convergence is given in Smith (1984). The algorithm is a refinement of that given in Smith (1983a).

3.1   Statement of algorithm (D)

For any non-equilibrium F in $\mathcal{G}$ let

$$\bar{\lambda}(F) = \left(\max_{1 \leqslant k \leqslant K} \left\{ \sum_{i=1}^{N_k} (U(F) \cdot (G_i^k - F))_+ \right\} \right)^{-1} .$$

Remark. This $\bar{\lambda}(F)$ has the property that $F + \lambda\Delta(F) \in \mathcal{G}$ if $F \in \mathcal{G}$. and $0 \leqslant \lambda \leqslant \bar{\lambda}(F)$. It is the largest number with this property.

Notation. In this section, and in 7.3, F_n will denote a route flow vector (and not the nth co-ordinate of such a vector).

To start the algorithm, choose any $\underline{F}_1$ in $\mathcal{D}$, and calculate $V(\underline{F}_1)$. If $V(\underline{F}_1) = 0$ then $\underline{F}_1$ is an equilibrium and so we terminate at $\underline{F}_1$. If $V(\underline{F}_1) \neq 0$, calculate $\bar{\lambda}(\underline{F}_1)$ and choose λ_1 so that $0 < \lambda_1 \leq \bar{\lambda}(\underline{F}_1)$. ($\bar{\lambda}(\underline{F}_1)$ would be a natural choice here perhaps.)

Suppose now that $\underline{F}_n$, λ_n have been determined already, and that $V(\underline{F}_n) \neq 0$. Putting $\Delta_n = \Delta(\underline{F}_n)$, calculate $V(\underline{F}_n + \lambda_n \Delta_n)$. If $V(\underline{F}_n + \lambda_n \Delta_n) = 0$, terminate at $\underline{F}_n + \lambda_n \Delta_n$ which is an equilibrium. If $V(\underline{F}_n + \lambda_n \Delta_n) > 0$, calculate

$$Q_n = \frac{V(\underline{F}_n + \lambda_n \Delta_n) - V(\underline{F}_n)}{\lambda_n W(\underline{F}_n)}$$

Then determine $\underline{F}_{n+1}$ and λ_{n+1} as follows:

If $Q_n > 0$ put
$$\underline{F}_{n+1} = \underline{F}_n \text{ and } \tilde{\lambda}_{n+1} = \frac{1}{2}\lambda_n.$$

If $0 \geq Q_n > -\frac{1}{2}$, put
$$\underline{F}_{n+1} = \underline{F}_n + \lambda_n \Delta_n \text{ and } \tilde{\lambda}_{n+1} = \frac{1}{2}\lambda_n.$$

If $-\frac{1}{2} \geq Q_n \geq -\frac{3}{2}$, put
$$\underline{F}_{n+1} = \underline{F}_n + \lambda_n \Delta_n \text{ and } \tilde{\lambda}_{n+1} = \lambda_n.$$

If $-\frac{3}{4} > Q_n$, put
$$\underline{F}_{n+1} = \underline{F}_n + \lambda_n \Delta_n \text{ and } \tilde{\lambda}_{n+1} = 2\lambda_n.$$

Finally put $\lambda_{n+1} = \min \left\{ \tilde{\lambda}_{n+1}, \bar{\lambda}(F_{n+1}) \right\}$.

Now $\underline{F}_{n+1}$, λ_{n+1} have been determined and $V(\underline{F}_{n+1}) \neq 0$. Thus the iterative step above may be repeated indefinitely, unless termination occurs.

Let $\{\underline{F}_n\}$ be an infinite sequence of route-flows generated by the above algorithm. The Euclidean distance $d(\underline{F}_n, \mathcal{E})$ between the route-flow $\underline{F}_n$ and the set $\mathcal{E}$ of equilibria tends to zero as $n \to \infty$ provided $C(\cdot)$ is continuously differentiable and monotone on $\mathcal{D}$. This result is proved in Smith (1984).

4. AN ELASTIC DEMAND EQUILIBRIUM PROBLEM

Suppose given a cost function $C : \mathcal{S} \to R_+^M$ as before, and a demand function

$$D : R_+^K \to R_+^K.$$

(If the (least) cost of travel between the kth origin-destination pair is L_k then the travel distribution will be D(L): $D_k(L)$ will be the travel volume between the kth origin-destination pair.)

In order to define an equilibrium it is useful to have two further functions defined in a purely mathematical way.

The <u>total flow function</u>

$$T : R_+^M \rightarrow R_+^K$$

is defined, for each $F \in R_+^M$, by

$$T_k(F) = \sum_{R_r \in \mathscr{R}_k} F_r . \qquad (k=1,2,\ldots,K.)$$

$T_k(F)$ gives the total travel between the kth origin-destination pair, in terms of the route-flow F, and T(F) is the vector of these origin-destination flows.

The <u>route price function</u>

$$Q : R_+^K \rightarrow R_+^M$$

is defined, for each $L \in R_+^K$, by

$$Q_r(L) = L_k \text{ if } R_r \in \mathscr{R}_k .$$

This function assigns the single price L_k of travel between the kth origin-destination pair to every route joining that origin destination pair.

4.1   Definition of equilibrium

(F,L) is an equilibrium if and only if:

$$\left. \begin{array}{l} F \in \mathscr{S}, \ L \in R_+^K, \\ T(F) = D(L) \text{ and} \\ Q(L) - C(F) \text{ is normal, at F, to } R_+^M \end{array} \right\} \qquad (6)$$

Now we have the following theorem due to Aashtiani and Magnanti (1983).

<u>Theorem 1</u>. Let $\mathcal{S} = \mathbf{R}_+^M$; $D(\cdot)$ be continuous, bounded and non-negative; and $C(\cdot)$ be continuous and always positive. Then there is an equilibrium (F,L).

4.2 Expressing the equilibrium problem as a problem (E)

The basic ideas in the proof of theorem 1 may be used to show that the equilibrium problem (6) is, under the hypotheses of theorem 1, <u>equivalent</u> to a problem (E) in that the set of equilibria coincides with the set of solutions of the problem (E).

Let $X = (F,L)$ and

$$U(X) = (Q(L) - C(F), D(L) - T(F))$$

for each $F \in \mathcal{S}$, $L \in \mathbf{R}_+^K$.

Suppose that the hypotheses of theorem 1 hold. Let, for each $k = 1,2,\ldots,K$, the constant A_k be chosen so that

$$D_k(L) < A_k \text{ for all } L \in \mathbf{R}_+^K$$

and now let, for each $r = 1,2,\ldots,M$, the constants B_r satisfy:

$$C_r(F) < B_r \text{ for all } F \in \prod_{r=1}^{M} [0,Q_r(A)].$$

Put

$$\mathcal{D} = \left(\prod_{r=1}^{M} [0,B_r] \right) \times \left(\prod_{k=1}^{K} [0,A_k] \right).$$

Then $X = (F,L)$ is an equilibrium if and only if

$$U(X) \text{ is normal, at } X, \text{ to } \mathcal{D}.$$

where $\qquad\qquad\qquad\qquad\qquad\qquad\qquad\qquad\qquad\qquad\qquad$ (7)

$$\mathcal{D} = \left(\prod_{r=1}^{M} [0,B_r] \right) \times \left(\prod_{k=1}^{K} [0,A_k] \right).$$

Since $\mathcal{D}$ has the right form (3), and $U(\cdot)$ is defined throughout $\mathcal{D}$, we have expressed the equilibrium problem (6) as a problem (E). (See (4).)

4.3   Some observations on the ensuing problem (E)

(i)   The result shows that, under the conditions stated in the theorem, there is no equilibrium outside $\mathscr{D}$.

(ii)   If $C(\cdot)$ and $-D(\cdot)$ are monotone then $-U(\cdot)$ is monotone.

(iii) It follows from (ii) that if $C(\cdot)$ and $-D(\cdot)$ are continuously differentiable and monotone then algorithm (D) solves the problem (E) given in (7).

(iv)   If $C(\cdot)$ and $-D(\cdot)$ are differentiable and monotone then it is easy to show that $-U(\cdot)$ is monotone. To do this let $J_1(\cdot)$, $J_2(\cdot)$ be the Jacobians of $C(\cdot)$ and $-D(\cdot)$. Then $-U(\cdot)$ has Jacobian

$$
J = \left(\begin{array}{c|c} \overset{\longleftarrow M \longrightarrow}{J_1} & \overset{\leftarrow K \rightarrow}{-H} \\ \hline H^T & J_2 \end{array} \right) \begin{array}{l} \updownarrow M \\[1em] \updownarrow K \end{array}
$$

where

$$
H_{rk} = \begin{cases} 1 \text{ if } R_r \in \mathscr{D}_k \\ 0 \text{ otherwise.} \end{cases}
$$

Hence

$$
J + J^T = \left(\begin{array}{c|c} J_1 + J_1^T & 0 \\ \hline 0 & J_2 + J_2^T \end{array} \right)
$$

and $J + J^T$ is positive semi-definite if $J_1 + J_1^T$ and $J_2 + J_2^T$ are. Thus $-U(\cdot)$ is monotone if $C(\cdot)$ and $-D(\cdot)$ are monotone and differentiable. This proof is similar to that given by Aashtiani and Magnanti (1982).

5.   ANOTHER ELASTIC DEMAND PROBLEM

This model has been considered by many in an economic context. Within transportation, the equilibrium conditions considered here have been given by Dafermos (1982b), by Florian (1979) and in a simple but generalisable model, by Smith (1979a).

5.1   The model

Suppose that, corresponding to each vector $T \in \mathbb{R}_+^K$ of origin-destination flows (T_k being the flow between the kth origin-destination pair) there is a unique vector W(T) of origin-destination travel costs which are consistent with $T(W_k(T)$ being the relevant cost for travel between the kth origin-destination pair).

The cost vector W(T) may be thought of as bringing forth the origin-destination flow vector T. The function $W : R_+^K \to R_+^K$ may be thought of as the inverse of an invertible demand function $D : R_+^K \to R_+^K$. But neither of these ways of thinking of $W(\cdot)$ admits the greatest generality in this model.

Now regard T as a function of the route-flow pattern F, as in section 4 of this paper, and define $Q : R_+^K \to R_+^K$ as in section 4. Let

$$\tilde{W}(F) = Q(W(T(F)))$$

for each $F \in R_+^M$; then $\tilde{W}$ takes values in R_+^M.

Put

$$U(F) = \tilde{W}(F) - C(F),$$

for each F in R_+^M.

5.2 Definition of equilibrium

The route-flow $F \in \mathcal{G}$ is an equilibrium if and only if

$$U(F) \text{ is normal, at } F, \text{ to } R_+^M. \tag{8}$$

5.3 This equilibrium problem expressed as a problem (E)

The condition (8) is not of the form E because the set R_+^M is not a product of bounded sets. However, if we suppose that

$$W(\cdot) \text{ is bounded} \tag{9}$$

and

$$C_r(F) \to +\infty \text{ as } F_r \to +\infty \tag{10}$$

then we may choose constants A_r so that, for $r = 1,2,\dots,M$;

$$U_r(F) < 0 \quad \text{if} \quad F_r > A_r.$$

In this case if we put

$$\mathcal{D} = \prod_{r=0}^{M} [0, A_r],$$

then F is an equilibrium if and only if $F \in \mathcal{D}$ and

$$U(F) \text{ is normal, at } F, \text{ to } \mathcal{D}. \tag{11}$$

This is now in the form of a problem (E).

It follows that there is an equilibrium if, in addition to conditions (9) and (10), $-W(\cdot)$ and $C(\cdot)$ are continuous.

It also follows that algorithm (D) solves the equilibrium problem, expressed in the form (E) in (11), provided $-W$ and C are monotone and smooth.

6.   ELASTIC DEMAND, QUEUES AND CONTROLS

Suppose that demand is elastic (in the manner of section 5), that queues impose delays on traffic, and that controls resolve conflicts at junctions.

In order to be as simple as possible let the average queueing delay (or cost) d_i on link i be given by:

$$d_i = q_i/g_i s_i \text{ (seconds)} \tag{12}$$

where q_i is the (average) number of vehicles queueing on link i, g_i is the proportion of time for which the signal is green for link i and s_i is the saturation flow of link i.

Let $R_r(F)$ be the running cost of travel along route r (excluding the queueing delay) when the flow pattern is F.

Let a_{ri} be 1 if link i lies on route r and 0 otherwise, and

$$D_r(q,g) = \sum_{i=1}^{m} a_{ri} d_i(q,g) \tag{13}$$

be the queueing delay on route r. $((a_{ri})$ is the route-link incidence matrix.)

The capacity limitations of the network are defined by a set $\mathscr{S}$ of supply-feasible route-flow vectors. We assume that $\mathscr{S}$ has the following form:

$$\mathscr{S} = \{F \in \mathbb{R}_+^M;\ F.N^k \le H_k \text{ for } k = 1,2,\ldots,K\}$$

where each N^k is a vector in $\mathbb{R}_+^M$ and each H_k is a positive real constant.

6.1   Definition of equilibrium

(F,D) is an _equilibrium_ if and only if $F \in \mathscr{S}$, $D \ge 0$ and

$$\tilde{W}(F) - R(F) - D \text{ is normal, at } F, \text{ to } \mathbb{R}_+^M.$$

This definition accords with that in section 5.2.

6.2   Control policy $\tilde{P}_o$ and equilibrium

Policy $\tilde{P}_o$ is defined in Smith (1981c, 1983c). In fact (F,D) is an equilibrium consistent with $\tilde{P}_o$ if and only if (F,D) is an equilibrium and

$$D = D(q,g) \text{ is normal, at } F, \text{ to } \mathscr{S}. \tag{14}$$

In this section we shall show that, under natural conditions, there is an equilibrium (F,D) satisfying (14), and that such an equilibrium is, under natural conditions, easy to calculate.

It is then a straightforward matter to calculate q, g which give rise to the delay vector D via (12) and (13), provided the set $\mathscr{S}$ arises in a natural way from constraints only at the junctions of the network. Some of the details are given explicitly in Smith (1983c). It is also easy to check that this vector q will, under natural conditions then be in equilibrium too in the sense that inflow will equal outflow for each link. All the details of this argument are given in Smith (1983c).

The following theorem is the basic result of this section.

Theorem 2. (F,D) is an equilibrium for some D such that

$$D \text{ is normal, at } F, \text{ to } \mathscr{S}$$

if and only if

$$W(F) - R(F) \text{ is normal, at } F, \text{ to } \mathscr{S}. \tag{15}$$

Proof.   (i)   Let E_r be the M-vector with 1 in the rth place and zeros elsewhere. Let

$$W(F) - R(F) \text{ be normal, at } F, \text{ to } \mathscr{S}.$$

Suppose that the active capacity constraints at F are

$$F \cdot N^k \leq H_k \quad \text{for } k \in J = J(F)$$

and that the active non-negativity constraints on F are

$$F \cdot E_r \geq 0 \quad \text{for} \quad r \in I = I(F).$$

Then any normal at F to $\mathscr{S}$ is a non-negative linear combination of these N^k and E_r. So, for some Λ_k, $\mu_r \geq 0$,

$$W(F) - R(F) = \sum_{k \in J} \Lambda_k N^k + \sum_{r \in I} \mu_r E_r.$$

It follows that if we put

$$D = \sum_{k \in J} \Lambda_k N^k$$

then D is normal, at F, to $\mathscr{S}$, $D \geq 0$ and

$$W(F) - R(F) - D = \sum_{r \in I} \mu_r E_r$$

is normal, at F, to R_+^M; or (F,D) is an equilibrium satisfying (15).

(ii)   Conversely suppose that $F \in \mathscr{S}$; $W(F) - R(F) - D$ is normal, at F, to R_+^M and that D is normal, at F, to $\mathscr{S}$. Then $W(F) - R(F) - D$ is also normal, at F, to $\mathscr{S} \subset R_+^M$ and

$$W(F) - R(F) = \{W(F) - R(F) - D\} + D$$

is the sum of two normals to $\mathscr{S}$. Hence $W(F) - R(F)$ is normal, at F, to $\mathscr{S}$.

The proof is complete.

Thus the calculation of an equilibrium F consistent with $\tilde{P}_0$ reduces to the solution of a variational inequality. This has the form (E), with $\mathscr{Q}_1 = \mathscr{Q} = \mathscr{S}$, but the vertices of $\mathscr{S}$ are numerous and not readily accessible. So we give an alternative formulation, still of the form, with a more convenient $\mathscr{Q}$.

6.3 An alternative formulation not involving the vertices of $\mathscr{S}$.

We need to suppose that there is a function

$$U : \mathbb{R}_+^M \to \mathbb{R}^M$$

such that, for $F \in \mathscr{S}$,

$$U(F) = W(F) - R(F).$$

The function U extends $W - R$. We shall also need $-U$ to be monotone and smooth, if $-W + R$ is monotone and smooth on $\mathscr{S}$. There should be little difficulty in practice here, as R is bounded on $\mathscr{S}$.

Initially, then, suppose that U is a continuous extension of $W - R$, so that $U(F)$ is defined for any F in $\mathbb{R}_+^M$.

Let F be a solution of (15), and let Λ_k be the "optimal" Lagrange multiplier corresponding to the constraint

$$F \cdot N^k \leq H_k \quad (k = 1, 2, \ldots, K).$$

Consider any route r such that $F_r > 0$. Then, resolving along E_r,

$$\left(\sum_k \Lambda_k N^k \right) \cdot E_r = U(F) \cdot E_r \tag{16}$$

since F is a solution and the Λ_k are "optimal". Since $N^k \geq 0$, each $N^k \cdot E_r \geq 0$ and so (16) implies that

$$\Lambda_k N^k \cdot E_r \leq U(F) \cdot E_r$$

for all k. Now suppose that

$$N^k \cdot E_r > 0.$$

Then

$$\Lambda_k < \frac{U(F) \cdot E_r}{N^k \cdot E_r} = \frac{U_r(F)}{N_r^k}$$

Let

$$I_k = \{ r; \quad N_r^k = N^k \cdot E_r > 0 \}$$

and

$$A_k = \max_{r \in I_k} \left\{ \frac{\max\limits_{F \in \mathscr{S}} U_r(F)}{N_r^k} \right\} .$$

The above argument shows that $\Lambda_k \leq A_k$.

Because we have found upper bounds A_k for each optimal Λ_k, it becomes possible to express (15) as a problem of form (E), without involving the vertices of $\mathscr{S}$, as follows.

Consider the function

$$Z = (X,Y) : \ \mathbb{R}_+^M \times \mathbb{R}_+^K \to \mathbb{R}^M \times \mathbb{R}^K$$

defined by letting

$$X(F,\Lambda) = U(F) - \sum_k A_k N^k$$

and

$$Y(F,\Lambda) = F \cdot N^k - H_k .$$

Then F is a solution of (15) and Λ is the corresponding vector of optimal multipliers if and only if $(F,\Lambda) \in \mathbb{R}_+^M \times \mathbb{R}_+^K$ and

$$Z(F,\Lambda) \text{ is normal, at } (F,\Lambda), \text{ to } \mathbb{R}_+^M \times \mathbb{R}_+^K \qquad (17)$$

Choose B_r so that

$$F \in \mathscr{S} \Rightarrow F_r < B_r$$

(this is possible since $\mathscr{S}$ is bounded) and let

$$\mathscr{D} = \prod_{r=1}^{M} [0,B_r] \times \prod_{k=1}^{K} [0,A_k]. \qquad (18)$$

Then (F,Λ) solves (17) if and only if

$$Z(F,\Lambda) \text{ is normal, at } (F,\Lambda), \text{ to } \mathscr{D}. \qquad (19)$$

This is of form (E) and the vertices of the $\mathscr{D}_j$ are clearly available.

Remark. (i) (18), (19) are essentially identical to (7).

(ii) $-Z$ is monotone if $-U$ is, so algorithm D may again be used if $-U$ is monotone and smooth.

(iii) The existence of an equilibrium, as in previous cases, follows from the continuity of U and the fact that $\mathscr{D}$ is closed, bounded and convex.

7.   A SPECTRUM OF SIMPLE COMPUTATIONAL TESTS

7.1   The network and a continuous spectrum of cost functions

In this section we consider a network with three routes joining one origin-destination pair. The fixed demand for travel will be 1 (vehicle per second).

If the flows along the routes are F_1, F_2, F_3, then (for $\mu \geqslant 0$) the costs c_1^μ, c_2^μ, c_3^μ of travel along these routes are to be as follows:

$$c_1^\mu(F) = 2F_1 + 3\mu F_2 + \mu F_3$$

$$c_2^\mu(F) = \mu F_1 + 2F_2 + 3\mu F_3$$

$$c_3^\mu(F) = 3\mu F_1 + \mu F_2 + 2F_3$$

7.2   Properties of the cost-flow function as μ varies

First, $c^0 = c^0(F)$ is clearly the gradient of the strictly convex function Z where

$$Z(F) = F_1^2 + F_2^2 + F_3^2.$$

Now let $\mu > 0$. Then c^μ has Jacobian

$$J_\mu = \begin{pmatrix} 2 & 3\mu & 1 \\ \mu & 2 & 3\mu \\ 3\mu & \mu & 2 \end{pmatrix}$$

This is asymmetric and so c^μ is not a gradient. However

$$J_\mu + J_\mu^T = \begin{pmatrix} 4 & 4\mu & 3\mu+1 \\ 4\mu & 4 & 4\mu \\ 3\mu+1 & 4\mu & 4 \end{pmatrix}$$

and the leading minors of $J_\mu + J_\mu^T$ are:

$$\begin{vmatrix} 4 \end{vmatrix} , \quad \begin{vmatrix} 4 & 4\mu \\ 4\mu & 4 \end{vmatrix}, \quad \begin{vmatrix} 4 & 4\mu & 3\mu+1 \\ 4\mu & 4 & 4\mu \\ 3\mu+1 & 4\mu & 4 \end{vmatrix}$$

or

4, $16 - 16\mu^2$, $(\mu-1)^2(96\mu+60)$.

These are all positive if $0 \le \mu < 1$ and are all non-negative if $\mu = 1$. Also these leading minors are not all non-negative if $\mu > 1$. It follows that C^μ is strictly monotone if $0 \le \mu < 1$ and monotone if $\mu = 1$. It also follows that C^μ is not monotone if $\mu > 1$.

Thus algorithm (D) converges for $0 \le \mu \le 1$.

7.3   Computational results

The starting route-flow $\underline{F}_1 = (1,0,0)$ was taken in each test. Also, λ_1 was always chosen to be $\bar{\lambda}_1(\underline{F}_1)$. In each problem considered there is a single equilibrium route-flow and this is $(\frac{1}{3}, \frac{1}{3}, \frac{1}{3})$.

Table 1 gives the distance d_n between $\underline{F}_{n+1}$, the route-flow after n iterations, and the equilibrium route-flow $(\frac{1}{3}, \frac{1}{3}, \frac{1}{3})$ for various values of n and various values of μ. Blank spaces occur when the accuracy limits of the computer were exceeded.

n \ μ	0·0	0·1	0·2	0·3	0·4	0·5	0·6	0·7	0·8	0·9	1·0	1·1
0	0·82	0·82	0·82	0·82	0·82	0·82	0·82	0·82	0·82	0·82	0·82	0·82
5	$1·9\times10^{-3}$	$1·8\times10^{-3}$	$3·6\times10^{-3}$	$2·2\times10^{-3}$	$5·1\times10^{-3}$	0·02	0·079	0·14	0·18	0·29	0·25	0·27
10	$7·7\times10^{-6}$	$9·7\times10^{-6}$	$1·9\times10^{-5}$	$3·1\times10^{-5}$	$9·1\times10^{-4}$	$1·7\times10^{-3}$	0·011	0·031	0·10	0·16	0·23	0·28
15	$3·2\times10^{-8}$	$5·5\times10^{-8}$	$1·0\times10^{-7}$	$5·3\times10^{-7}$	$1·3\times10^{-8}$	$9·0\times10^{-5}$	$1·9\times10^{-3}$	0·019	0·067	0·12	0·24	0·19
20	$5·8\times10^{-10}$	$8·9\times10^{-10}$	$8·3\times10^{-10}$	$8·8\times10^{-9}$		$3·5\times10^{-6}$	$5·8\times10^{-4}$	$8·6\times10^{-3}$	0·044	0·096	0·17	0·17
50							$7·7\times10^{-8}$	$1·6\times10^{-4}$	0·0091	0·044	0·079	0·15

Table I. The distance d_n from equilibrium after n iterations of algorithm (D), for various values of n and various values of μ.

Naturally, convergence is in general slower when C^μ is "less monotone".

The above results were obtained using an 8K PET.

8.   CONCLUSION

The paper has shown how algorithm (D) may be applied to a variety of equilibrium problems, and gives a few computational results.

Acknowledgement   I am grateful to Dr G R Walsh for the computational results and to Dr D W Hearn for sending me the paper by Payne and Thompson, and his own work on bounded flow equilibrium problems. I also acknowledge two invigorating conversations with Mr J Polak.

9.   REFERENCES

Aashtiani H and Magnanti T (1983) Equilibrium on a congested transportation network. SIAM Journal of Algebraic and discrete Methods, 2, 213 - 216.

Aashtiani H and Magnanti T (1982) A linearization and decomposition algorithm for computing urban traffic equilibrium. Proceedings of the 1982 IEEE Large scale systems symposium.

Dafermos S C (1980) Traffic equilibrium and variational inequalities. Transportation Science, 14, 42 - 54.

Dafermos S C (1982a) Relaxation algorithms for the general asymmetric traffic equilibrium problem. Transportation Science, 16, 231 - 240.

Dafermos S C (1982b) The general multimodal network equilibrium problem with elastic demand. Networks, 12.

Daganzo C F (1977) On the traffic assignment problem with flow dependent costs I and II. Transportation Research, 11, 433 - 441.

Florian M (1979) Asymmetrical variable demand multi-mode traffic equilibrium problems: existence and uniqueness of solutions and a solution algorithm. Divicao de Intercambio de Edicoes, INDO/O/O, Pontificia Universidad Catolica de Rio de Janeiro,

Hearn D W (1982) The gap function of a convex program. Operations Research Letters, 1, 67 - 71.

Hearn D W and Lawphongpanich S (1982) Simplicial decomposition of the asymmetric assignment problem. ISE Department Research Report 82 - 12, University of Florida (October).

Hearn D W and Ribera J (1980) Bounded flow equilibrium problems by penalty methods. Proceedings of the 1980 IEEE International Conference on Circuits and Computers.

Leventhal T L, Nemhauser G L and Trotter L E (1973) A column generating technique for traffic assignment. Transportation Science, 7, 168 - 172.

Nguyen S and Dupuis C (1981) An efficient method for computing traffic equilibria in networks with asymmetric transportation costs. Publication No. 404-A, Department d'Informatique et de Recherche Operationnelle, Université de Montreal.

Payne H J and Thompson W A (1975) Traffic assignment on transportation networks with capacity constraints and queueing. Paper presented at the 47th National ORSA/TIMS North American Meeting.

Polak J (1983) Some methodological aspects of equilibrium assignment algorithm. Paper presented at the Annual Conference of the Universities' Transport Study Group (January).

Smith M J (1979a) The marginal cost taxation of a transportation network. Transportation Research, 13B, 237 - 242.

Smith, M J (1979b) A local traffic control policy which automatically maximises the overall travel capacity of an urban road network. Paper presented at the Conference on Urban Traffic Control Systems, University of California, Berkeley (August), printed in the Proceedings and in Traffic Engineering and Control, 21 (1980), 289 - 302.

Smith M J (1979c) The existence, uniqueness and stability of traffic equilibria Transportation Research, 13B, 295 - 305.

Smith M J (1981a) The existence of an equilibrium solution to the traffic assignment problem when there are junction interactions. Transportation Research, 15B, 443 - 451.

Smith M J (1981b) Properties of a traffic control policy which ensure the existence of a traffic equilibrium consistent with the policy. Transportation Research, 15B, 453 - 462.

Smith M J (1981c) A theoretical study of traffic equilibrium and traffic control. Paper presented at the Eighth International Symposium of Transportation and Traffic Theory, University of Toronto (June) and printed in the proceedings.

Smith M J (1983a) The existence and calculation of traffic equilibria. Transportation Research, 17B, 291 - 303.

Smith M J (1983b) An algorithm for solving asymmetric equilibrium problems with a continuous cost-flow function. Transportation Research, 17B, 365 - 371.

Smith M J (1983c) Traffic assignment and traffic control on a capacity constrained network with queueing. Paper presented at the Annual Conference of the Italian Operations Research Society, Naples (June).

Smith M J (1984) A descent algorithm for solving monotone variational inequalities and monotone complementarity problems. Journal of Optimization Theory and Applications, to appear.

Wardrop J G (1952) Some theoretical aspects of road traffic research. Proceedings of the Institute of Civil Engineers, Part II, 1, 325 - 378.

Ninth International Symposium on
Transportation and Traffic Theory
© 1984 VNU Science Press, pp. 299-330

MODELLING INTER URBAN ROUTE CHOICE BEHAVIOUR

M. BEN-AKIVA,[1] M. J. BERGMAN,[2] A. J. DALY[3] and R. RAMASWAMY[3]

[1] Department of Civil Engineering, Massachusetts Institute of Technology, Cambridge, MA, U.S.A.
[2] Ministry of Transport, Rijkswaterstaat, The Hague, The Netherlands
[3] Cambridge Systematics Europe, The Hague, The Netherlands

ABSTRACT

This paper develops and tests innovative models of driver's route choice behavior. The approach taken in this paper is to define choice sets of "labelled" paths. The idea is to transform a large number of physical routes into a smaller number of routes each representing a specific "label". A labelled path is defined as the optimal physical path with respect to some criterion function. The criteria that might be relevant to route choice include travel time, distance, scenery, congestion, signposting, etc. The label functions are estimated by maximizing the proportion of observed routes included in the sets of labelled paths. This paper describes the analysis of a set of labels and their estimated functions. This modelling approach assumes a two stage decision process. In the first stage, a choice set of alternative routes between origin and destination are generated; these are the labelled paths. In the second stage the driver is selecting one of these routes on the basis of route attributes and personal and trip characteristics. A random utility theory is employed to derive the choice model for the second stage of this decision process. The general model has a nested-logit form that represents alternative assumptions about the unobserved route choice utilities; some unobserved attributes may be link-related and others label-related. The estimation results demonstrate the important effects of factors other than time and distance on driver's route choice behavior.

INTRODUCTION

This paper develops and tests models of route choice behavior with the objective of gaining better understanding of drivers' route-choice preferences and ultimately improving traffic assignment procedures.

Previous research into route choice behavior can best be summarized by the following list of hypotheses that have been classified into three broad categories: drivers' knowledge about alternative routes, route choice decision processes, and route attributes and drivers' route choice preferences.

(1.) Route and Network Information and Knowledge:

- Long distance drivers tend to restrict their travel to highways, due to their lack of knowledge about local street networks and the easy availability of information (signs, maps) about highways (Ueberschauer, 1971 and Tagliacozzo and Pirzio, 1973).

- Drivers are often unable to evaluate simple characteristics of paths, such as time or distance, due to lack of knowledge and thus are unable to meet their objectives in route choice; e.g., minimizing time or distance (Burrell, 1968; Dial, 1970; Aashtiani and Powell, 1978).

- Travellers may drive on a particular route because they are unaware of any other alternatives (Wright, 1976).

- Drivers may plan some trips in a hierarchical fashion; they start at the lowest level in the road network on local streets, proceed up the hierarchy to expressways for the bulk of their travel and then go down the hierarchy in the vicinity of their destination (Aashtiani and Powell, 1978 and Hidano, 1983). Bovy (1979) has shown with a sample of urban trips that about 60 percent of travellers follow this type of hierarchical route.

(2.) Decision Processes:

- Preplanning: Drivers are assumed to plot out their route before beginning a trip.

- Markov Process: Drivers' decisions about routes are made at intersections as they are encountered during a trip, and are independent of prior decisions.

o Intermediate Process: Both preplanning and some immediate decisions take place (Hamerslag, 1977).

In general, these hypotheses are not specific enough to be testable. However, one aspect of the Markov process type of behavior could be the result of inadequate information about the road network, in which case the driver would follow the posted signs.

(3.) Route Attributes and Route Choice Preferences--The factors affecting route choice can be divided into two types: (a) factors that are the attributes of the routes themselves and (b) personal characteristics of the driver:

- Travel time (Morisugi et. al., 1981; Outram, 1976; Lessieu and Zupan, 1970 and Trueblood, 1952).

- Distance (Trueblood, 1952).

- Number of stops or traffic signals (Ueberschaer, 1971).

- Aesthetic appeal--Drivers may be more concerned with scenery on optional trips (e.g., visit a friend) than on obligatory ones (e.g., work purpose) (Wachs, 1967; Tagliacozzo and Pirzio, 1973).

- Time or Distance on Limited Access Highways--Preference for travel on limited access highway routes has been postulated to increase (Wachs, 1967) as: distance to work increases; age of the driver decreases; education level of the driver decreases (probably because of the level of skills required in map reading); and peak hour travellers prefer limited access routes more than off-peak travellers.

- Safety--Wach (1967) showed that safety becomes more impor- tant to drivers as the length of the trip increase. He also shows an interrelationship with class of employment (blue collar, white collar), length of job tenure (people new to their jobs are more concerned with safety), and fre- quency of use of other modes (people who use other modes besides driving are less concerned with safety). Morisugi et al (1981) found that safety is more important to female drivers. Safe routes are obviously important considera- tions from a driver's viewpoint. Drivers, especially those who drive frequently, may know about the existence of "black spots" in a network. However, a driver does not usually have an idea of how safe a route is.

- Commercial Development--When drivers are interested in reaching their destination quickly, such as on the work trip, they may avoid commercial development along their path, because it causes traffic conflicts. On trips to shopping destinations, commercial development may provide additional shopping opportunities.

- Congestion--Wachs' (1967) research indicated that drivers
  avoidance of congestion was related to: length of residence (as a proxy for knowledge of the network), location
  of residence (Central Business District residents were less
  concerned with congestion), and social status (as measured
  by income and education). Michaels (1962) speculated that
  travel impedance associated with congestion was linear up
  to some level and thereafter exponential. The ability of
  frequent drivers to avoid congestion is also a reflection
  of the driver's knowledge about a route.

- Road quality.

- Road signs--For some types of trips, like long-distance
  ones, drivers rely on signs along the road to choose their
  routes (Wootton, Ness and Burton, 1981). This hypothesis
  is com- patible with those about knowledge of alternative
  routes and the Markovian decision process.

These hypotheses are derived from a wide range of sources (see, for
example, Carpenter, 1979) and most of them are not reflected in the
existing traffic assignment models. Significant progress has been
achieved in recent years in modelling traveller choices using discrete
choice models estimated with individual data. The direct application
of these modelling methods to route choice behavior, however, is not
feasible. The nature of route choice poses a unique problem because
of:

- the very large number of practically feasible routes between
  most origins and destinations; and

- the complex overlapping of these routes.

In this paper, we develop a route choice model in which these difficulties are treated through a model structure consisting of two
stages: choice set generation and choice from a choice set (Manski,
1977). We refer to the method as the "labelling" approach because in
the first stage model, we assign descriptive labels to physical paths
in the network. The essential idea of the labelling approach is to
replace consideration of the huge number of feasible routes by consideration of a much smaller number of routes, each embodying a specific criterion that might be relevant to route choice. These criteria
(e.g., time, scenery, etc.) are the labels. For each label, a criterion (or a generalized impedance) function is defined so that a

network minimum path algorithm can be used to build trees that are minimal with respect to the criterion and which represent paths emphasizing the label characteristic. For example, for the scenery label, time spent on roads with poor scenery would be much more highly weighted (have greater impedance) than time on scenic roads.

Thus, this model views route choice as essentially a two-stage process. In the first stage, labels are used to define the range of alternatives considered. In the second stage, a model of choice from the set of labels is applied to predict the chosen route. For the first stage of the model, a method was developed of generating labels that maximize the coverage of observed behavior; the method selected is described in the next section of the paper. It is followed by a section that presents the discrete choice model formulated for the second stage of the model. It is a form of the nested logit model which is derived from specific assumptions about the nature of the choice set. Estimation results are then presented and analyzed and the paper concludes with directions for further work.

The data set used for model estimation was collected in the area between Utrecht and Amersfoort in the Netherlands. In this area, many drivers are confronted with a choice from a range of feasible paths, rather than being "captive" to one clearly superior path. Data on the routes chosen by drivers were taken from a survey that was conducted during 1979. Automobile trips were surveyed at a number of roadside survey stations which constituted a cordon. At some cordon crossings, cars were stopped and mail-back questionnaires were handed to drivers. At others, car number plates were recorded and mail-back survey questionnaires were sent to their owners.

A network was set up for this analysis which was very detailed inside the study area and much coarser outside. The network attributes of distance and speed class were supplemented within the study area by a wide range of other link attributes and information on the nodes at the ends of the links.

1.   THE CHOICE SET GENERATION MODEL

This section presents the choice set generation model. It includes the definitions and specifications of the label impedance functions, the method of estimating the unknown parameters and the estimation results.

1.1 The Definition of Labels

The first step is to determine a comprehensive set of labels that
could be potential candidates for the choice set available to a driver
travelling between a given origin and destination. Existing literature on route choice and postulated hypotheses about drivers' route
choice behavior are useful for this purpose. Table 1 lists possible
criteria for drivers' consideration of a particular route. It is
clear that not all the criteria apply to all drivers, but it is reasonable to assume that drivers make their actual choice from some subset of these criteria.

1.2 The Specification of the Label Impedance Functions

The quantification of the above criteria involved two steps. The
first was to find a quantitative descriptor of each of the criteria.
This was limited by the availability of data and only the first 10
criteria in Table 1 were implemented. The second step was to define
an impedance function for each label. Potentially, it was possible to
generate optimum paths using the attribute itself, e.g., a maximum
scenic or a minimum signals path; however, such a path could deviate
appreciably from the minimum time path to an extent that made it unreasonable. To avoid this, for all labels other than the time and
distance labels, a function was defined for link impedances which was
a weighted sum of travel time and the link attribute describing the
label. The link attributes and the corresponding form of each impedance function are shown in Table 2.

1) Minimize time

2) Minimize distance

3) Maximize scenery along route

4) Minimize the number of traffic signals

5) Maximize the use of highways

6) Minimize the use of congested routes

7) Maximize shopping and commercial areas along route

8) Travel on roads with high pavement quality

9) Travel on high capacity roads

10) Follow a "hierarchical" route; i.e., start from a
    lower level road and hierarchically attempt to reach
    a higher level road at the origin end; follow the
    reverse process at the destination end.

11) Follow the sign posted route

12) Travel on safe roads

<u>Table 1</u>: Drivers' Criteria for Route Choice

Label Criterion	Attribute	Impedance Factor for Each Link
Min time	Travel time on a link	Time
Min distance	Link length	Distance
Max scenic	% of link length through scenic areas	Time $(1 + \beta_1$ % non-scenic)
Min traffic lights	Number of traffic lights on link	Time $+ \beta_2$ no. of lights
Min congestion	Volume/Capacity ratio	Time $(1 + \beta_3$ high V/C dummy)
Max highways	% of link length on highway	Time $(1 + \beta_4$ % non-highway)
Max capacity	% of link length of high capacity road	Time $(1 + \beta_5$ low capacity dummy)
Max commercial	% of link length through commercial areas	Time $(1 + \beta_6$ % low commercial)
Max road quality	Dummy variable for quality of pavement	Time $(1 + \beta_7$ low quality dummy)
Hierarchical travel pattern	Link "level" in network (3 levels considered)	Time $(1 + \beta_{81}$ (level 1 dummy) $+ \beta_{82}$ (level 2 dummy))

Table 2: Label Attributes and Impedance Functions

1.3 Label Parameter Estimation

In designing and selecting labels, the objective is to generate a reasonable set of paths that include the actual paths chosen by the drivers. The objective function of the entire label set, therefore, is to maximize the coverage by the label set of the set of chosen paths, and the optimal values of the parameters of the impedance functions are the values that maximize this coverage. This estimation procedure of a deterministic choice set generation model is analogous to the maximum score estimator that was developed by Manski (1973) for discrete choice models.

The objective is to maximize the number of observed paths in a sample of size N which are covered by a set of J labels. Let C_{in} be a dummy variable that indicates coverage of a chosen path in observation n by label path i, and D_{ijn} be a dummy variable that indicates overlap between label path i and label path j for the O/D pair of observation n; i, j = 1, ..., J and n = 1, ..., N. If P_{in} represents the path for label i and path P_{cn} the chosen path for O/D pair n = 1, ..., N; then we have as our objective function:

$$\underset{\beta}{\text{Max}} \quad \sum_{i=1}^{J} \sum_{n=1}^{N} (C_{in} - Q_{in}) \tag{1}$$

where

$$C_{in} = \begin{cases} 1 & \text{if } P_{in} = P_{cn} \\ 0 & \text{if } P_{in} \neq P_{cn} \end{cases} \tag{2}$$

$$i=1, ..., J, n=1, ..., N;$$

$$Q_{in} = \begin{cases} 1 & \text{if } \sum_{j=1}^{i-1} D_{ijn} > 0 \\ 0 & \text{otherwise} \end{cases} \tag{3}$$

$$i=1, ..., J, n=1, ..., N;$$

and for $i \neq j$

$$D_{ijn} = \begin{cases} 1 & \text{if } P_{in} = P_{jn} \\ 0 & \text{if } P_{in} \neq P_{jn} \end{cases} \tag{4}$$

$$i, j=1, ..., J, n=1, ..., N.$$

The paths P_{in}, i=1, ..., J and n=1, ..., N are conditional on the values of the unknown parameters β used to generate the label paths in the network. The estimation problem is to find values for the unknown parameters ($β_1$, $β_2$, ..., $β_J$) that maximize the objective function defined above, subject to sign constraints on the unknown parameters, i.e.,

$$β_1, β_2, β_3, ..., β_J \geq 0, \tag{5}$$

and to the condition that P_{in} exists for all i=1, ..., J and n=1, ..., N.

It is clear that dealing with the problem as defined and seeking a global optimum is too complicated given the complexity of the objective function. There is also the issue of the limits of β. While it is intuitively obvious that $β \geq 0$, there is no upper bound on β except that it should be reasonable; for example, the notion that a non-scenic link could be 200 or 300 times as unattractive as a scenic appears to be unreasonable. With these in mind, a heuristic approach to the estimation of the β's based on a limited range of parameter values was followed.

The labels were loaded one at a time onto the network, and the number of new (i.e., previously unmatched) paths matched by this label was ascertained. The first two labels loaded were time and distance since these did not have any parameters to be estimated. The remaining labels were then loaded in turn, keeping the β values within a relatively narrow range of intuitively acceptable values. These values of 'β' were then optimized separately for each label as it was loaded. The score at each step is conditional on the score obtained at the previous step. Obviously, for a given set of parameter values the same total score is obtained from all possible label orderings. Table 3 shows the set of scores for a certain order obtained for the optimal parameter values. In general, the total score was insensitive to the values of the parameters within the limited range of values that was investigated.

Label	Paths Matched	β
Time	1505	–
Distance	34	–
Scenic	123	2.0
Signals	59	30 sec.
Capacity	53	1.5
Hierarchy	72	5.0, 100.0
Quality	8	2.0
Commercial Dev.	0	1.5
Highway	3	3.0
Congestion	2	3.0

TOTAL MATCHED PATHS:	1859
TOTAL CHOSEN PATHS:	2152
TOTAL UNMATCHED PATHS:	293

Table 3: Number of New Paths Matched by Labels

1.4 Number of Labels

As Table 3 indicates, the time and the distance labels in themselves
matched over 70 percent of the total number of chosen paths, and all
the labels together covered almost 90 percent of the chosen paths.
The last four labels contributed very little by way of new matches.
However, these match rates depend heavily on the order in which the
labels are loaded, and further investigations were necessary to establish that the subset of the first six labels was the most effective.

These further investigations compared the new matches given by each
label for various different orders of label loading with the total
matches given by the label. The total matches, which are of course
not order dependent, are shown in Table 4. The criterion applied was
to include a label if the ratio of new matches to total matches was
generally high for different loading orders. Labels for which this
ratio was generally low were considered to be ineffective and were
eliminated, since the addition of purely overlapping coverage only
confuses the explanation of behavior rather than adding new information. By this process a set of six labels was derived.

1.5 Robustness of Label Parameter Estimates

The last set of tests, displayed in Table 5, investigated the robustness of estimated parameter values. This involved changing the value
of the estimated β and testing the stability of the coverage scores.
These "partial derivatives" of coverage with respect to β were very
close to zero, indicating robustness of the parameter estimates.

Label	Paths Matched
Time	1505
Distance	462
Scenic	770
Signals	851
Capacity	1058
Hierarchy	712
Quality	1477
Commercial Dev.	1508
Highway	501
Congestion	425
TOTAL PATHS:	2152

Table 4: Total Number of Paths Matched by Labels

Label	Label Estimation Results		Sensitivity Tests																		
			1		2		3		4		5		6		7		8		9	
	Optimal β	Paths Matched	β	Paths Matched	β	Paths Matched	β	Paths Matched	β	Paths Matched	β	Paths Matched	β	Paths Matched	β	Paths Matched	β	Paths Matched	β	Paths Matched
Time	-	1505		1505		1505		1505		1505		1505		1505		1505		1505		1505
Distance	-	34		34		34		34		34		34		34		34		34		34
Scenic	2.0	123	1.75	113	2.25	124		123		123		123		123		123		123		123
Signals	30 sec	59		59		58	20 sec	53	40 sec	57		59		59		59		59		59
Hierarchy	5,100	72		82		72		72		71		72		72		72	4.6*	72	5.4*	71
Capacity	1.5	53		53		53		54		54	1.35	56	1.25	53	1.7	53		53		53
Unmatched		306		306		306		311		308		303		306		306		306		307

* Hierarchy was found to be insensitive to β_{82} values between 50 and 100.

Table 5: Robustness of Estimated β Values

2. CHOICE MODEL STRUCTURE AND ESTIMATION PROCEDURE

For any traveller observed in the survey, the coded "chosen" path may overlap with the estimated set of labelled paths:

- not at all;

- entirely (i.e. all the labelled paths are the same and the same as the chosen);

- with some labels but not with others.

In the first two cases, we can deduce nothing from modelling the traveller's behavior. In the first case, the route chosen is not explicable by the labels, perhaps because of irrational behavior or some coding error in the data. In the second case, it appears that the traveller has no alternative: all the reasonable paths are identical. Thus the modelling is based on cases where the chosen route is the same as some label(s) and not the same as some other(s): the labels may all be distinct or there may be as few as two distinct paths.

In the following subsection the model structure is presented. The second subsection deals with alternative forms of the utility function for the paths and labels, and shows how the different models may be estimated.

2.1 Model Structure

Let A_n be the set of label paths for individual n with J_n elements. Subsets of label paths in A_n may represent the same physical path on the network. Partition the set A_n into subsets of physically identical label paths as follows

$$A_n = \left\{ I_1, \ldots, I_r, \ldots, I_{R_n} \right\} \tag{6}$$

where R_n is the number of physically distinct paths in the set A_n (such that $2 \leqslant R_n \leqslant J_n$) and I_r is a set of label paths that are represented by the rth physical path.

Denote by U_{in} the utility of the label path $i \in A_n$ for individual n. Treat U_{in} as a random variable with a mean of V_{in} and a disturbance ε_{in}. The attributes, and as a consequence the utilities, of the label paths are defined by the minima of the label impedance functions over all the paths in the network. Thus, a random label utility can be assumed to have an extreme value distribution, and the joint cumulative distribution of the ε_{in}'s to have the following form of the multivariate extreme value distribution (McFadden, 1978):

$$F(\varepsilon_{1n}, \ldots, \varepsilon_{in}, \ldots, \varepsilon_{J_n n}) =$$

$$= \exp \left[- \sum_{r=1}^{R_n} \left(\sum_{i \in I_r} \exp(-\varepsilon_{in}/\mu) \right)^{\mu} \right] \qquad (7)$$

where μ is an unknown parameter satisfying the condition

$$0 < \mu \leq 1 \qquad (8)$$

This distribution assumes that the random utilities of label path i and label path j ($j \neq i$) are independently distributed if they belong to different physical paths. If two labels i and j ($i \neq j$) belong to the same physical path then their utilities are not independently distributed and their correlation coefficient is equal to $(1 - \mu^2)$ (see Ben-Akiva and Lerman, 1983). The distribution assumption given by (7) and (8) results in the following nested logit model for the choice probability of label path i by the nth observation:

$$P_n(i) = P_n(i|I_r) \cdot P_n(I_r), \text{ for } i \in I_r \subseteq A_n \qquad (9)$$

where

$$P_n(i|I_r) = \frac{e^{V_{in}/\mu}}{\sum_{j \in I_r} e^{V_{jn}/\mu}} \qquad (10)$$

$$P_n(I_r) = \frac{e^{V_{I_r n}}}{\sum\limits_{r'=1}^{R_n} e^{V_{I_{r'} n}}} \; ;$$ (11)

and

$$V_{I_r n} = \mu \log \left[\sum_{j \in I_r} e^{V_{jn}/\mu} \right]$$ (12)

Two limiting cases are particularly interesting.

(i)   For $\mu = 1$ the label utilities of two labels that belong to the
      same physical path are assumed to be independently
      distributed and the model reduces to a simple logit
      model as follows:

$$P_n(i) = \frac{e^{V_{in}}}{\sum\limits_{j \in A_n} e^{V_{jn}}}$$ (13)

However, the observed choice is only known to be included in a
subset of labels I_c and the probability of the observed choice
is

$$P_n(I_c) = \frac{\sum\limits_{i \in I_c} e^{V_{in}}}{\sum\limits_{j \in A_n} e^{V_{jn}}}$$ (14)

(ii) For $\mu \to 0$ the utilities of label paths that correspond to the
      same physical path are perfectly correlated and the
      following limit which is a maximum model is obtained

$$V_{I_r n} = \max_{i \in I_r} V_{in} \tag{15}$$

$$P_n(i|I_r) \begin{cases} > 0 & \text{for } V_{in} = V_{I_r n} \\ = 0 & \text{for } V_{in} < V_{I_r n} \end{cases} \tag{16}$$

and

$$P_n(I_r) = \frac{e^{V_{I_r n}}}{\sum_{r'=1}^{R_n} e^{V_{I_{r'} n}}} \tag{17}$$

These two simple forms may be used to determine limiting estimates for particular forms of V_{in}. In general, however, we are interested in estimating $0 < \mu \leqslant 1$ and this is the case considered in the following sub-section.

2.2   Estimation of the Choice Model

The general choice model, as described in the previous sub-section, has the functional form of a nested logit model with two levels. However, the choice is observable only at the upper level: i.e., the physical path that was chosen by the traveller. The lower level of the model gives the conditional choice probability among labels that correspond to the same physical path. Therefore, the observations of choice are made at the higher level and no direct estimation can be made of the utilities at the lower level. However, these utilities enter the upper level model and can be estimated at this level as will be shown below.

The model for the choice probability of a physical path I_r can be rewritten as follows, with the subscript n omitted to simplify the notation:

$$\log P(I_r) = V_{I_r} - \log \sum_{r'=1}^{R} \exp (V_{I_{r'}}) \qquad (18)$$

The average utility of a physical path I_r has the following form

$$V_{I_r} = x_{I_r} \gamma + \mu \log \sum_{j \in I_r} \exp (x_j \theta / \mu) + \sum_k z_{kI_r} \eta_k \qquad (19)$$

where x_{I_r} is a vector of variables describing the physical path I_r, and include generic path attributes like time and distance; x_j is a vector of variables associated with the label j (which coincides with path I_r); z_{kI_r} is a dummy variable which is equal to one if I_r contains the kth label combination; and γ, θ and η are, respectively, associated vectors of unknown parameters.

Note that for a total of J labels there are $2^J - 2$ possible label combinations, excluding the case of complete overlap which is omitted from the estimation data set. Thus, for a sufficiently large sample it is possible, in principle, to estimate the model with $2^J - 3$ label combination constants. In practice, however, a much smaller set of constants for label combinations that frequently appear in the data is included.

This physical path utility is based on the following specification of the utility for label path $i \in I_r$

$$v_i = x_{I_r} \gamma + x_i \theta + \sum_k z_{kI_r} \eta_k \qquad (20a)$$

and the conditional choice probability for a label path is

$$\log P (i|I_r) = x_i \theta / \mu - \log \sum_{j \in I_r} \exp (x_j \theta / \mu) \qquad (20b)$$

The kernel of the log likelihood function for this model for a random (or exogenously stratified) sample is

$$L = \sum_{n=1}^{N} \log P_n(I_c) \tag{21}$$

where I_c denotes the chosen route. For a fixed value of μ the first order conditions for the maximum likelihood estimator obtained by differentiating with respect to Θ are

$$\sum_{n=1}^{N} [\sum_{i \in I_{cn}} x_{in} P_n(i|I_c) - \sum_{r=1}^{R_n} P_n(I_r) \sum_{j \in I_r} P_n(j|I_r) x_{jn}] = 0 \tag{22}$$

The equations obtained by differentiating with respect to γ are

$$\sum_{n=1}^{N} [x_{I_c n} - \sum_{r=1}^{R_n} x_{I_r n} P_n(I_r)] = 0 \tag{23}$$

Equations (23) and similar conditions obtained for the η parameters are identical to the first order conditions for a linear logit model and equation (22) is a modification in which the chosen alternative is generalized to a set of alternatives. Therefore, an algorithm for this non-linear logit model can be developed as an extension of an existing linear logit maximum likelihood algorithm. However, joint estimation of γ, Θ, η and μ of the above model in its most general form presents a difficulty. First, the model is of a form for which software does not exist, and although software could be prepared for maximum likelihood estimation, the form of the likelihood function is not necessarily globally concave, so that difficulty in finding a global optimum could be anticipated. More fundamentally, the parameter μ gives the ratio of variance of the random components of utility at the two levels, and because of lack of information about the lower level utilities it cannot be well estimated from the observed data.

Difficulties were encountered in estimating the general model and in particular the parameters θ and μ. Estimations were therefore made with a series of successively less severe restrictions imposed on the general model.

(i): $\cancel{\mu = 1}$

As explained in the previous sub-section, this restriction leads to the model in (18) with

$$V_{I_r} = \log \sum_{j \in I_r} \exp (x_{I_r} \gamma + x_j \theta + \sum_k z_{kI_r} \eta_k) \tag{24}$$

This is a simple logit model except that the chosen alternative in this case is the set of elementary alternatives (i.e. label paths) coinciding with the chosen path I_c. This formulation assumes no correlation between the error terms of the utilities of the elementary alternatives belonging to sets I_r, $r=1,\ldots,R$.

For this model, software was prepared to calculate the derivatives of the likelihood function to allow maximum likelihood estimation of γ, θ and η.

(ii): Restricted Dimensionality ($0 < \mu \leq 1$)

A restriction which provides the possibility of making joint estimates of μ with existing software is to take x_j as a vector of length J, where J is the total number of labels, having only one non-zero component, which is the value 1 in position j. The utility in (19) then becomes

$$V_{I_r} = x_{I_r} \gamma + \mu \log \sum_{j \in I_r} \exp (\theta_j / \mu) + \sum_k z_{kI_r} \eta_k$$

$$\tag{25}$$

$$= x_{I_r} \gamma + \mu \log \sum_{j=1}^{J} \exp (\theta_j / \mu) y_{jI_r} + \sum_k z_{kI_r} \eta_k$$

where $Y_{jI_r} = \begin{cases} 1 \text{ if } j \in I_r \\ 0 \text{ otherwise} \end{cases}$

This non-linear specification can be estimated with the logit estimation software developed by Daly (1982) for spatial choice with multiple
attraction variables. Note that only $J-1$ of the θ_j parameters,
$j=1,\ldots,J$; can be identified and therefore estimating this model would
require that one of these parameters, say θ_J, is set equal to zero.

(iii): $\underline{\mu \nrightarrow 0}$

This restriction is the second limiting case of the general model.
The assumption is that the lower level choice is deterministic. In
other words, the conditional choice probability is equal to one for
the label in the set I_r that has the largest utility. In this case

$$V_{I_r} = x_{I_r}\gamma + \underset{j \in I_r}{\text{Max}} \ x_j\theta + \sum_k z_{kI_r}\eta_k \tag{26a}$$

Under the restricted dimensionality assumption about x_j made above
it becomes

$$V_{I_r} = x_{I_r}\gamma + \underset{j \in I_r}{\text{Max}} \ \theta_j + \sum_k z_{kI_r}\eta_k \tag{26b}$$

It is possible to estimate this model by repeated application of a linear
logit estimation program with different ranking of labels to establish a
unique self-consistent ranking of the parameters θ_j, $j=1,\ldots,J$.

(iv): $\underline{\theta = 0}$

This restriction implies that the lower level utilities for labels
that correspond to the same physical path are exactly identical. In
other words, the conditional probabilities are equal to $1/|I_r|$ for

all i∈I_r and for r=1,...,R. The utility function is reduced to the form

$$V_{I_r} = x_{I_r} \gamma + \mu \log|I_r| + \sum_k z_{kI_r} \eta_k \qquad (27)$$

where $|I_r|$ denotes the number of label paths included in I_r. This specification is a linear logit model that can be estimated with standard logit estimation software. This restriction implies that any information on the specific combination of labels that are contained in a given physical path can be included in the model only through the z_{kI_r} variables. The simplest specification of this information consists of the following variables: for j=1,...,J define

$$Y_{jI_r} = \begin{cases} 1 & \text{if } j \in I_r \\ 0 & \text{otherwise} \end{cases}$$

The utility function for this model has the form

$$V_{I_r} = x_{I_r} \gamma + \sum_{j=1}^{J} Y_{jI_r} \tilde{\theta}_j + \mu \log |I_r| \qquad (28)$$

where $\tilde{\theta}$ is a vector of J unknown parameters.

3. CHOICE MODEL ESTIMATION RESULTS

The previous section described the structure of the model and the restrictions that could be imposed on the general model to facilitate estimation. This section reports the results obtained from these estimations.

3.1 Alternatives and Attributes

Section 1 of this paper describes the initial set of ten labels that
were generated as possible descriptors of route choice behavior and
the experimentation which resulted in the use of six labels for the
modelling. The selected labels are time, distance, scenic, signals,
hierarchy and capacity. A variety of generic path attributes were
tried and from the ten label attributes listed in Table 2, seven were
chosen to form the vector X_I described in Section 2.2. These
were the time, distance, scenic time, no. of traffic lights, length
which is highway, length which is on high capacity road and length
which is on high quality road surface. The precise specification of
the model is described below.

3.2 Label Combinations

An initial analysis was made of all the ways in which label combinations occurred in the sample. Out of the 64 (2^6) possible combinations, 36 were actually present, two of which were the all chosen and
none chosen combinations leaving 34 useful combinations. The distribution of the label combinations of the chosen paths is given in Table
6. The specification of the label combination constants is shown in
Table 7. All pairs of combinations were treated individually, while
the triples, quadruples and sets of five were differentiated into sets
that contained and did not contain the hierarchy label. Hierarchy was
chosen because in preliminary estimations without label combination
constants the hierarchy constant made the largest positive contribution to the utility and the inclusion or exclusion of the hierarchy
label was likely to have a signficant effect.

3.3 Estimation Results

The most general model that was estimated is the specification with
restricted dimensionality of the label variables (x_j) given in equation (25). However, the estimation algorithm failed to converge

		LABEL				
TIME	DIST	SCEN	SIGN	HIER	CAPC	TOTAL OBSERVATIONS
*					*	29
		*				3
			*			54
*			*			70
*		*	*		*	40
*			*	*	*	76
*						542
		*			*	99
*			*			56
*	*		*	*	*	28
	*			*		1
			*	*		2
	*					25
*		*		*	*	48
*		*	*	*	*	110
*	*				*	7
*	*	*	*		*	72
*				*	*	39
*	*	*	*	*	*	306
*	*	*		*	*	1
		*		*		11
*		*			*	51
				*		71
	*		*			2
					*	53
			*	*	*	1
			*		*	1
	*	*			*	1
*	*	*			*	3
*	*		*		*	6
		*	*	*		5
	*	*	*	*		2
		*	*			4
	*	*	*			2
*		*	*			1

'*' indicates chosen labels
' ' indicates unchosen labels

Table 6: Label Combinations

Combination Constants	Explanation
1	Time and Capacity combined
2	Distance and Hierarchy combined
3	Distance and Signals combined
4	Scenic and Hierarchy combined
5	Scenic and Capacity combined
6	Signals and Hierarchy combined
7	Time and Signals combined
8	Distance and Scenic combined
9	Scenic and Signals combined
10	Triples that include Hierarchy
11	Triples that exclude Hierarchy
12	Quadruples that include Hierarchy
13	Quadruples that exclude Hierarchy
14	Sets of 5 labels that include Hierarchy
15	Sets of 5 labels that exclude Hierarchy

Table 7: Definition of Label Combination Constants (η's)

on a value for the parameter μ. Therefore, estimations were performed for fixed values of the parameter μ between zero and one. The model estimation results for $\mu = .3$ are presented in Table 8.

The log likelihood values at convergence for different values of μ were as follows:

μ	Log Likelihood
0	−793.61
.3	−792.15
.5	−792.32
1	−792.84

Although $\mu = .3$ may be selected as the optimal value, it is evident that the effect of the value of μ on the choice probabilities in the estimation sample is not significant. It indicates that the non-linear part of the utility function

$$\mu \log \sum_{j \in I_r} \exp (\theta_j/\mu),$$

can be approximated by the linear form

$$\sum_{j \in I_r} \widetilde{\theta}_j$$

that was included in the simplified model specification in equation (28). This linear approximation results in a linear Logit Model without the parameter μ. However, preliminary model estimation results that excluded the label combination constants indicated a significantly better fit for the non-linear specification. The optimal value of μ for the model without the label combination constants was zero. This result implies that without the label combination constants the correlation of the random utilities of overlapping labelled paths approaches one. The inclusion of the label combination constants has significantly improved the fit of the model; from a log likelihood of −849 to −792.

Parameter		Estimate	Standard Error	t Ratio
Attributes γ				
Time	1	-3.21	1.35	-2.38
Distance	2	-8.72	2.04	-4.27
Scenic Time	3	2.55	0.59	4.35
# of Signals	4	-0.0307	-0.04	-0.77
Xway Distance	5	3.90	0.73	5.32
High Cap. Dist.	6	2.40	0.89	2.69
High Qual. Dist.	7	0.977	1.06	0.92
Label Specific Constants θ / μ				
Time	1	1.76	1.39	1.26
Distance	2	-6.44	2.00	-3.23
Scenic	3	-16.70	2.24	-7.42
Signals	4	-2.94	1.92	-1.53
Hierarchy	5	4.53	1.69	2.68
Label Combination Constants η				
Time & Capacity	1	-2.71	0.47	-5.72
Dist & Hierarchy	2	-1.55	1.53	-1.01
Dist & Signals	3	-0.737	1.05	-0.70
Scenic & Hier	4	-0.841	0.54	-1.56
Scenic & Cap	5	0.0533	0.19	0.27
Sign & Hierarchy	6	-3.19	1.02	-3.11
Time & Signals	7	-0.548	0.26	-2.08
Dist & Scenic	8	-0.688	1.28	-0.54
Scenic & Signals	9	-1.33	0.69	-1.99
Triples with Hier	10	-2.72	0.62	-4.38
Triples w/o Hier	11	-1.26	0.42	-2.96
Quads w/Hier	12	-2.23	0.63	-3.56
Quads w/o Hier	13	-1.39	0.49	-2.82
5's w/ Hier	14	-0.327	0.78	-0.42
5's w/o Hier	15	-0.744	0.53	-1.40

No. of Observations	1515
Log Likelihood	-792.15
ρ^2 w.r.t. constants	.31
ρ^2 w.r.t. zero	.62

Table 8:   Estimation Results (μ = .3)

The estimation results support some of the hypotheses presented in the introduction to this paper. Time and distance play major roles as determinants of route choice and have significant coefficients. The other attributes that appear in the hypotheses, for example scenic time, high capacity and highway distance are also demonstrated to contribute to the utility function of route choice. The signals variable fails to appear significantly as a generic level of service variable. The calculation of time along a path as done in this study constitutes both travel time and waiting time, and it is probable that the estimated waiting time at signals have already included this penalty effect. Hierarchial route choice does appear to be significant as is seen by the fact that the hierarchy label constant makes the largest positive contribution to the utility function.

The model specification includes two travel time variables: total travel time and travel time on scenic roads. The effective travel time coefficient for a non-scenic road is -3.21 and for a scenic road it is (-3.21 + 2.55 =) -.66. Thus, the disutility of travel time on a non-scenic road is about five times the disutility of scenic travel time. The model also includes four distance variables: total distance, highway distance, high capacity distance and high quality distance. The effective coefficient of distance for a low capacity, low quality arterial road is -8.72 and for a high capacity, high quality expressway it is (-8.72 + 3.90 + 2.40 + .977=) -1.44. Thus, the disutility of distance on a minor road may be as high as six times the disutility of distance travelled on a high standard expressway.

The value of time implied by the model estimation results depends on the road characteristics. For a non-scenic minor road the trade-off between travel time and travel cost; assuming car operating costs of .2 fl/km; is 4.4 fl/hr. For a non-scenic major road this value may be as high as 26.8 fl/hr.

The small value of the coefficient of the number of traffic signals implies a penalty of .1 minutes per signal on a non-scenic road. In

terms of road distance one traffic signal on a high quality major road
is perceived to be equivalent to a distance of about .2 km. As indicated above, this penalty is in addition to the wait time at signals
that is included in the network estimate of the total travel time.

4. CONCLUSION

The empirical results clearly indicate that factors other than time
and distance play a significant role in inter-urban route choice behavior. The estimation results of alternative model specifications
that were discussed in the previous section and the market segmentation and partial choice set tests that have been performed (but not
reported in this paper) have demonstrated the feasibility of the labelling route choice model.

Further developments should initially be focused on improvements to
the model specification. The market segmentation tests that were conducted demonstrated the importance of driver and trip characteristics.
It was shown that route choice preferences vary significantly by trip
length, trip purpose, and trip frequency. Thus, a useful extension
would be to include these systematic preference variations directly
in the model specification. On the other hand, it would also be useful to test a range of simpler specifications based on a limited set
of attributes which are commonly available from transportation network
models.

Further indications of the validity of the modeling approach can be
gained by applications to similar data sets. Our knowledge of route
choice behavior and our confidence in the new route choice model would
be greatly enhanced with comparable estimation results from samples
taken at different locations and different points in time.

ACKNOWLEDGEMENTS

Much of the work was performed and sponsored by the Rijkswaterstaat Dienst Verkeerskunde of the Netherlands Ministry of Transport. Staff of that agency have contributed significantly to the study, as have Steve Pitschke and Lionel Silman of Cambridge Systematics who participated in earlier phases of this study and contributed to the developments that are reported in this paper. The data used in this study was principally collected by the Adviesgroep Verkeer en Vervoer, of Utrecht.

REFERENCES

Aashtiani, H. and W. Powell (1978). "Route Choice Models in the Traffic Assignment Process", unpublished paper for M.I.T. course 1.202.

Ben-Akiva, M. and S.R. Lerman (1983) Travel Behavior: Theories, Models and Prediction Methods (forthcoming).

Bergman, M.J.; M. Ben-Akiva, L.A. Silman, S.B. Pitschke (1982). "An Analysis of Interurban Route Choice in The Netherlands", Proceedings of the PTRC Summer Annual Meeting.

Bovy, P.H.L. (1979). "Het Kortste - Tijd Routekeuzecriterium: Een Empirische Toetsing", Colloquium Vervoersplanologisch Speurwerk, The Netherlands.

Burrell, J.E. (1968). "Multiple Route Assigment and Its Application to Capacity Restraint", in W. Leutzbach and P. Baron, eds., Fourth International Symposium on the Theory of Traffic Flow, Karlsruhe, Germany.

Carpenter, S.M. (1979). Driver's Route Choice Project - Pilot Study, Research Report, Transport Studies Unit, Oxford University.

Daly, A.J. (1982). "Estimating Choice Models Containing Attraction Variables", Transportation Research.

Dial, R.B. (1970). "Probabilistic Assignment: A Multipath Traffic Assignment Model which Obviates Path Enumeration", Ph.D. dissertation, University of Washington, Seattle.

Hamerslag, R. (1979). "Onderzoek naar Routekeuze met Behulp van een Gedisaggregeerd Logitmodel", Verkeerskunde, No. 8, pp. 377-382.

Hidano, N. (1983). "Driver's Route Choice Model: An Assessment of Residential Traffic Management", paper presented at the WCTR, Hamburg, Germany.

Lessieu, E.J. and J.M. Zupan (1970). "River Crossing Travel Choice: The Hudson River Experience", Highway Research Record 322, pp. 54-67.

Manski, C.F. (1973). The Analysis of Qualitative Choice, Ph.D. dissertation, Department of Economics. MIT.

Manski, C.F. (1977). "The Structure of Random Utility Models", Theory and Decision, Vol. 8, pp. 229-254.

McFadden, D. (1978). "Modeling the Choice of Residential Location", Transportation Research Record 673, pp. 72-77.

Michaels, R.M. (1962). "The Effect of Expressway Design on Driver Tension Responses", Public Roads, Vol. 32, No. 5.

Morisugi, H., N. Miyatake, A. Katoh (1981). "Measurement of Road User Benefits By Means of a Multi-Attribute Utility Function", Papers of the Reigional Science Association, Vol 46, pp. 31-43.

Outram, V.E. (1976). "Route Choice", Proceedings of the PTRC Summer Annual Meeting.

Tversky, A. (1972). "Elimination by Aspects: A Theory of Choice", Psychological Review, Vol 79, pp. 281-299.

Taglizcozzo, F., F. Pirzio (1973). "Assignment Models and Urban Path Selection Criteria: Results of a Study of the Behavior of Road Users", Transportation Research, Vol 7, pp. 313-329.

Trueblood, D.L. (1952). "Effect of Travel Time and Distance on Freeway Usage", Highway Research Bulletin 61, pp. 18-35.

Ueberschaer, M.H. (1971), "Choice of Routes on Urban Networks for the Journey to Work", Highway Research Record 369.

Wachs, M. (1967). "Relationship Between Driver's Attitudes Toward Alternative Routes and Driver and Route Charactertistics", Highway Research Record 197, pp. 70-87.

Wootton, H.J., M.P. Ness, R.S. Burton (1981). "Improved Direction Signs and the Benefits for Road Users", Traffic Engineering and Control.

Wright, C.C. (1976). "Some Characteristics of Driver's Route Choice in Westminster", Proceedings of the PTRC Summer Annual Meeting.

Ninth International Symposium on
Transportation and Traffic Theory
© 1984 VNU Science Press, pp. 331–355

OPTIMAL TRANSIT TIMETABLES FOR A FIXED VEHICLE FLEET

AVISHAI CEDER[1] and HELMAN I. STERN[2]
[1] *Department of Civil Engineering, Transportation Research Institute, Technion-Israel Institute of Technology, Haifa, Israel*
[2] *Department of Industrial Engineering and Management, Ben Gurion University of the Negev, Beersheva, Israel*

ABSTRACT

The problem addressed in this paper is that of optimal multi-terminal timetable construction for a transit property. The motivation for the study came from a large-scale bus company — as its schedulers face a challenging task in reconstructing timetables when the available vehicle fleet is reduced. The purpose of this paper is to crystalize this identified problem in terms of an integer programming formulation and an heuristic approach designed for a person-computer interactive procedure. Both the mathematical programming formulation and the heuristic algorithm are interpreted by an example in a step-by-step fashion. These developments provide the integration of two major components in the transit operational planning process that have heretofore traditionally been treated in a serial manner. These two components are the generation of timetables followed by a vehicle scheduling procedure.

1.   INTRODUCTION & BACKGROUND

        This paper presents an extension of a continuing research
effort to provide the integration between transit timetable
construction and transit vehicle scheduling procedures.   The
motivation and background of this effort are presented in this section.

1.1. Introduction

        The transit planner working on creating schedules is in charge
of allotting vast resources, and naturally his aim is to allocate the
vehicles in an optimal and feasible manner.   The motivation for this
study comes from Egged - The Israel National Bus Carrier - which is
responsible for scheduling an average of about 54,000 daily trips
using a fleet of about 5,000 buses.   The schedulers in Egged face
frequent changes in the schedule, particularly, due to a limited
available vehicle fleet.   The reconstruction of timetables is then
an extremely cumbersome and time consuming task.   This is the
problem addressed in the paper which, in other words, is that of
optimal multi-terminal timetable construction for different fixed
sizes of vehicle fleet.

        It is the purpose of this paper to crystallize this problem
in terms of an integer programming formulation, and a heuristic
approach designed for a person-computer interactive procedure.
The heuristic procedure is considered due to (i) the inherent
limitation of integer programming for tackling large practical problems;
and (ii) the opportunity to include in the analysis practical
constraints such as deadheading and shifting of departure times.
The following section provides the background of a graphical bus
scheduling approach to be used both in the mathematical
programming formulation and the heuristic procedure.

Section 2 provides an integer programming formulation of the problem, and illustrates its use through a two terminal - two time period example. Section 2 also includes a fine tuning procedure which, when combined with the integer programming, helps to reduce the value of objective for the given fleet. Section 3 presents the heuristic algorithm along with a step-by-step solution of the example. Section 4 concludes the paper with remarks about the transit operational planning process.

1.2. Background on the deficit function

An early development of the heuristic approach used in this paper is based on the deficit function theory. A deficit function is simply a step function which increases by one at the time of each trip departure, and decreases by one at the time of each trip arrival. Such a function may be constructed for each terminal in a multi-terminal transit system. To construct a set of deficit functions, the only information needed is a timetable of required trips. The main advantage of the deficit function is its visual nature. Let $d(k,t)$ denote the deficit function for terminal k at time t. The value of $d(k,t)$ represents the total number of departures less the total number of trip arrivals at terminal k up to and including time t. The maximal value of $d(k,t)$ over the schedule horizon $[T_1, T_2]$ is designated $D(k)$.

It is possible to partition the schedule horizon of $d(k,t)$ into a sequence of alternating hollow and maximal intervals. The maximal intervals $M_r^k = [s_r^k, e_r^k]$, $r=1,2...n(k)$, define the interval of time over which $d(k,t)$ takes on its maximum value. The index r represents the rth maximal interval from the left, and $n(k)$ the total number of maximal intervals in $d(k,t)$. A hollow interval is defined as the interval between two maximal intervals. Hollows may consist of only one point. In case a hollow consists of one point not on the schedule horizon boundaries, the graphical representation of $d(k,t)$ is emphasized by a clear dot.

If we denote the set of all terminals as T, the sum of
D(k) ∀ k∊T is equal to the minimum number of buses required to
service the set T. This is known as the Fleet Size Formula and
was independently derived by Bartlett (1957),
Gertsbach and Gurevich (1977), and Salzborn (1972, 1974). Mathematically, for a given fixed schedule:

$$N = \sum_{k \in T} D(k) = \sum_{k \in T} \max_{t \in [T_1, T_2]} d(k, t) \tag{1}$$

where N is the minimum number of buses to service the set T.

When deadheading (DH) trips are allowed, the fleet size may
be reduced below the level described in Eq. (1). Ceder and Stern (1981)
and Stern and Ceder (1981a) describe this procedure based on the
construction of a Unit Reduction Deadheading Chain (URDHC). Such a
chain is comprised of a set of non-overlapping DH trips which, when
inserted into the schedule, reduces the fleet size by one. The
procedure continues inserting URDHC's until no more can be inserted
or a lower bound on the minimum fleet is reached. Determination of
the lower bound is detailed in Stern and Ceder (1983). The deficit
function theory for bus scheduling is extended by Ceder and Stern
(1982, 1983) to include possible shifting in departure times within
bounded tolerances. Finally, a chain construction method for assigning
vehicles to trips can be carried out by the first in - first out (FIFO)
rule or by a chain extraction procedure described by Gertsbach and
Gurevich (1977).

The above-mentioned deficit function developments have been
implemented on a microcomputer vehicle scheduler named AUTOBUS - an
AUTOmated BUs Scheduling graphic interactive program. AUTOBUS,
detailed in Ceder and Stern (1983), has the ability to handle multi-

garage systems, variable trip departure and travel times, deadheading trips (including those for interlining pull-in and pull-out) and timetable construction. A timetable editor is included which allows the addition, deletion and modification of trips.

2.   MATHEMATICAL FORMULATION AND EXAMPLE

There exists a feedback relationship between vehicle scheduling and timetable construction. However, the task to find this relationship for a multi-terminal schedule is extremely cumbersome and time consuming to carry out manually. The purpose of this section is to present an approach to automate this task. Recently, Koutsopoulos et al., (1983) used a similar approach but for independent routes.

2.1. Problem description

More concretely the problem is: given a fixed fleet size (number of vehicles) and a known passenger demand between set of terminals (predefined routes), it is desired to determine the frequency of vehicle departures such that some measure of passenger satisfaction is maximized. A fixed set of terminals and a fixed planning horizon are given. The time horizon for each terminal is partitioned to a fixed number of time periods (not necessarily equal). For each time period and route, the demand is represented by the total number of passengers demanding service on vehicles travelling between a given pair of terminals. The number of vehicle departures for a given terminal and for each time period must lie within a bounded range. The lower bound is based on a predetermined minimum service policy requirement or vehicle ridership measure. The upper bound is based on a minimum passenger loading factor.

For each set of vehicle departures there exists a frequency rate from which one can establish the departure times and ridership levels of each trip. The service measure for each interval is a function of the difference between the number of riders and the number of seats (plus

Allowable standees). In addition to crowding, this service measure can also include passenger waiting time (e.g., aircraft operations do not allow crowding). These and other service measures are discussed in Amar (1983) and Koutsopoulos et al., (1983). The objective is to minimize the sum of all service measures which might be interpreted as a minimum passenger disservice cost or crowding.

2.2. Formulation

Let K be the set of all terminals and T_k be the set of all departure times from terminal k. Let $(j, k_1, k_2) \equiv (\cdot)$ represent a possible trip bundle departing during period j from terminal k_1 to terminal k_2.

Define a 0-1 variable as:

$$x^q(\cdot) = \begin{cases} 1 \text{ , if } q \text{ departures are selected during} \\ \quad \text{period } j \text{ from terminal } k_1 \text{ to } k_2 \\ 0 \text{ , otherwise} \end{cases}$$

where q is an index running from $L(\cdot)$ to $U(\cdot)$, i.e.,

$$q = L(\cdot), L(\cdot)+1, L(\cdot)+2, \ldots, U(\cdot)-1, U(\cdot).$$

Define $P(\cdot)$ as the total number of passengers usually observed at the route maximum load point and $R(\cdot)$ as the desirable amount of riders in terms of number of seats plus allowable standees -- both definitions are associated with period j and the route from terminal k_1 to k_2. Both $P(\cdot)$ and $R(\cdot)$ are common measures used by transit properties world-wide, as described by Ceder (1983). The cost associated with $x^q(\cdot)$ is defined as:

$$C^q(\cdot) = \max [P(\cdot) - q R(\cdot), 0] \qquad (2)$$

and the total cost defines the objective function, Z :

$$Z = \sum_{\forall(\cdot)} \sum_{q=L(\cdot)}^{U(\cdot)} C^q(\cdot) x^q(\cdot)$$

The mathematical programming formulation which is inspired from Salzborn's note (1972) on Fleet Routing Models for Transportation Systems is stated below:

$$\min \sum_{\forall (j,k_1,k_2)} \sum_{q=L(j,k_1,k_2)}^{U(j,k_1,k_2)} c^q(j,k_1,k_2) \, x^q(j,k_1,k_2)$$

s.t.

(i) <u>Bundle departure constraints</u>

$$\sum_{q=L(\cdot)}^{U(\cdot)} x^q(\cdot) = 1 , \quad \forall \ (\cdot)$$

(ii) <u>Deficit function bounds</u>

$$\left\{ \begin{array}{l} \text{the net number of departures less arrivals} \\ \text{that occur before or at } t \text{ at terminal} \\ k \text{ as determined by the value of } x^q(\cdot) \end{array} \right\} \leq D(k), \ t \in T_k, k \in K$$

where $D(k)$ is the maximal value of the deficit function or the number of vehicles assigned to terminal k.

(iii) <u>Resource constraint</u> (total fleet size)

$$\sum_{k \in T} D(k) \leq N_0$$

where N_0 is the total (fixed and given) fleet size.

(iv) $x^q(\cdot) = 0,1 \qquad \forall \ q,(\cdot)$

$D(k) \geq 0 \qquad \forall \ k \in K$

Note that $D(k)$ will be integer in any feasible solution.

Constraint (i) ensures that only one bundle of departures is selected for a given terminal pair (route) and time period. Constraint (ii) ensures that the number of vehicles used at a given terminal k up to time t does not exceed the number of vehicles, D(k) assigned to terminal k. The left hand side of constraint (ii) can be represented as a linear function of the $X^q(\cdot)$ variable (as demonstrated in the example in the next section). The value of this function for a given solution is that of an associated deficit function for terminal k at time t. Not all departure times need be considered, as some lead to redundant equations which occur when there exists a sequence of departure times unbroken by intervening arrivals. Constraint (iii) indicates that the sum of vehicles assigned to all terminals is not greater than the given fleet size N_o.

The optimal solution $\{\overline{X}^q(\cdot), \overline{D}(k)\}$ indicates both the assignment of vehicles to terminals, $\overline{D}(k)$, and the optimal number of departures within each time period. From $\overline{X}^q(\cdot)$ an optimal timetable may be constructed. Given this timetable, a vehicle schedule can easily be derived using, for example, the FIFO rule. The vehicle schedule indicates the sequence of departures (trips) assigned to each vehicle in the fleet.

2.3. Example

In order to understand further the underlying structure of the mathematical formulation, an example comprised of two terminals and two time periods is presented. The basic input data of the example problem along with the 0-1 variables and their associated cost, $C^q(\cdot)$ are indicated in Table 1. For each trip bundle (·) the upper and lower bounds on the number of departures are shown in Table 1. For example, L(1,b,a) = 2 and U(1,b,a) = 3. The indicated cost for each (·) is calculated in accordance with Eq. (2). That is, max(p-q·R, 0). There are three 0-1 variables for the first trip bundle (1,a,b) and two variables for each other trip bundle. Altogether there are 11 variables -- 9 X's and D(a) and D(b).

In order to construct the deficit function bounds for constraint (ii), the arrival and departure times of each of the 9 X's need to be determined. Table 2 contains this information where the headways are equally spaced for each set of departures. This assumes an arrival rate among the passengers, as indicated in Table 1.

Table 1. The basic input data for the example

terminals (k_1, k_2)	travel time	period j	time span	no. of passengers P	desired occupancy R	Number of departures					
						q=1		q=2		q=3	
						cost	variable	cost	variable	cost	variable
(a,b)	4	1	0-4	145	65	80	X_1	15	X_2	0	X_3
		2	4-8	75	47	28	X_4	0	X_5	—	—
(b,a)	3	1	0-4	160	65	—	—	30	X_6	0	X_7
		2	4-8	70	47	23	X_8	0	X_9	—	—

Table 2. Departure and arrival times for each set of departures

Departure terminal	a									b							
Arrival terminal	b									a							
Variable	X_1	X_2		X_3			X_4	X_5		X_6		X_7				X_8	X_9
Departure time	4	2	4	1 1/3	2 2/3	4	8	6	8	2	4	1 1/3	2 2/3	4	8	6	8
Arrival time	8	6	8	5 1/3	6 2/3	8	12	10	12	5	7	4 1/3	5 2/3	7	11	9	11

The example problem can now be constructed in terms of integer programming.

$$\min \{Z = 80X_1 + 15X_2 + 28X_4 + 30X_6 + 23X_8\}$$

s.t.

$$X_1 + X_2 + X_3 = 1 \qquad\qquad (i)$$
$$X_4 + X_5 = 1$$
$$X_6 + X_7 = 1$$
$$X_8 + X_9 = 1$$

$$X_1 + 2X_2 + 3X_3 \leq D(a) \qquad\qquad (ii)$$
$$X_1 + 2X_2 + 3X_3 + X_5 - X_6 - 2X_7 \leq D(a)$$
$$X_1 + 2X_2 + 3X_3 + X_4 + 2X_5 - 2X_6 - 3X_7 \leq D(a)$$
$$2X_6 + 3X_7 \leq D(b)$$
$$-X_2 - X_3 + 2X_6 + 3X_7 + X_9 \leq D(b)$$
$$-X_1 - 2X_2 - 3X_3 + 2X_6 + 3X_7 + X_8 + 2X_9 \leq D(b)$$

$$D(a) + D(b) \leq N_0 \qquad\qquad (iii)$$

$$X_i = 0,1 \quad ; \quad i = 1,2,\ldots, 9 \qquad\qquad (iv)$$

$$D(a), D(b) \geq 0$$

The constraints in (ii) are based on the information given in
Table 2. That is, each possible combination of the net number of
departures for a given terminal is restricted not to exceed the number
of vehicles (maximal value of the deficit function) assigned to that
terminal. For example, the first constraint in (ii) refers to
$0 \leq t \leq 4$ and the second-to $0 \leq t \leq 6$ regarding the net number of
departures in terminal a. For $0 \leq t \leq 6$ in terminal a, one considers
three possible departures of X_3, two departures of X_2 and one departure
of X_1 and X_5 as opposed to two arrivals of X_7 and one arrival of X_6.

The MPSX package has been used to solve this simple example for
N_o = 7, 6, 5, 4, 3, 2. It is possible to attain the solutions by relaxing
the integrality constraint on the X's, and if necessary, round off any
fractions to the nearest integer. The results are presented in Table 3.
Note that right hand side on N_o may be used to obtain directly a fleet size
vs. minimum cost curve in a single computer run. This trade-off is shown
as the solid line in Fig. 1. In the next section, the problem is resolved
when small departure time shifts are allowed.

Table 3. Optimal results for different fleet sizes

Fleet size N_o	sets of departures in the solution, $X_i = 1$	D(a)	D(b)	min. cost Z
7	X_3, X_5, X_7, X_9	3	3	0
6	X_3, X_5, X_7, X_9	3	3	0
5	X_2, X_5, X_7, X_9	2	3	15
4	X_2, X_5, X_6, X_9	2	2	45
3	X_1, X_5, X_6, X_8	1	2	133
2	infeasible			--

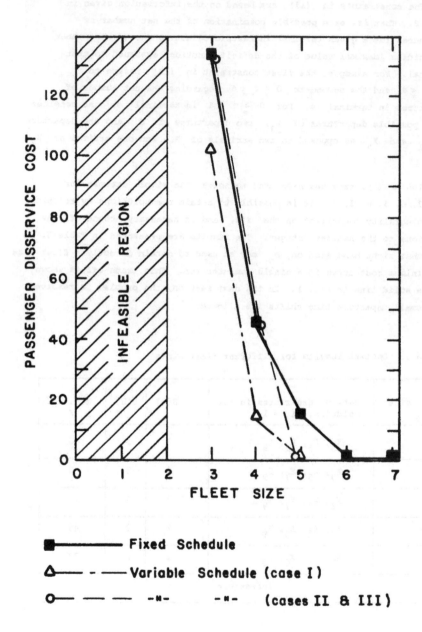

Fig. 1:   The optimal results of the example problem.

2.4. Extension: fine tuning and variable scheduling

Practical vehicle scheduling often involves shifting departure
times to better match vehicle assignment with a given set of trips.
In this case, the departure times are allowed to vary over prespecified limits, as explained in detail by Ceder and Stern (1982,
1983). The interpretation of vehicle scheduling through the deficit
function approach makes it possible to fine tune the solution
procedure.

A pre-computation analysis can be carried out regarding the
bounds on the number of departures. The steps of the analysis are
shown in a flow chart form in Fig. 2. The deficit functions are first
constructed for the minimum and maximum number of departures $(L(\cdot)$
and $U(\cdot))$ for all the trip bundles $(\cdot)$. The lower and upper bounds
on the fleet size are then determined as N_L and N_U, respectively.
Before executing the integer program, it is possible to perform two tests.
First, to check feasibility and second to check if the $U(\cdot)$ solution
is appropriate. These two tests are shown in Fig. 2. By applying this
pre-computation stage (without shifting departure times) to the example
problem, there is no need to execute the integer programming for three
fleet size values: $N_o = 7, 6, 2$ since $N_L = 3$ and $N_U = 6$.

Assume that the departure time tolerance for the example problem,
described in Tables 1 and 2, is $\Delta(\cdot) = \pm 1/3$ time units for all the
trip bundles. The graphical representation of the deficit functions
reveals that by varying individual departure times, N_U can be reduced
by one whereas N_L remains unchanged. The deficit functions for the
case $q = U(\cdot), \forall (\cdot)$ are shown in the next section in Fig. 3. The
value of N_U can be reduced from 6 to 5 through three alternative
shift procedures:

(I) decrease the first departure time of X_7 from 1 1/3 to 1.0;

(II) increase the third departure time of X_3 from 4.0 to 4 1/3;

(III) shift both departure times in (I) and (II) in opposite direction
by 1/6; i.e., from 1 1/3 to 1 1/6 and from 4.0 to 4 1/6.

Note that in all three cases, it is possible to maintain equal headways
by shifting all departures in each bundle by the same amount, but this is
not true in general.

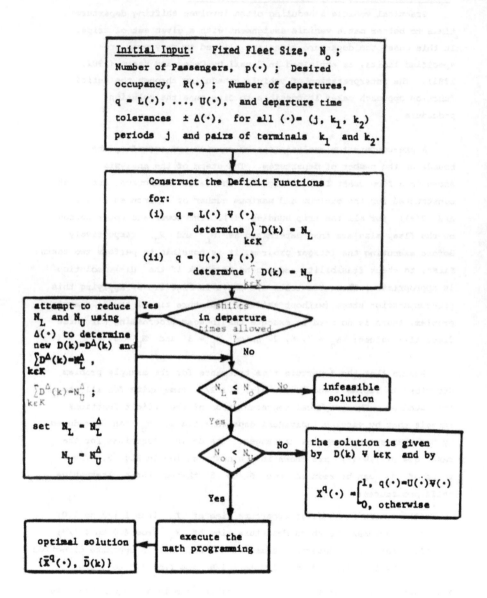

Fig. 2 : A pre-computation analysis.

The integer programming for these three possibilities is similar to that indicated in section 2.3., except the following:

For case (I) the first constraint in (ii) is changed into:

$$3 X_3 + 2 X_2 + X_1 - X_7 \leq D(a) \quad .$$

For cases (II) and (III) the first and last constraints in (ii) are changed into:

$$2 X_3 + 2 X_2 + X_1 \leq D(a) \quad ,$$

$$3 X_7 + 2 X_6 - 2 X_3 - 2 X_2 - X_1 + X_8 + 2 X_9 \leq D(b) \quad ,$$

and an additional constraint is added to (ii) :

$$3 X_3 + 2 X_2 + X_1 - X_7 \leq D(a) \quad .$$

The results of the modified problem are presented in Table 4.

Table 4. Optimal results for the variable scheduling example problem

Case	Fleet size N_o	Sets of departures in the solution, $X_i = 1$	D(a)	D(b)	min.cost Z
(I)	≥ 5	X_3, X_5, X_7, X_9	2	3	0
	4	X_2, X_5, X_7, X_9	1	3	15
	3	X_1, X_5, X_7, X_8	0	3	103
(II) and (III)	≥ 5	X_3, X_5, X_7, X_9	2	3	0
	4	X_2, X_5, X_6, X_9	2	2	45
	3	X_1, X_5, X_6, X_8	1	2	133

The pre-computation analysis shown in Fig. 1 has been applied to this modified problem. Consequently, the MPSX package has been used (while relaxing the integrality constraints) only for $N_o = 4,3$ in all the shifting cases. The optimal results exhibited in Tables 3 (without shifting) and 4 (after shifting) are compared in Fig. 1. This figure demonstrates the trade-off between the passenger disservice cost and the fleet size. Such a graphical representation can be used as an evaluation tool by the transit planner.

3. DEFICIT FUNCTION HEURISTIC PROCEDURE

The graphical representation of the deficit function allows both a visual and computerized examination of different sets of departure times in the attempt to construct an optimal timetable for a fixed vehicle fleet. It also allows the design of a pre and post computation fine tuning procedure to be implemented in an interactive manner. The conversational person-machine mode gives schedulers an opportunity to consider practical constraints such as deadheading and further shifting of departure times.

3.1. Observations based on the deficit function

The deficit function theory provides the basis for constructing a heuristic procedure to change specific sets of departure times. Two propositions can be derived for establishing heuristic rules.

Proposition 1

If the number of departures from terminal k that occur at $t \leq s_1^k$ is reduced by Q, and only the remaining departure times at $t \leq s_1^k$ are rearranged, then the required fleet at k is reduced by Q.

Proof:

Based on the fleet size formula in Eq. (1), the maximal net number of departures in terminal k across the schedule horizon, $D(k)$ is the minimum required fleet at k. Since $d(k,t)$ first approach $D(k)$ at s_1^k, the reduction of Q departures before and at s_1^k result in $d(k,s_1^k) = D(k) - Q$. It is also given that the departure times after s_1^k remain the same. Hence, the minimum fleet size at k is reduced by Q.

Proposition 2

The changes mentioned in proposition 1 lead to the reduction of
the overall multiterminal fleet size if, after the rearrangement of the
remaining departure times,

$$\sum_{u \varepsilon \{K-k\}} [D'(u) - D(u)] > 0 \quad ,$$

where $D'(u)$ and $D(u)$ are the maximal value of the deficit function
at terminal u before and after the changes, respectively.

Proof:

Consider a deleted departure from terminal k to u. If its
departure time occurs at $t \leq s_1^k$, then based on proposition 1, $D(k)$
is reduced by one. However, if the arrival time of this deleted trip
occurs at $t \leq s_1^u$, $D(u)$ is increased by one. The net reduction of the
fleet size is then zero. This argument suffices to show that for reducing
the overall fleet size, the condition $\sum_{u \varepsilon \{K-k\}} [D'(u) - D(u)] > 0$ must be
satisfied.

3.2. Heuristic algorithm

In order to construct the trade-off curve (see Fig. 1) without pre-
shifting and solving the integer program, a heuristic algorithm is offered.
This heuristic approach is designed to be used in either interactive mode
or automated mode. The usefulness of this approach can be seen from two
aspects. First, it is well known that there is inherent limitation of
integer programming for tackling large practical problems. Second, and more
important, practical scheduling tasks involve many constraints which cannot
be considered in the mathematical formulation. Usually, the scheduler who
seeks assistance prefers to receive it through an interactive computer
technique rather than to manipulate a complete optimal solution which does
not fully fulfil his practical constraints.

The heuristic algorithm is based on the deficit function properties,
among them the characteristics mentioned in propositions 1 and 2. It allows
examination of different fleet sizes, starting from the upper bound value, N_{TJ}
and reducing the fleet size to N_o. The following is a description of the
algorithm in a step-wise manner.

step 1: initialization with a timetable, in which $q(\cdot) = U(\cdot) \; \forall(\cdot); \; N=N_U$.

step 2: construct the deficit functions for all $k\epsilon K$ and find $s_1^k \; \forall \; k\epsilon K$.

step 3: determine in each terminal $k\epsilon K$ all sets of departures that terminate at $t \leqq s_1^k$; if there are no such sets - go to **step 10**.

step 4: Reduce the number of departures in each bundle by one, and compute the penalty cost, C, as defined by Eq. (3) below. This cost involves setting $q(\cdot) = q(\cdot) - \Delta q$, where $\Delta q = 1$ if not otherwise specified. Disregard sets in which $q(\cdot) = L(\cdot)$ and alert the user for cases in which $q(\cdot) - \Delta q < L(\cdot)$.

step 5: insert the determined sets into the list L arranged in increasing order of C.

step 6: examine for the first departure set in L the condition that $D(k_1)$ is decreased and $D(k_2)$ is not increasing when $q(j,k_1,k_2)$ is reduced; if the condition is not satisfied -- continue, if yes -- delete this set from L and go to **step 8**; if $L = \phi$ go to **step 10**.

step 7: delete the considered departure set from L and go to **step 6**.

step 8: set $q(j,k_1,k_2) = q(j,k_1,k_2) - \Delta q$ and update the deficit functions $d(k_1,t)$ and $d(k_2,t)$ with the new set of departure times; check if $\sum_{k\epsilon K} D(k) = N_o$; if yes - STOP; if not - continue.

step 9: determine in k_1 and k_2 the new sets of departures that terminate at $t \leqq s_1^{k_1}$ and at $t \leqq s_1^{k_2}$, respectively; go to **step 4**.

step 10: determine in each terminal $k\epsilon K$ all sets of departures, such that their q can be reduced; if there are not such sets in all $k\epsilon K$ - STOP, otherwise - continue.

step 11: list all sets of departures that terminate at $t \leqq s_1^k \; \forall \; k\epsilon K$ in increasing order of their C values and call this list $\bar{L}_1$; list the remaining sets in increasing order of their C values and call this list $\bar{L}_2$.

step 12: find for the first set in $\bar{L}_1$ all the sets in $\bar{L}_2$ such that the arrival terminal of the $\bar{L}_1$ set coincides with the departure terminal of the $\bar{L}_2$ set; if there are no such pairs, delete the $\bar{L}_1$ set from its list and start this step again; if $\bar{L}_1 = \phi$ go to **step 15**; if $\bar{L}_2 = \phi$ - STOP.

step 13: given that the $\bar{L}_1$ set is referred to (j,k_1,k_2), examine that $D(k_2)$ is not increasing when the number of departures of the pair of sets is reduced simultaneously; if $D(k_2)$ remains same - delete the $\bar{L}_1$ and $\bar{L}_2$ sets from their lists and go to **step 14**. If $D(k_2)$ increases - delete the $\bar{L}_2$ set (temporarily - to be reconsidered after the $\bar{L}_1$ set is deleted) from its list and go to **step 12**.

step 14: set the appropriate q values for the considered pair of sets and update their associated two or three deficit functions; check if $\sum_{k \in K} D(k) = N_0$; if yes - STOP; otherwise go to **step 12**.

step 15: enter the first set in $\bar{L}_2$ into $\bar{L}_1$, delete it from $\bar{L}_2$ and go to **step 12**.

In step 4 of the algorithm, the scheduler can examine different settings of sets of departure times. If the scheduler knows apriori that $N_0 \ll N_U$ then he may set $\Delta q > 1.0$ in order to expedite the process to reduce the fleet size requirement. In an interactive mode, the scheduler may determine Δq in each iteration, bearing in mind that for $\Delta q > 1.0$ the process skips some alternative solutions. The penalty cost is defined as the change in disservice between two alternative sets of departures. This expression is based on Eq. (2):

$$C = \max\left[P(\cdot) - q(\cdot)R(\cdot) + \Delta q R(\cdot),\ 0\right] - \max\left[P(\cdot) - q(\cdot)R(\cdot),\ 0\right] = \min\left\{\max\left[P(\cdot) - q(\cdot)R(\cdot) + \Delta q R(\cdot),\ 0\right],\ \Delta q R(\cdot)\right\} \quad (3)$$

Step 3 of the algorithm is based on proposition 1 and step 6 on proposition 2. After examination of all possible sets that terminate at $t \le s_1^k \ \forall\ k \in K$, the heuristic procedure continues in step 10. Then an attempt is made to examine simultaneously the reduction of q in two sets of departures. The algorithm at present does not consider simultaneous examination of more than two sets of departures.

3.3. <u>Heuristic solution for the example problem</u>

The example problem in section 2.3. is characterized by $N_U = 6$ and $N_L = 3$. The heuristic algorithm is applied to solve the problem for $N_o = 3$. In the process, the optimal solutions for $N_o = 5$ and $N_o = 4$ are obtained. Deficit functions, generated by the AUTOBUS microcomputer program, are used to enhance the understanding of the algorithm. The solution of the example problem is attained through the following steps.

<u>Start with fleet size of 6.</u>

(a) insert the initial data with $q(\cdot) = U(\cdot)$ into AUTOBUS
     to construct $d(a,t)$ and $d(b,t)$ shown in Fig. 3 where
     $D(a) = 3$, $D(b) = 3$, $N = N_U = \sum_{k=a,b} D(k) = 6$ and $s_1^a = 4$, $s_1^b = 4$.

(b) the sets of departures that terminate at $t \le s_1^a$ and
     $t \le s_1^b$ are X_3 and X_7, respectively.

(c) penalty costs of X_3 and X_7 are $C(X_3) = 15-0=15$ and
     $C(X_7) = 30-0=30$.

(d) the list L includes X_3 and thereafter X_7.

(e) examination: if $q(1,a,b)$ is reduced by $\Delta q = 1$, then
     $X_3 = 0$ and $X_2 = 1$; the value of $D(a)$ is decreased and
     $D(b) = 3$ remains same as is shown in Fig. 4; hence, the procedure
     continues with $X_2 = 1$.

(f) the set X_3 is deleted from L; the deficit functions $d(a,t)$
     and $d(b,t)$ are updated with $q(1,a,b) = 2$ and are presented in
     Fig. 4; this figure demonstrates the optimal solution for
     $N_o = \sum_{k=a,b} D(k) = 5$ in which the minimal cost is equal to 15.

<u>Now the fleet size is 5.</u>

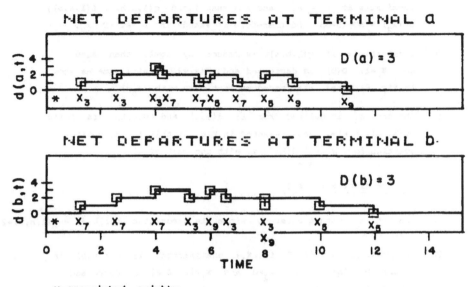

* associated variables

Fig. 3: AUTOBUS's deficit functions for N=6

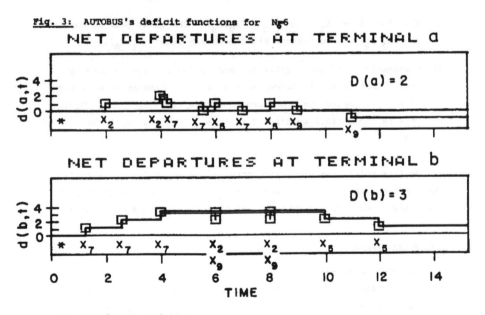

* associated variables

Fig. 4: AUTOBUS's deficit functions for N=5

(a) the new set of departures to enter L is X_2 since it
     terminates at $t \leq s_1^a$, and its associated $q(1,a,b) > L(1,a,b)$;
     $C(X_2) = 80-15=65$ and therefore it stands in L after X_7.

(b) examination: if $q(1,b,a)$ is reduced by $\Delta q=1$, then $X_7=0$
     and $X_6=1$; $D(b)$ is decreased and $D(a)=2$ remains same as shown
     in Fig. 5; hence, the procedure continues with $X_6=1$.

(c) the set X_7 is deleted from L; $d(a,t)$ and $d(b,t)$ are updated
     with $q(1,b,a)=2$ and presented in Fig. 5; this is the optimal
     solution for $N_o = \sum\limits_{k=a,b} D(k) = 4$ with the cost of 45.

Now the fleet size is 4.
--- -- ----- ---- -- -

(a) the set X_6 cannot enter L since its associated $q(1,b,a) = L(1,b,a)=2$.

(b) the remaining set in L is X_2; examination: if $q(1,a,b)$ is
     reduced by $\Delta q=1$, then $X_2=0$ and $X_1=1$; $D(a)$ is decreased
     but $D(b)$ is increased; hence, X_2 is deleted from L and $L=\phi$,
     and the procedure continues in steps 10-15 of the algorithm.

(c) the list $\bar{L}_1 = \{X_2\}$ and $\bar{L}_2 = \{X_9,X_5\}$; the first pair to examine
     is (X_2,X_9) to be changed simultaneously into (X_1,X_8).

(d) examination: if both $q(1,a,b)$ and $q(2,b,a)$ are reduced by
     $\Delta q=1$, then $D(a)$ is decreased and $D(b)=2$ remains same as shown
     in Fig. 6; hence the procedure continues with $X_1=1$ and $X_8=1$.
     This is the optimal solution for $N_o = \sum\limits_{k=a,b} D(k) = 3$ with the cost of
     133.

At this point, the algorithm stops, since $N_o = 3$. In case of
continuation, the list $\bar{L}_1=\phi$ and $\bar{L}_2=\{X_5\}$, and according to step 15
of the algorithm, X_5 enters $\bar{L}_1$ and then $\bar{L}_2=\phi$ stops the procedure.

4. CONCLUDING REMARKS

The problem addressed in this paper is that of optimal multi-
terminal timetable construction -- a problem receiving to our best
knowledge limited coverage in the literature. Four basic components are
involved in the transit planning process: (i) planning routes,

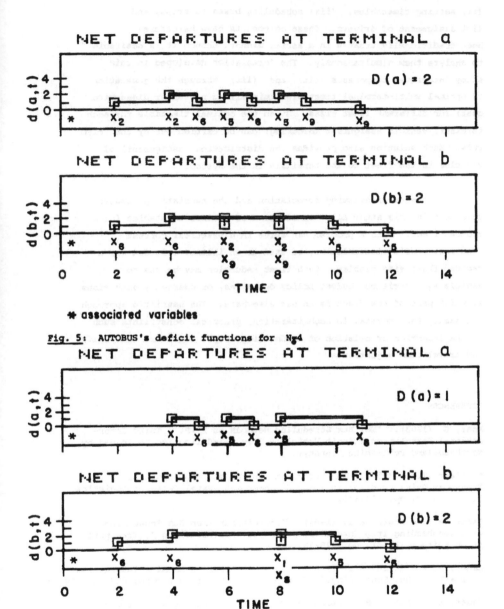

Fig. 5: AUTOBUS's deficit functions for $N_o = 4$

Fig. 6: AUTOBUS's deficit functions for $N_o = 3$

(ii) setting timetables, (iii) scheduling buses to trips, and
(iv) assignment of drivers. These components have heretfore,
been traditionally treated in a serial manner, though it is desirable
to analyze them simultaneously. The formulation developed in this
study integrates components (ii) and (iii) through the generation
of optimal multi-terminal timetables (minimizing passenger disservice
cost) for different fleet sizes. Given the optimal timetable for each
terminal, vehicle assignment (blocking) can be carried out by the FIFO
rule. Each solution also provides the distribution (assignment) of
vehicles in the fleet to the terminals in the system.

The integer programming formulation and the heuristic procedure
developed in this study allow examination of optimal timetables for a
reduced fleet. Such a problem can arise in a practical context
when last year's timetables may be used as an initial solution for a
reduced fleet size problem. Such fleet reduction may be due to
vehicle age attrition, budget policy decisions, or emergency situations
in which part of the fleet is in use elsewhere. The heuristic approach
can easily incorporate, in each iteration, practical constraints such
as the insertion or deletion of deadheading trips, departure time shifts
and travel time adjustments.

REFERENCES

Amar, G. (1983). "New Bus Scheduling Methods at RATP." Third International Workshop on Transit Vehicle & Crew Scheduling (to be issued by
North-Holland Publishing Company).

Bartlett, T.E. (1957). "An Algorithm for the Minimum Number of
Transport Units to Maintain a Fixed Schedule." Naval Research Logistics
Quarterly, 4, pp. 139-149.

Ceder, A. and Stern, H.I. (1981). "Deficit Function Bus Scheduling
with Deadheading Trip Insertions for Fleet Size Reduction." Transportation Science, 15(4), pp. 338-363.

Ceder, A. and Stern, H.I. (1982). "Graphical Person-Machine Interactive
Approach for Bus Scheduling." Transportation Research Record, 857, pp. 69-72.

Ceder, A. (1983). "Bus Frequency Determination Using Passenger Count
Data." To appear in Transportation Research (part A).

Ceder, A. and Stern, H.I. (1983). "The Variable Trip Procedure Used in the AUTOBUS Vehicle Scheduler." Third International Workshop on Transit Vehicle and Crew Scheduling (to be issued by North-Holland Publishing Company).

Gertsbach, I. and Gurevich, Y. (1977). "Constructing an Optimal Fleet for a Transportation Schedule." Transportation Science, 11, pp. 20-36.

Koutsopoulos, H.N., Odoni, A., Wilson, N.H.M. (1983). "Determination of Headways as a Function of Time Varying Characteristics on a Transit Network." Third International Workshop on Transit Vehicle and Crew Scheduling (to be issued by North-Holland Publishing Company).

Stern, H.I. and Ceder, A. (1983). "An Improved Lower Bound to the Minimum Fleet Size Problem." Transportation Science, 17, pp. 471-477.

Salzborn, F.J.M. (1972). "Optimum Bus Scheduling." Transportation Science, 6, pp. 137-148.

Salzborn, F.J.M. (1972). "A Note on the Fleet Routing Models for Transportation Systems." Transportation Science, 6,(3).

Salzborn, F.J.M. (1974). "Minimum Fleet Size Models for Transportation Systems in Transportation and Traffic Theory." D.J. Buckley (Ed.), North-Holland/Elsevier, N.Y., pp. 607-624.

Ninth International Symposium on
Transportation and Traffic Theory
© 1984 VNU Science Press, pp. 357–374

THE DYNAMIC VEHICLE ALLOCATION
PROBLEM WITH UNCERTAIN DEMANDS

WARREN B. POWELL,[1] YOSEF SHEFFI[2] and
SEBASTIEN THIRIEZ[2]

[1] *Department of Civil Engineering, Princeton University, Princeton, NJ 08544, U.S.A.*
[2] *Department of Civil Engineering, Massachusetts Institute of Technology,
Cambridge, MA 02139, U.S.A.*

ABSTRACT

A stochastic formulation of the dynamic vehicle allocation problem is
proposed. Two solution algorithms are developed, the first being a
direct extension of the deterministic formulation and the second using
the space-time structure of the network to describe the impact of decisions made now on future profits. The latter algorithm is shown to be
significantly faster. Experiments using an actual network are used to
help quantify the potential benefits from using a stochastic model
over a deterministic one.

The dynamic vehicle allocation problem arises frequently in transportation, as common carriers must determine how to use their fleet to maximize profits over time. Decisions made today must anticipate their consequences in the future. Such problems are particularly pronounced for carriers that are not bound by published time schedules, but rather have a relatively free-floating fleet of vehicles that must be managed on a day-to-day basis. Examples include truck-load (TL) motor carriers, railroads and sea container companies.

The decisions faced by these companies can be stated simply. At each point in time, each vehicle that is not in transit (moving full or empty) must be either a) assigned to an available demand, which will carry the vehicle to another point in space at some future point in time, b) moved empty or c) left where it is (a decision which can be viewed as a type of empty movement). The problem can be formulated mathematically by first discretizing the time axis and letting each point in time and space be a node. A network is then used to describe permissible movements between nodes, and costs and revenues are assigned to quantify the impacts of different decisions on a firm's profits.

The most difficult step in formulating such a model is estimating future demands for vehicles. The standard approach is to forecast these demands over a given planning horizon (say a week) and then to assume that the forecasted demands are known with certainty. In this case the problem can be easily formulated as a linear network optimization problem. To see this, assume that the vehicles are being run between a given set of cities C and that time has been discretized into days. Let (i,n) represent city $i \epsilon C$ on day n, n=1,...,N, where N is the length of the planning horizon, and define

$E_{ij,n}$ = the flow of empties going from i to j, leaving i on day n

$F_{ij,n}$ = the flow of full vehicles going from i to j, leaving on day n

T_{ij} = travel time to go from i to j, where T_{ij} = 1,2,...

c_{ij} = cost to send an empty from i to j

r_{ij} = net revenue received from sending a full vehicle from i to j

$D_{ij,n}$ = demand for vehicles going from i to j, leaving on day n

$R_{i,n}$ = supply of vehicles entering the system for the first time at city i on day n

The deterministic vehicles allocation problem can be stated as follows:

$$\text{(DVA)} \quad \max \Pi = \sum_{i \in C} \sum_{j \in C} \sum_{n=1}^{N} [r_{ij}F_{ij,n} - c_{ij}E_{ij,n}] \tag{1}$$

subject to:

$$\sum_{i \in C} [F_{ij,n-T_{ij}} + E_{ij,n-T_{ij}}] + R_{i,n} = \sum_{k \in C} [F_{jk,n} + E_{jk,n}] \quad \forall \, j,n \tag{2}$$

$$0 \leq F_{ij,n} \leq D_{ij,n} \quad \forall \, i,j,n \tag{3}$$

$$E_{ij,n} \geq 0 \tag{4}$$

DVA can be efficiently solved using a network simplex algorithm (see, for example, Glover et al., 1979).

Typically, the solution of DVA would be used only to determine what should be done on a given day. The problem would be solved again the next day as more up to date information becomes available. The limitation to this approach is that decisions are being made today assuming perfect information on future demands. Errors in these predictions can cause a company to position trucks for future demands that never materialize. This situation is exacerbated by the tendency of deterministic formulations to seek extreme solutions which in effect put all the eggs in one basket. The purpose of this research is to propose a stochastic formulation which provides a more realistic description of future demands. This is done by fitting a probability distribution around forecasted demands and then maximizing expected net revenues over the planning horizons. Such distributions can be developed by comparing actual versus forecasted demands.

In view of the importance of the problem, the literature in this area is relatively sparse. Misra (1972) formulates the empty rail car allocation problem as a Hitchcock transportation problem, thereby ignoring the time dependency. White (1972) uses an out-of-kilter algorithm to minimize costs over a time-space network, thereby introducing the time element into the problem. Demands, however, are still assumed to be known deterministically. Cooper and LeBlanc (1977) present a stochastic

version of the transportation problem by modelling the demands as random
variables, a situation that might arise in inventory problems. The
formulation, however, uses only one time period and is not easily extended to the vehicle allocation problem. The most in depth investigation to date is provided by Jordan and Turnquist (1983) who look at
empty rail car allocation over time. Their model incorporates randomness in demands and in travel times, but does not track the movements
of full cars. In addition, they assume that an empty car can be moved
only once during the planning horizon.

The purpose of this paper is to present a stochastic formulation of the
dynamic vehicle allocation problem. Two equivalent formulations are
provided which suggest alternative solution algorithms. These algorithms are compared, and numerical experiments are presented which support the use of a stochastic model.

1.0 PROBLEM DESCRIPTION

The dynamic vehicle allocation problem is best represented using a space-
time network such as that shown in Figure 1. Links between nodes represent movements in space forward in time. Supplies to a node can be
either external, representing movements begun before the beginning of
the planning horizon, or internal, representing flows on links into a
node as assigned by the model. Vehicles may move full or empty, but any
vehicle coming into a node is assumed to become empty and hence available for assignment to a new demand. Assignments made which terminate
after the end of a planning period are connected to a supersink node.

Normally, in a deterministic model, movements between two nodes would be
represented using two links, one for full moves with a positive "cost",
representing revenue, and an upper bound equal to the demand. The other
link would be for empties with a negative cost and no upper bound. For
the purposes here, however, every movement is represented using only one
link which carries all the flow. The mixture of fulls and empties is
reflected in the cost function on the link which is now a function of
the flow.

Two approaches are used to solve the stochastic vehicle allocation
problem. The first is a direct extension of DVA which fully exploits
the network structure of the underlying problem. The second approach

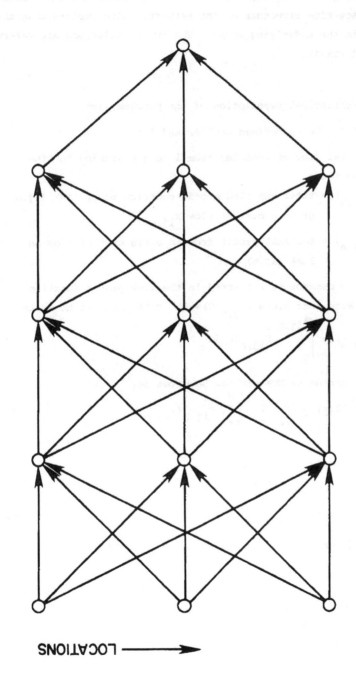

LOCATIONS

DAYS

Figure 1

Illustration of space-time network representation

does not require the use of any network flow algorithms and takes advantage of the space-time structure of the network, which implies that there are no cycles in the underlying graph. The two formulations are referred to as SVA-I and SVA-II.

Formulation I

To provide a mathematical description of the problem, let

$D_{ij,n}$, r_{ij} and c_{ij} be as defined earlier and let

$\quad x_{ij,n}$ = total flow of vehicles from i to j beginning in time
$\qquad$ period n

$\quad P_{ij,n}(x_{ij,n})$ = total expected profit contributed by link (ij,n)
$\qquad$ given a current flow $x_{ij,n}$

$\quad \hat{P}_{ij,n}(x_{ij,n})$ = marginal profit from an extra unit of flow on
$\qquad$ link (ij,n)

The randomness in demands is reflected in the link profit function $P_{ij,n}(x_{ij,n})$. Note that since $P_{ij,n}(0) = 0$, it is possible to write

$$P_{ij,n}(x_{ij,n}) = \int_0^{x_{ij,n}} \hat{P}_{ij,n}(\omega)\,d\omega \tag{5}$$

The stochastic version of DVA can now be given by:

$$(\text{SVA-I})\max \Pi(\underline{x}) = \sum_{i \in C} \sum_{j \in C} \sum_{n=1}^{N} P_{ij,n}(x_{ij,n}) \tag{6}$$

subject to

$$\sum_{i \in C} x_{ij,n-T_{ij}} + R_{j,n} - \sum_{k \in C} x_{jk,n} = 0 \quad \forall\, j,n \tag{6a}$$

$$x_{ij,n} \geq 0 \tag{6b}$$

If demands are deterministic, where $D_{ij,n} = d_{ij,n}$, then

$$\hat{P}_{ij,n}(x_{ij,n}) = \begin{cases} r_{ij} & 0 \leq x_{ij,n} \leq d_{ij,n} \\ -c_{ij} & x_{ij,n} > d_{ij,n} \end{cases} \tag{7}$$

If demands are stochastic with density function $f_{ij,n}(d)$, then

$$P_{ij,n}(x_{ij,n}) = r_{ij}x_{ij,n} - (r_{ij} + c_{ij}) \int_0^{x_{ij,n}} (x_{ij}-\omega)f_{ij,n}(\omega)\,d\omega \qquad (8)$$

The marginal revenue is easily verified to be:

$$\hat{P}_{ij,n}(x_{ij,n}) = r_{ij} - (r_{ij}+c_{ij}) \int_0^{x_{ij,n}} f_{ij,n}(\omega)\,d\omega \qquad (9)$$

The solution of SVA requires specifying the density function of demand for each link (ij,n) in the network. The simplest density function which best characterizes the possible range of demands that can occur is the Erlang distribution, given by

$$f(t) = \lambda^\alpha t^{\alpha-1} e^{-\lambda t}/(\alpha-1)! \quad t \geq 0 \qquad (10)$$

where α is a positive integer and $\lambda \geq 0$. For a given link (ij,n), the density function of demands is characterized by $\lambda_{ij,n}$ and $\alpha_{ij,n}$. The use of a continuous distribution to approximate a discrete random variable (demands) should not significantly affect the results.

The problem SVA-I can be easily solved using a standard nonlinear programming algorithm. First, it is useful to establish the global concavity of the objective function. This result is summarized in the following proposition:

Proposition 1: The optimization problem SVA-I is globally concave.

Proof: SVA-I is separable, thus we have only to provide that $P_{ij,n}(x_{ij,n})$ is concave with respect to $x_{ij,n}$. Differentiating (9) gives

$$\frac{d^2 P_{ij,n}}{d\,x_{ij,n}^2} = - (r_{ij} + c_{ij})f_{ij,n}(x_{ij,n}) \leq 0 \qquad (11)$$

Since $f_{ij,n}(x_{ij,n}) > 0$ for $x_{ij,n} > 0$, $P_{ij,n}(x_{ij,n})$ is strictly concave everywhere. □

The direct implication of proposition 1 is that $\Pi(\underline{x})$ has a unique global maximum.

Solution algorithm for SVA-I

SVA-I can be solved relatively simply using the Frank-Wolfe algorithm, which requires solving the following linearized subproblem:

$$\max_{\underline{x}} \sum_{i\epsilon C} \sum_{j\epsilon C} \sum_{n=1}^{N} \hat{P}_{ij,n}(x_{ij,n}^o) \cdot x_{ij,n} \qquad (12)$$

subject to (6a) and (6b). This is a standard linear transshipment problem which can be efficiently solved using a network simplex code.

The complete algorithm is given as follows:

Step 1: Solve DVA to obtain an initial flow vector $\underline{x}^{(1)}$.
         Set k=1.

Step 2: Solve the linear transshipment problem (12) using the current
         solution $\underline{x}^{(k)}$ to find a new set of flows $\underline{y}^{(k)}$.

Step 3: Solve the one dimensional search

$$\max_{0 \leq \gamma \leq 1} \Pi(\underline{x}^k + \gamma(\underline{y}^{(k)} - \underline{x}^{(k)})) \tag{13}$$

         and let γ^* be the optimum stepsize.

Step 4: Set $\underline{x}^{(k+1)} = \underline{x}^{(k)} + \gamma^*(\underline{y}^{(k)} - \underline{x}^{(k)})$.

         If $|\Pi(\underline{x}^{(k+1)}) - \Pi(\underline{x}^{(k)})| < \epsilon$, stop. Otherwise, set k=k+1
         and go to step 2.

This algorithm is guaranteed to converge and, in addition, provides
bounds on the objective function at every iteration (see Frank and
Wolfe (1956) and LeBlanc et al. (1975)). The rate of convergence near
the optimum, however, is known to be quite slow. An alternative procedure is the scaled reduced gradient algorithm developed by Dembo and
Klincewicz (1981), but the implementation of this approach is considerably more complex. An alternative solution algorithm is found by re-
casting the problem in a somewhat different format.

Formulation II

The second formulation first requires the following additional variables:

   $\theta_{ij,n}$ = fraction of the supply of vehicles at city i at time n to
            be sent to city j

   $s_{i,n}$ = supply of vehicles at city i at time period n

Clearly, the flow on link (ij,n) is $x_{ij,n} = \theta_{ij,n} \cdot s_{i,n}$. The optimization problem is now given as follows:

$$(\text{SVA-II}) \max \Pi(\underline{\theta}) = \sum_{i \epsilon C} \sum_{j \epsilon C} \sum_{n=1}^{N} P_{ij,n}(\theta_{ij,n} \cdot s_{i,n}) \tag{14}$$

subject to

$$\sum_j \theta_{ij,n} = 1 \quad \forall\ i,n \tag{14a}$$

$$\theta_{ij,n} \geq 0 \quad \forall\ ij,n \tag{14b}$$

where

$$s_{j,n} = \sum_{i \in C} \theta_{ij,n-T_{ij}} \cdot s_{i,n-T_{ij}} \tag{14c}$$

Note that (14c) is simply a definitional constraint and is incorporated implicitly in (14). Thus the decision variables are the vector of flow fractions ($\underline{\theta}$) and the only constraints are (14a) and (14b). The advantage of this problem is that the constraints (14a) are very simple, which simplifies the design of a solution algorithm. The disadvantage is that the implicit incorporation of (14c) in (14) implies that the objective function is no longer separable. This fact does not, however, pose any major problems in the design of a solution algorithm. This fact is best illustrated in the calculation of the derivatives

$$\hat{\Pi}_{ij,n}(\underline{\theta}) = \frac{\partial \Pi(\underline{\theta})}{\partial \theta_{ij,n}} \tag{15}$$

For SVA-I, these derivatives were found easily by taking advantage of the separability of the objective function. The calculation of (15), however, is easily accomplished by using the following result:

Theorem 1: The derivatives $\hat{\Pi}_{ij,n}(\underline{\theta})$ can be calculated using the following recursion:

$$\hat{\Pi}_{ij,n}(\underline{\theta}) = \hat{P}_{ij,n}(\theta_{ij,n} \cdot s_{i,n}) + s_{i,n} \tilde{\Pi}_{k,n+T_{ij}}(\underline{\theta}) \tag{16}$$

where:

$$\hat{P}_{ij,n}(\theta_{ij,n} s_{i,n}) = \frac{\partial P_{ij,n}(\theta_{ij,n} \cdot s_{i,n})}{\partial \theta_{ij,n}} \tag{17}$$

$$\tilde{\Pi}_{i,n}(\underline{\theta}) = \frac{\partial \Pi(\underline{\theta})}{\partial s_{i,n}} \tag{18}$$

The derivatives $\tilde{\Pi}_{i,n}(\underline{\theta})$ are calculated using the recursion:

$$\tilde{\Pi}_{i,n}(\underline{\theta}) = \sum_{j \in C} [\tilde{P}_{ij,n}(\theta_{ij,n} s_{i,n}) + \theta_{ij,n} \tilde{\Pi}_{j,n+T_{ij}}(\underline{\theta})] \tag{19}$$

where

$$\tilde{P}_{ij,n}(\theta_{ij,n} \cdot s_{i,n}) = \frac{\partial P_{ij,n}(\theta_{ij,n} \cdot s_{i,n})}{\partial s_{i,n}} \tag{20}$$

Proof: Equation (16) is a direct application of the chain rule. The first term, $\hat{P}_{ij,n}(\cdot)$ is the impact on profits on link (ij,n) due to a small increase in $\theta_{ij,n}$. The second term is the impact of an increase in $\theta_{ij,n}$ on all other links in the network. It is here that the space-time structure of the network is utilized, since $\theta_{ij,n}$ can only impact

other links at future points in time. This term can be found by looking
at the impact of $\theta_{ij,n}$ on the supply of vehicles at node $(j,n+T_{ij})$, and
multiplying this times the impact of an increase in $s_{j,n+T_{ij}}$ on all future
profits, given by $\tilde{\Pi}_{j,n+T_{ij}}(\cdot)$. This term, then, would be written

$$\frac{ds_{j,n+T_{ij}}}{d\theta_{ij,n}} \cdot \tilde{\Pi}_{j,n+T_{ij}}(\underline{\theta}) \tag{21}$$

Referring to (14c) gives the result that $ds_{j,n+T_{ij}}/d\theta_{ij,n} = s_{i,n}$. The
proof of (19) follows the identical logic. $\square$

The application of the recursions in theorem 1 require the partial deri-
vatives $\hat{P}_{ij,n}(\cdot)$ and $\tilde{P}_{ij,n}(\cdot)$. These are easily verified to be given by

$$\hat{\tilde{P}}_{ij,n}(\cdot) = \frac{dx_{ij,n}}{d\theta_{ij,n}} \cdot \frac{dP_{ij,n}(x_{ij,n})}{dx_{ij,n}}$$

$$= s_{i,n}\hat{P}_{ij,n}(\theta_{ij,n}s_{ij,n}) \tag{22}$$

$$\tilde{P}_{ij,n}(\cdot) = \theta_{ij,n}\hat{P}_{ij,n}(\theta_{ij,n} \cdot s_{i,n}) \tag{23}$$

where $\hat{P}_{ij,n}(\theta_{ij,n} \cdot s_{i,n})$ is given by (9). When the distribution of
demands are described by an Erlang, then it is easily verified that

$$\int_0^x \frac{\lambda^k t^{k-1} e^{-\lambda t}}{(k-1)!} \, dt = 1 - \sum_{n=0}^{k-1} \frac{(\lambda x)^{k-n-1} e^{-\lambda x}}{(k-n-1)!} \tag{24}$$

and

$$\int_0^x t \left(\frac{\lambda^k t^{k-1} e^{-\lambda t}}{(k-1)!}\right) \, dt = \frac{k}{\lambda}\left(1 - \sum_{n=0}^{k} \frac{(\lambda x)^{k-n} e^{-\lambda x}}{(k-n)!}\right) \tag{25}$$

Solution algorithm for SVA-II

Once the derivatives $\hat{\Pi}_{ij,n}(\underline{\theta})$ are known, it is possible to find an improved solution by again using the Frank-Wolfe algorithm, which would
involve solving the following linearized subproblem:

$$\max \sum_{i\epsilon C} \sum_{j\epsilon C} \sum_{n=1}^{N} \hat{\Pi}_{ij,n}(\underline{\theta}^\circ) \cdot \theta_{ij,n} \tag{26}$$

subject to (14a) and (14b). This problem decomposes into a series of
trivial subproblems for each node (k,n). If the vector $(\underline{\beta})$ is the solution of (26), then for each node (i,n),

$$\beta_{ij,n} = \begin{cases} 1 & \text{if } \hat{\pi}_{ij,n} = \max_{k \in C} \{\hat{\pi}_{ik,n}\} \\ 0 & \text{otherwise} \end{cases} \tag{27}$$

The overall algorithm requires procedures for calculating all the supplies $\{s_{i,n}\}$, $\forall i,n$, and the derivatives $\{\hat{\pi}_{ij,n}\}$, $\forall i,j,n$, for a given value of $\underline{\theta}$. The set $\{s_{i,n}\}$ can be found using a forward pass over the network, starting with time $n=1$ and proceeding forward in time. This procedure is summarized as follows:

FORWARD PASS:

Step 1: Set $n=1$ and initially set $s_{i,n} = R_{i,n}$, $\forall i,n$.

Step 2: For all cities $j \in C$ set $s_{j,n+T_{ij}} = s_{j,n+T_{ij}} + \theta_{ij,n} \cdot s_{i,n}$.

Step 3: If $n < N-1$, set $n = n+1$ and go to step 2.

The set of derivatives $\{\hat{\pi}_{ij,n}(\underline{\theta})\}$ can be found using a backward pass over the network, starting with time $n=N$ and proceeding backward in time. This procedure is summarized as follows:

BACKWARD PASS

Step 1: Set $n=N-1$

Set $\hat{\pi}_{i,N} = 0$, $\bar{\pi}_{i,N} = 0$ $\forall i \in C$

Step 2: For all cities $i \in C$ calculate $\hat{\pi}_{ij,n}$ using (16) and $\bar{\pi}_{i,n}$ using (19) by looping over all links emanating from node (i,n).

Step 3: If $n > 1$, set $n = n-1$ and go to step 2.

The FORWARD PASS and BACKWARD PASS are both very easy to implement and are very fast. The complete solution algorithm is given as follows:

Step 1: Solve DVA to find an initial set of link flows and set $\underline{\theta}^{(1)}$ using

$$\theta_{ij,n}^{(1)} = x_{ij,n}/s_{i,n}$$

where $x_{ij,n}$ and $s_{i,n}$ are determined using the solution of DVA.

Step 2: If $k > 1$, use FORWARD PASS to find $\{s_{i,n}\}$ (if $k=1$, these are already known).

Step 3: Use BACKWARD PASS to find $\{\hat{\pi}_{ij,n}(\theta^{(k)})\}$

Step 4: Solve linearized subproblem; for each node (i,n), determine the linearized solution $\underline{\beta}^{(k)}$ using (27).

Step 5: Find step size γ by finding

$$\max_{0 \le \gamma \le 1} \quad \Pi(\underline{\theta}^{(k)} + \gamma(\underline{\beta}^{(k)} - \underline{\theta}^{(k)}))$$

and let γ^* be the optimum stepsize.

Step 6: Set $\underline{\theta}^{(k+1)} = \underline{\theta}^{(k)} + \gamma^*(\underline{\beta}^{(k)} - \underline{\theta}^{(k)})$

If $\left| \Pi(\underline{\theta}^{(k+1)} - \Pi(\underline{\theta}^{(k)}) \right| < \epsilon$, stop. Otherwise,

set k=k+1 and go to step 2.

The simplicity of the constraint set for $\underline{\theta}$ allows the use of more flexible strategies for finding ascent directions than the simple all-or-nothing approach dictated by the Frank-Wolfe algorithm, as given by (27). For example, it is possible to split the flow among competing "good" links. For example, for each node (i,n) let $\beta_{ij,n}$ be given by

$$\beta_{ij,n} = \begin{cases} \dfrac{\exp(\xi \hat{\Pi}_{ij,n})}{\sum\limits_{k \in B} \exp(\xi \hat{\Pi}_{ik,n})} & j \in B \\ 0 & j \notin B \end{cases} \tag{28}$$

where $B = \{j \,|\, \hat{\Pi}_{ij,n} > 0\}$. If $B=\emptyset$, then equation (27) is used. The parameter ξ is an empirically determined parameter that may be increased with each iteration, producing a procedure that converges to the Frank-Wolfe algorithm.

An alternative approach to determining $\underline{\beta}$ is to use a second order algorithm. Such an approach is very difficult with SVA-I since it is difficult to find a feasible descent direction. With SVA-II, it is relatively easy to take any descent direction and then project it onto the feasible region, taking advantage of the simple structure of the constraint set.

The next section compares the algorithms SVA-I and SVA-II as well as comparisons between the solutions provided by stochastic and deterministic formulations.

3.0 NUMERICAL ALGORITHMS

The algorithms SVA-I and SVA-II were tested on the network of a medium sized truckload trucking company in the United States with national coverage. The country was divided into approximately 40 regions, and

flows were optimized over a seven day planning horizon using one day planning periods. Demands for trucks were forecasted over this period, and
possible empty movements were determined using travel times between regions.
The network required to represent this system consisted of 295 nodes and
4840 links. Demands were typically small, generally ranging between zero
and five loads per day. Variances of demands were not available at the
time of the study; instead, the coefficient of variation of demands,
given by $C_D = \sigma_D/\mu_D$, was assumed to be the same for all links and was
used as an input parameter. Before any numerical tests were run, some experiments were run to determine the best choice of step size. Two approaches
were compared. The first used a golden section search to determine the
optimal stepsize to within .01. The second began with a stepsize
$\gamma^{(0)} = .5$ and then determined $\gamma^{(k)}$ at the kth iteration, k=1,2,3,...,
using the following logic:

$$\gamma^{(k)} = \begin{cases} \gamma^{(k-1)} & \text{if } \Pi(\underline{x}^{(k)} + \gamma^{(k-1)} (\underline{y}^{(k)} - \underline{x}^{(k)})) > \Pi(\underline{x}^{(k-1)}) \\ (.75)^M \gamma^{(k-1)} & \text{otherwise} \end{cases} \qquad (29)$$

where M is the smallest positive integer such that $\Pi(\underline{x}^{(k)} + (.75)^M \cdot \gamma^{(k-1)} (\underline{y}^{(k)} - \underline{x}^{(k)})) > \Pi(x^{(k-1)})$. Equation (29) proved to be significantly faster than the golden section search for both SVA-I and SVA-II,
and hence was adopted for all the remaining experiments.

The first experiment simply compared the convergence rates of SVA-I and
SVA-II using $C_D = 1.0$. The experiments were run on an IBM 3081, and all
the software was written in FORTRAN IV and compiled using the FORTRAN H
compiler. The execution times do not include any I/O or the development
of an initial solution. The results, shown in Fig. 2, clearly demonstrate that SVA-II is significantly faster than SVA-I, despite the fact
that an efficient network simplex code was used to solve the linearized
subproblems for SVA-I. Experiments were conducted to test the value of
the logit splitting function, Eq. (28), as an alternative to the all-or-
nothing rule that results from the straight use of the Frank-Wolfe algorithm for solving the linearized subproblems. The implementation of
Eq. (28) also included doubling the scaling parameter after each iteration to ensure that the algorithm converged to the all-or-nothing solution. Initial values of ξ between .0001 and .05 were tested. The tech-

nique did not, however, significantly improve the rate of convergence over that produced by the all-or-nothing solution resulting from the Frank-Wolfe algorithm.

Using SVA-II, a second series of experiments were conducted to test the value of using a stochastic model over a deterministic one, particularly in view of the significant increase in execution times (the deterministic solution required approximately 0.5 CPU seconds). The difference between the deterministic and stochastic objective functions is best illustrated by the graphs of $P_{ij}(x_{ij})$ for different variances in the demand distribution, shown in Fig. 3. Clearly, the difference between the two functions is greatest when the flow on the link is close to the expected demand. In many instances, a deterministic solution will return flows that are exactly equal to the expected demand, suggesting that significant differences in the solutions may result.

The experiments consisted of varying C_D from .1 to 1.0. In each case, the problem was first solved deterministically. Next, the deterministic solution was evaluated using the stochastic objective function. Finally, the search algorithm was used to optimize the stochastic objective function. The results, given in Table 1, indicate that the stochastic search algorithm could improve profits over that provided by the deterministic solution by 1.5 to 49 percent, depending on the value of C_D. For networks with small flows, the coefficient of variation can easily be greater than .5, suggesting that the improvements from using a stochastic model may be substantial.

4.0 SUMMARY

The results of this research demonstrate that the stochastic vehicle allocation problem can be solved reasonably efficiently by taking advantage of the structure of the network. In addition, the numerical results based on actual data suggest that the stochastic formulation may offer significantly better results than a deterministic model. At the same time, the research raises a number of interesting questions that could lead to improvements in the current model and extensions to more general problems.

o Taking advantage of the simple structure of the constraint set, it may be possible to develop improved descent directions using a second order algorithm.

o   For networks with small flows (as occurred for the example described
    in the paper) the loss of integrality can potentially exaggerate
    the benefits of a stochastic model. It would be useful to develop
    a method for either ensuring integrality during the search process
    or restoring integrality once an optimal continuous solution is found.
    Alternatively, a piecewise linear formulation could be used which
    could be solved using a standard network simplex code. The drawback
    of such an approach is it is not easily extended to more general
    formulations.

o   The model assumes that if the supply exceeds the demand that the
    vehicles move empty. Such an assumption overstates the cost of
    uncertainty since in reality dispatchers can sometimes solicit addi-
    tional loads (which generate lower revenues) from shippers in the area.

	Coefficient of Variation			
	1.0	0.5	0.2	0.1
Optimal deterministic objective function	84505	84505	84505	84505
Stochastic objective function with deterministic solution	31032	58269	76065	81981
Optimal stochastic objective function	46253	67098	80761	83203
Percent improvement	49.0	15.2	6.2	1.5

Table 1. Comparison of solutions from deterministic
         and stochastic models

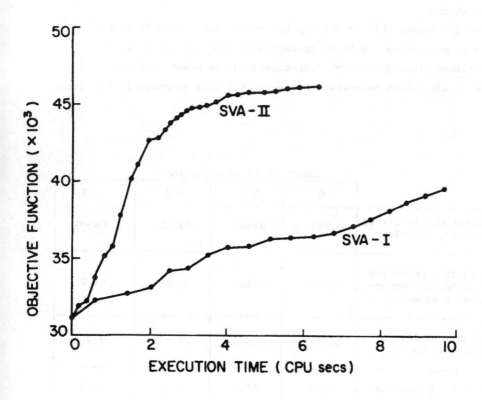

Figure 2

Comparison of rates of convergence between SVA-I and SVA-II

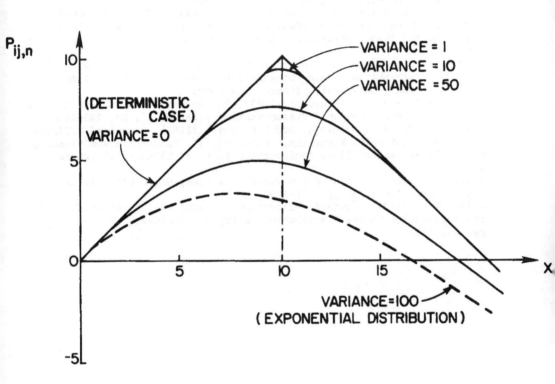

Figure 3

Expected profit on a link for alternative variances in demand

REFERENCES

Cooper, L. and L.J. LeBlanc (1977), "Stochastic Transportation Problems and Other Network-related Convex Problems," Naval Research Logistics Quarterly, 24, 2, pp. 327-337.

Dembo, R.S. and J.G. Klincewicz (1981), "A Scaled Reduced Gradient Algorithm for Network Flow Problems with Convex Separable Costs," Mathematical Programming Study, 15, pp. 125-147.

Frank, M. and P. Wolfe (1956), "An Algorithm for Quadratic Programming," Naval Research Logistics Quarterly, 3, pp. 95-110.

Glover, F., D. Karney, D. Klingman, and A. Napier (1979), "A Computational Study on Start Procedures, Basic Change Procedure and Solution Algorithm for Transportation Problems," Management Science, 20, pp. 793-819.

Jordan, W.C. (1982), "The Impact of Uncertain Demand and Supply on Empty Rail Car Distribution," Ph.D. Dissertation, Department of Civil Engineering, Cornell University, Ithaca, New York.

Jordan, W.C. and M.A. Turnquist (1983), "A Stochastic, Dynamic Model for Railroad Car Distribution," Transportation Science, 17, pp. 123-145.

LeBlanc, L.J., R.V. Helgason, and D.E. Boyce (1982), "Improved Efficiency of the Frank-Wolfe Algorithm," Publ. No. 13, Transportation Planning Group, Department of Civil Engineering, Univ. of Illinois at Urbana-Champaign.

Misra, S.C. (1972), "Linear Programming of Empty Wagon Disposition," Rail International, 3, pp. 151-158.

White, W.W. (1972), "Dynamic Transshipment Networks: An Algorithm and Its Application to the Distribution of Empty Containers," Networks, 2, pp. 211-236.

Ninth International Symposium on
Transportation and Traffic Theory
© 1984 VNU Science Press, pp. 375-396

THE PREDICTION OF INTERREGIONAL GOODS VEHICLE FLOWS: SOME NEW MODELLING CONCEPTS

HEINZ HAUTZINGER

Institute of Applied Transport and Tourism Research, Heilbronn, F.R.G.

ABSTRACT

This paper deals with the problem of forecasting interregional goods vehicle flows. In a variety of planning situations, especially in the context of transport planning at the national level, the necessity arises to predict the flows of goods vehicles on the basis of an exogeneous forecast of the interregional commodity flow pattern. The model presented in this contribution is designed to solve this type of forecasting problem.

After a brief discussion of previous work in the field of goods vehicle flow estimation the results of an empirical investigation into the nature of interregional highway truck flows are presented. In close correspondence with the empirical findings a new model of interregional goods vehicle flow forecasting is then developed. The endogeneous variables of the model are the origin-destination goods vehicle flows disaggregated by base zone of the vehicle and trip type (full or empty load). Starting from a number of explicite behavioural hypotheses these variables are shown to be functionally related to each other and to certain explanatory variables of the model. Commodity flow volumes, mean vehicle tonnage, and a factor reflecting the probability of a return trip without load are the ultimate determinants of the vehicle flow pattern. Various results of a practical application of the model are described.

1.   INTRODUCTION

Transport policy and investment decisions are normally
based on the evaluation of alternative plans or projects.
Perhaps the most important prerequisite for project evaluation is the availability of reliable forecasts of the traffic flows on the transport network under consideration.
This paper deals with the prediction of interregional goods
vehicle flows, an important but rather neglected area within the field of freight transport modelling.

Both aggregate and disaggregate models of freight transport
demand usually concentrate on the prediction of interregional commodity flow volumes by commodity type and transport
mode. In many cases origin-destination matrices of commodity flow volumes broken down by commodity type and transport
mode are obtained using really sophisticated model systems
whereas the crucial step of predicting interregional vehicle
flow volumes and network link loads is completed by applying
rather crude forecasting techniques. Therefore, it is felt
that increasing attention should be paid to the improvement
of existing methods for predicting interzonal goods vehicle
flows on the basis of observed or predicted commodity flow
matrices. The estimation of goods vehicle flow matrices
is by no means a simple or trivial problem and the accuracy
of the interregional vehicle flow estimates is perhaps the
most important determinant of the reliability of the final
network link load forecasts.

Subsequently, after a brief discussion of existing models
of goods vehicle flow prediction some empirical findings
relevant to interregional goods vehicle flow modelling will
be presented. These empirical findings form the starting
point for the development of a new model. The assumptions
of this new model are discussed and the results derived
from these assumptions are critically examined. It turns
out that the proposed forecasting methodology obviates some

detrimental inconsistencies and weaknesses of certain previous modelling approaches while at the same time remaining
a simple and analytically tractable concept.

2.   A BRIEF REVIEW OF PREVIOUS WORK ON GOODS VEHICLE
      FLOW PREDICTION

2.1 Some general remarks

It has already been pointed out that relatively little work
on goods vehicle flow prediction has been done in the past.
In some cases freight transport models have been developed
which do not consider at all vehicle movements from origin
to destination zones. For instance, the model of Hutchinson et al. (1975) produces as its final output the commodity volume in tons per annum on each link of each of the
modal transport networks.

The rather few models of interregional goods vehicle flows
reported in the literature can broadly be classified into
two different categories:

(i)   Direct approaches

(ii) Commodity flow based approaches

Direct approaches relate the volume of interregional goods
vehicle movements directly to certain explanatory variables
such as population size or certain measures of economic
activity. In contrast to this, commodity flow based approaches first predict origin-destination matrices of
commodity flows by commodity group and transport mode; in
a second step the resulting interregional vehicle flow
pattern is estimated for each mode of transport. As an
example of such a comprehensive approach see Kessel et al.

(1982) where origin-destination matrices of vehicle flows
are estimated for the highway, rail, and waterway mode.
Subsequently, the discussion is restricted to methods suited
to the analysis and forecasting of truck and trailer movements on the highway network.

2.2   Direct approaches to predict interregional goods vehicle flows

Direct approaches to predict interregional goods vehicle
flows may be suited in situations where only relatively
little amount of money and time is available for the study.
Also in cases of restricted availability of empirical data
such a direct approach may be useful. An approach of this
type which is both simple and theoretically sound has recently been proposed by Hutchinson and Wilkinson (1982).
The forecasting technique is based on the theory of log-
linear statistical models and consists of the following
sequence of activities:

(i)   A survey of truck origin-destination movements
        stratified by major commodity group;

(ii)   Estimation of a log-linear model for each origin-
        destination matrix in order to establish the in-
        teraction structure for each commodity group;

(iii)   Establish the relationship of truck trip genera-
        tion and attraction magnitudes for each commod-
        ity group to the intermediate and final demand
        characteristics of each spatial unit;

(iv)   Estimate the probable interaction structure for
        the forecast year by modifying the base year in-
        teraction matrix to account for potential
        changes in the economic and spatial structure;

(v)   Estimate the truck trip generation and attrac-
        tion magnitudes for each commodity group for
        each spatial unit in the forecast year using the
        relationships established in (iii) and fore-
        casts of economic activity;

(vi)   Adjust the interaction matrix established in
        (iv) to the new generation and attraction mag-
        nitudes estimated in (v) using bi-proportional
        balancing procedures, and

(vii) Assign the estimated origin-destination magnitudes to the highway network using an all-or-
nothing assignment procedure.

Although the approach described above has not actually been
applied in a major study it can be supposed that the method
will give useful results. A growth factor model of commercial vehicle flow forecasting which also represents a direct
approach can be found in DEPARTMENT OF TRANSPORT (1979) and
a refinement of this model has been described by EASTMAN (1981)

2.3 Commodity flow based approaches to predict interregional goods vehicle flows

If commodity flow matrices are available for the forecast
year, these matrices can serve as the main determining factors in a commodity flow based model to predict the inter-
regional goods vehicle flows. Let m_{ij} denote the commodity flow volume for a certain commodity group from zone i
to j measured in tons per year and let z_{ij} represent
the goods vehicle flow from i to j measured in vehicles
per year $(i,j=1,\dots,n)$.

In a number of practical freight transport planning studies
a very simple approach to predict z_{ij} has been applied.
This approach uses the relationship

$$z_{ij} = \frac{1}{\bar{m}} m_{ij} \qquad \text{('naive proportionality model')} \qquad (1)$$

where $\bar{m}$ is the mean transport volume per vehicle trip.
Naturally, $\bar{m}$ has to be determined empirically in advance
in a way that both trips with and without load are included.
Sometimes $\bar{m}$ is determined seperately for different distance
categories and/or seperately for for-hire and private truck
trips. In many cases, however, this approach will lead to
completely unreasonable results which is mainly due to the
fact that no explicite distinction is made between trips
with and without load (most trips without load are return
trips). If, for instance, m_{ij} increases, the vehicle flow

z_{ij} also increases according to (1) while the flow in the opposite direction, i.e. the quantity z_{ji} remains unchanged. This is not plausible since an increase in m_{ij} will lead to an increase in z_{ji} stemming from an increased number of return trips without load from j to i .

A model which is considerably more realistic has been described by Noortman and van Es (1978). For every pair (i,j) of zones the following model has been proposed

$$
\begin{aligned}
z_{ij} &= x_{ij} + y_{ij} \\
x_{ij} &= (1/a_{ij})m_{ij} \\
y_{ij} &= f(x_{ji}, d_{ij}) \\
a_{ij} &= g(d_{ij}) \qquad (a_{ij} = a_{ji})
\end{aligned}
\qquad (2)
$$

where m_{ij} and z_{ij} are defined as before and the other symbols have the following meaning

x_{ij} number of vehicle trips with load from i to j

y_{ij} number of vehicle trips without load from i to j

a_{ij} mean transport volume (tons per trip) of the vehicle trips with load from i to j

d_{ij} distance from i to j .

As can be seen, both directions of the relationship (i,j) are considered simultaneously and a distinction is made between vehicle trips with and without load. The crucial point is, of course, the appropriate specification of the functional form of $f(\cdot)$. The Noortman/van Es paper does not contain the function f actually used. In a working paper of Kessel et al. (1981) a model of the type described above has been investigated using the relationship

$$
y_{ij} = px_{ji} \quad \text{and} \quad y_{ji} = px_{ij} \quad (0<p<1) \qquad (3)
$$

where p is a constant - possibly dependent on distance class - which has to be determined empirically. In this case one obtains

$$z_{ij} = \frac{1}{a_{ij}}\left(m_{ij} + p m_{ji}\right) \tag{4}$$

From (4) we conclude that the total number of vehicle trips from i to j increases as m_{ij} and m_{ji} increases, respectively. In addition, the magnitude of the goods vehicle flow z_{ij} also depends on the mean transport volume per loaded trip (tons per trip) and on the factor p reflecting the probability of a (return) trip without load. Whereas these properties of the model are reasonable there is another property which seriously calls in doubt the usefulness of the model (2), (3) . We have

$$z_{ij} - z_{ji} = \frac{1-p}{a_{ij}}\left(m_{ij} - m_{ji}\right) \tag{5}$$

which means that the difference between the vehicle flows is directly proportional to the difference between the commodity flows. This result is, however, not consistent with empirical data which show that even in cases of extreme differences between m_{ij} and m_{ji} the vehicle flows z_{ij} and z_{ji} tend to be approximately equal.

There is another model due to Southworth (1982) which has to be mentioned in the context of commodity flow based approaches. Within the framework of an urban goods movement model logistic choice models form the theoretical basis for estimating both the daily interzonal commodity flows and the daily direct interzonal movements. Both spatial interaction models are calibrated using a disaggregated data base made up of a random sample of individual daily truck circuit activities. The approach of Southworth deserves attention particularly because the whole urban goods movement model contains a truck circuits sub-model in addition to the interzonal truck trips sub-model already mentioned.

The new model presented subsequently was developed by the author in the context of a research project financed by the German Ministry of Transport. See Kessel et al. (1982).

3.   A NEW APPROACH TO INTERZONAL GOODS VEHICLE
     FLOW FORECASTING

3.1 Some empirical evidence

In the Federal Republic of Germany currently about 10
million tons of goods are transported by truck each workday of
the week. The number of truck trips per day is approximately equal to 2 millions and about 30 percent of these trips
are made without load. It appears that the mean tonnage
per vehicle trip (trips with load) is equal to 65 percent
of the mean capacity (in tons) per trip. Trips originating
from the truck's base zone are to a very high percentage
made with load. In contrast to this, the probability of
a return trip without load is relatively high, especially,
if it is a return trip from a location close to the base
terminal of the truck. At an aggregate level it appears
that the proportion of interzonal vehicle trips made without load is high if there is a significant difference
between the commodity flows in both directions. Given a
division of the Federal Republic of Germany into 79 zones
(planning regions) it seems that more than 80 percent of
the total number of vehicle trips between two zones are
made by vehicles the base terminal of which is located in
either of the two zones considered.

Finally, it appears that the total number of vehicle trips
from zone i to j is approximately equal to the total
number of trips in the opposite direction even in the case
of a substantial disequilibrium of the zone-to-zone commodity flows. For instance, although 5.9 million tons have
been transported from the remainder of the FRG to Berlin
(West) in 1979 but only 3.2 million tons from Berlin (West)
to the remainder of the FRG, the total number of vehicle
trips is approximately equal (420,000 and 424,000 trips
per year, respectively). A more detailed analysis can be
found in Kessel et al. (1982).

3.2 Formulation of the basic model

Let us consider a pair (i,j) of zones or regions and let us introduce the following notation:

$x_{ij}(k)$ number of vehicle trips with load from i to j made by vehicles based in k $(k=i,j)$

$y_{ij}(k)$ number of vehicle trips without load from i to j made by vehicles based in k $(k=i,j)$

Consequently,

$$z_{ij} = x_{ij}(i) + x_{ij}(j) + y_{ij}(i) + y_{ij}(j) \qquad (6)$$

and

$$z_{ji} = x_{ji}(j) + x_{ji}(i) + y_{ji}(j) + y_{ji}(i) \qquad (7)$$

denotes the total goods vehicle flow from i to j and from j to i, respectively. Implicitely, it is assumed that only vehicles based in i and j have to be considered which is, of course, a simplifying assumption. From (6) and (7) we see that eight variables have to be determined if the vehicle flows z_{ij} and z_{ji} are to be estimated.

If we assume that all trips out of zone i made by vehicles based in i are trips with load and if we make a corresponding assumption for zone j we have

$$y_{ij}(i) = y_{ji}(j) = 0 \qquad (8)$$

and the number of variables reduces to six for each pair of zones. For the remaining six variables we formulate the following set of behavioural hypotheses:

(A)   The total interzonal freight transport demand is
      satisfied, i.e.

$$\left.\begin{aligned}
x_{ij}(i) + x_{ij}(j) &= \frac{1}{a}\, m_{ij} \\[2mm]
x_{ji}(j) + x_{ji}(i) &= \frac{1}{a}\, m_{ji}
\end{aligned}\right\} \tag{9}$$

where $a > 0$ denotes the mean tonnage of trips
with load.

(B)   The number of **trips** out of the base zone is equal
      to the number of return trips, i.e.

$$\left.\begin{aligned}
x_{ij}(i) &= x_{ji}(i) + y_{ji}(i) \\[2mm]
x_{ji}(j) &= x_{ij}(j) + y_{ij}(j)
\end{aligned}\right\} \tag{10}$$

(C)   The number of return trips without load is propor-
      tional to the number of trips leaving the base zone:

$$\left.\begin{aligned}
y_{ji}(i) &= p_i\, x_{ij}(i) \\[2mm]
y_{ij}(j) &= p_j\, x_{ji}(j)
\end{aligned}\right\} \tag{11}$$

where p_i and p_j are elements of the unit interval
$[0,1]$.

(D)   The six endogeneous variables of the model, i.e. the
      quantities $x_{ij}(i)$, $x_{ji}(i)$, $y_{ji}(i)$, $x_{ji}(j)$, $x_{ij}(j)$,
      and $y_{ij}(j)$ satisfy a nonnegativity condition.

From these assumptions it is clear that the six vehicle
flow variables depend on the exogeneous factors m_{ij}, m_{ji},
p_i, p_j, and a . However, given m_{ij}, m_{ji}, and a , the
proportionality factors p_i and p_j which can be interpreted as the probabilities of return trips without load
for vehicles based in zone i and j , respectively,
have to be chosen such that the nonnegativity condition (D)
is satisfied.   The admissible range of p_i and p_j can
easily be determined as follows:

Ratio of commodity flow volumes	Admissible ranges for the probabilities of return trips without load	
$m_{ij}/m_{ji} = 1$	$0 \leq p_i \leq 1$	$0 \leq p_j \leq 1$
$m_{ij}/m_{ji} < 1$	$0 \leq p_i \leq 1$	$\left(1 - \dfrac{m_{ij}}{m_{ji}}\right) \leq p_j \leq 1$
$m_{ij}/m_{ji} > 1$	$\left(1 - \dfrac{m_{ji}}{m_{ij}}\right) \leq p_i \leq 1$	$0 \leq p_j \leq 1$

Provided that admissible values of p_i and p_j have been chosen, the six unknown vehicle flow variables can be determined by solving the system of equations (9), (10), (11). Inserting these solutions into equations (6) and (7) yields the desired estimates of the goods vehicle flows between the two zones under consideration.

The model estimates of the interzonal vehicle flow volumes are easily shown to be given by

$$z_{ij} = \frac{p_i m_{ij} + p_j m_{ji}}{a\{1 - (1-p_i)(1-p_j)\}} \tag{12}$$

From (12) it is obvious that our model has the desired property

$$z_{ij} = z_{ji} \quad \text{irrespective of } m_{ij} \text{ and } m_{ji}. \tag{13}$$

Moreover, we have the following properties:

(a) The sum of all vehicle trips originating from i equals the sum of trips having i as the destination.

(b) The total vehicle flow z_{ij} increases as the commodity flows m_{ij} and m_{ji} increase, respectively.

(c) The total vehicle flow z_{ij} decreases as the mean tonnage of trips with load, i.e. the quantity a, increases.

(d) If the probability of a return trip without load de-

creases, the expected number of vehicle trips also decreases and vice versa. More specifically, if p_i and
p_j are changed to $p_i^* = \alpha p_i$ and $p_j^* = \alpha p_j$, respectively, the vehicle flow changes from z_{ij} to z_{ij}^* where

$$z_{ij}^* = \frac{p_i m_{ij} + p_j m_{ji}}{\alpha\{1 - (1-p_i)(1-p_j) + (1-\alpha)p_i p_j\}} \qquad (\alpha > 0)$$

Consequently, if $0 < \alpha < 1$ we have $z_{ij}^* < z_{ij}$ and we
have $z_{ij}^* > z_{ij}$ for $\alpha > 1$.

Summarizing it can be said that the model is capable to
simulate the effects of changes in the commodity flow
pattern and the aggregate effect of improved vehicle scheduling and more efficient cooperation between shippers,
vehicle operators, and receivers.

In view of empirical results it is reasonable to assume
that the factors p_i and p_j are not simply constants but
rather depend on the degree of noncorrespondence between
the interzonal commodity flows m_{ij} and m_{ji} . Moreover,
empirical findings suggest that p_i and p_j decrease as
the distance between i and j increases. In the next
section it will be demonstrated how the model (12) can be
generalized to capture these effects.

3.3 A model of return trips without load

In the preceeding section it has been shown that the range
of feasible values of p_i (and p_j) depends on the ratio
m_{ij}/m_{ji} of the two commodity flows. Only in the symmetric
case $m_{ij}/m_{ji} = 1$ and in the case where m_{ji} exceeds m_{ij}
$(m_{ij}/m_{ji} < 1)$ the factor p_i can vary from 0 to 1. If,
however, m_{ij} exceeds m_{ji} $(m_{ij}/m_{ji} > 1)$ at least
$(1 - m_{ji}/m_{ij})\cdot100$ percent of the return trips of vehicles
based in zone i have to be made without load. Empirical
evidence, in addition, suggests that even if the commodity
flows are equal $(m_{ij}/m_{ji} = 1)$ return trips without load

are of considerable importance and cannot be neglected.
Thus, we need a model relating the probability p_i of return trips to zone i without load to the ratio m_{ij}/m_{ji}
and - in view of the remarks at the end of Section 3.2 -
to the distance between the two zones under consideration.

Let d_{ij} denote the distance from i to j . Then, it is
reasonable to assume

$$p_i = f(m_{ji}/m_{ij}, d_{ij}) \tag{14}$$

where f is a monotonically decreasing function of the
commodity flow ratio m_{ji}/m_{ij} and also decreasing in d_{ij} .
Naturally, the function f has to be specified such that
the condition

$$1 - m_{ji}/m_{ij} \le p_i \le 1 \quad \text{whenever} \quad m_{ji}/m_{ij} < 1$$

is satisfied for every value of distance d_{ij} .

A useful model specification is the following:

$$p_i = \exp\{-\lambda(m_{ji}/m_{ij})^2\} \tag{15}$$

where $\lambda > 0$ is a parameter depending on distance class.
See Figure 1 where the function (15) is depicted for
$\lambda = 1.4$. It appears that return trips to zone i without
load become less likely as the ratio m_{ji}/m_{ij} increases
which is consistent with empirical observations. If
$m_{ij} = m_{ji}$ we have $p_i = \exp(-\lambda) > 0$ i.e. the model can
reflect the consequences of limited information and non-
optimal vehicle scheduling as well as effects of seasonal
disequilibria which are the main causes of return trips
without load in the case of symmetric commodity flows.
If m_{ji} becomes small as compared to m_{ij} the probability
of return trips without load to i increases which is also
a reasonable property of the model (15).

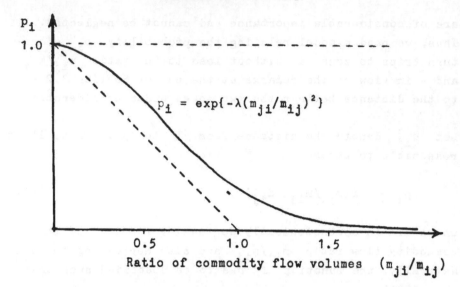

Figure 1: Probability of return trips to zone i without load as a function of the ratio m_{ji}/m_{ij}

From Figure 1 it becomes clear that p_i must satisfy the relation

$$\max\left\{0 \, , \, 1 - \frac{m_{ji}}{m_{ij}}\right\} \; \leq \; p_i \; \leq \; 1 \tag{16}$$

From the definiton of p_i it is clear that $0 \leq p_i \leq 1$ whenever $\lambda \geq 0$. On the other hand it can be shown that for arbitrary values of the ratio m_{ji}/m_{ij} the inequality $p_i \geq 1 - m_{ji}/m_{ij}$ is satisfied if $\lambda \leq \lambda_0$ where $\lambda_0 \approx 2.46$. Thus, the necessary condition (16) is always satisfied provided that $0 \leq \lambda \leq \lambda_0$.

Empirically relevant values of λ proved to lie in the interval $[0.1 , 2.3]$ indicating that the specification (15) is useful and does not cause any consistency problems. In the case of equal commodity flows in both directions $(m_{ij} = m_{ji})$ a value of $\lambda = 0.1$ yields $p_i = 0.9$ and a value of $\lambda = 2.3$ leads to $p_i = 0.1$. Consequently, all

empirically relevant situations can be considered within
the model (15).

3.4 A numerical example

In order to demonstrate the properties of the model we consider the following example. Let m_{ij} = 1,000 (tons/year)
and m_{ji} = 1,500 (tons/year). Moreover, let a =10 (tons
per vehicle trip with load) and let us assume that $\lambda = 1.5$.
Since m_{ji}/m_{ij} = 1.50 and m_{ij}/m_{ji} = 0.67 we obtain

$$p_i = \exp\{-1.5(1.50)^2\} = 0.034 \quad (= 3.4\%)$$

$$p_j = \exp\{-1.5(0.67)^2\} = 0.513 \quad (=51.3\%)$$

As can be seen, the factors p_i and p_j differ considerably due to the inequality of the interzonal commodity
flows.

The various vehicle flow variables of interest, i.e. the
solutions of the simultaneous equations (9) to (11) can be
determined recursively as follows:

$$x_{ji}(i) = \frac{(p_i-1)\{(p_j-1)m_{ji} + m_{ij}\}}{a\{(p_i-1)(p_j-1) - 1\}} = 49.2$$

$$x_{ji}(j) = (1/a)m_{ji} - x_{ji}(i) = 100.8$$

$$x_{ij}(j) = (1 - p_j)x_{ji}(j) = 49.1$$

$$x_{ij}(i) = (1/a)m_{ij} - x_{ij}(j) = 50.9$$

$$y_{ji}(i) = p_i x_{ij}(i) = 1.7$$

$$y_{ij}(j) = p_j x_{ji}(j) = 51.7$$

As a result, we have the total vehicle flow

$$z_{ij} = x_{ij}(i) + x_{ij}(j) + y_{ij}(j)$$

$$= 50.9 + 49.1 + 51.7$$

$$= 151.7 \quad \text{(vehicle trips per year)}$$

Similarly, we can determine z_{ji} which is, of course, equal to z_{ij}:

$$z_{ji} = z_{ij} = 151.7 \quad \text{(vehicle trips per year)}$$

Naturally, we would have obtained the same values of z_{ij} and z_{ji}, respectively, by direct insertion into equation (12). See Figure 2.

Obviously, since the mean tonnage is $a = 10$ (tons per trip with load) and $m_{ij} = 1,000$ (tons per year) have to be transported from zone i to zone j the total number $x_{ij} = x_{ij}(i) + x_{ij}(j)$ of trips with load from i to j is $x_{ij} = 100$. Similarly, we have $x_{ji} = 100.8 + 49.2 = 150$.

The overall proportion of vehicle trips without load is

$$q_{ij} = y_{ij}(j)/z_{ij} = 0.341 \quad (=34.1\%)$$

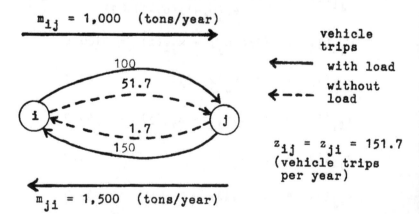

$m_{ij} = 1,000$ (tons/year)

vehicle trips

with load

without load

100
51.7

i

j

1.7
150

$z_{ij} = z_{ji} = 151.7$
(vehicle trips per year)

$m_{ji} = 1,500$ (tons/year)

<u>Figure 2:</u> Numerical example

and

$$q_{ji} = y_{ji}(j)/z_{ji} = 0.011 \quad (= 1.1\%)$$

respectively. If we consider both directions, we find that
$51.7 + 1.7 = 53.4$ vehicle trips are made without load corresponding to a proportion of $53.4/(151.7 + 151.7) = 0.176$
or 17.6% idle vehicle trips. As can be seen, the 'productivity' of road transport may be quite different for the
different directions of an origin-destination relationship.

3.4 Estimation of model parameters

We can write the complete model (12), (15) as follows:

$$z_{ij} = \frac{p_i(\lambda)m_{ij} + p_j(\lambda)m_{ji}}{a[1 - \{1 - p_i(\lambda)\}\{1 - p_j(\lambda)\}]} \qquad (17)$$

where

$$p_i(\lambda) = \exp\{-\lambda(m_{ji}/m_{ij})^2\}$$

$$p_j(\lambda) = \exp\{-\lambda(m_{ij}/m_{ji})^2\} . \qquad (18)$$

Using this representation it evident that our model has two
explanatory variables m_{ij} and m_{ji} whereas the quantities
a and λ are parameters which have to be determined empirically. The method of parameter estimation depends on
the type and quality of available data.

If, for instance, estimates of the commodity flows m_{ij}, m_{ji}
and counts or estimates of the vehicle flows z_{ij}^0, z_{ji}^0 are
available for a certain base year and if, in addition, the

mean tonnage of trips with load, i.e. the quantity a , is
known from some appropriate source, then the parameter λ
can be estimated using the method of least squares. In this
case it would be useful to group the origin-destination-
pairs (i,j) , which constitute the observational units,
according to distance and estimate the parameter λ seperately for the different distance classes.

In view of data limitations it can be necessary to develop
other - perhaps more heuristic or ad hoc - techniques of
parameter estimation. From a practical point of view this
needs not to be a serious drawback. For instance, in the
study of Kessel et al. (1982) the following categories of
information were available for parameter estimation purposes:

- mean values of the proportion of actually used vehicle
  capacity (tonnage of all trips as a percentage of the
  total capacity) for different vehicle types

- maximum capacity of different vehicle types

- proportion of trips without load averaged over all
  origin-destination pairs

Based on this information the mean tonnage of trips with
load, i.e. the quantity a , and the parameter λ have
been determined iteratively such that these two values were
consistent with the input information mentioned above.

4.   PRACTICAL APPLICATION OF THE MODEL

The new modelling approach described in Section 3 has been
tested empirically in several different applications. These
are described in detail in Kessel et al. (1982). Subsequently, some of the results are presented briefly. All empirical
results refer to the Federal Republic of Germany.

For the mean tonnage of vehicle trips with load the following estimates have been obtained.

Commodity group	Mean tonnage (tons/trip)				
For-hire trucks	under 100 km	100-200 km	200-300 km	300-500 km	over 500 km
Agricultural products	5.2	9.8	11.9	15.2	20.6
Coal	10.7	14.1	15.4	16.7	21.0
Iron, steel	3.8	7.6	10.0	12.7	17.8
Chemical products	4.2	8.5	10.7	13.3	19.5
Private trucks					
Agricultural products	2.1	4.9	6.4	8.8	13.5
Coal	8.4	12.0	13.5	15.0	20.5
Iron, steel	1.4	3.6	5.2	7.2	11.7
Chemical products	1.6	4.1	5.6	7.5	12.7

It appears that there are significant differences between
different distance classes and between for-hire and private
trucks stressing the necessity of an appropriate segmentation. The estimates of λ are given below.

Distance class	Estimate of λ		Overall proportion (%) of trips without load	
	for-hire	private	for-hire	private
under 100 km	1.4	0.1	19.0	43.8
100 - 200 km	2.1	0.2	20.8	39.7
200 - 300 km	2.0	0.3	21.1	37.4
300 - 500 km	2.1	0.4	21.8	37.6
over 500 km	2.3	0.5	20.4	40.2
Total	-	-	20.7	41.3

The parameter estimates indicate that return trips without
load become less likely as distance increases. The proportion of trips without load, however, also depends on the
degree of noncorrespondence between the commodity flows in
both directions of an origin-destination relationship.
Therefore, the overall proportion of trips without load
does not necessarily decrease monotonically as distance increases. In the Federal Republic the proportion of trips
of this kind is high. This is especially true for private
trucks (trucks owned by manufacturers) because of certain
policy regulations. Consequently, an explicite distinction
between trips with and without load is to be recommended
for models of interzonal goods vehicle flows.

The goods vehicle flow prediction model has been used to
simulate the likely consequences of various alternative
scenarios of the future economic situation as far as interzonal truck movements are concerned. The results obtained
can serve as an input into a traffic assignment model for
long distance goods transport.

5.   FINAL REMARKS

The model described in this paper is a contribution to the
rather neglected field of modelling interregional commercial
vehicle trips in a competitive environment. Some of the
model's most important features are that an explicite distinction is made between full and empty runnings and that
the empirically most relevant case of interregional imbal-
ances in commodity flows is treated. Practical application
of the model requires exogeneous forecasts of commodity flow
matrices to be available. This is, of course, not always
the case but at least goods transport planning at the national level is often based on this type of commodity flow data

(e.g. in the Federal Republic of Germany).

Although both empirical and logical support of the model has
already been given more detailed tests of its forecasting
capacity in the context of really comprehensive transport
studies are required. Similarly, the model needs to be generalized in various directions. For instance, in its present
form the model treats the different origin-destination pairs
of zones seperately; it would be a natural extension of the
model to allow for more complex trip patterns instead of
being restricted to simple base-destination-base cycles.
It should also be considered which modifications of the model are required if it is to be applied in other fields such
as international goods transport or in the case of other
transport modes operating in intramodal competition (e.g.
waterway transport). The basic concept of the model seems
flexible enough for such generalizations and modifications.

Finally, the author wishes to thank three anonymous referees
for their helpful critical comments.

6.   REFERENCES

Department of Transport (1979). Regional Highway Traffic
Model, London.

Eastman, C.R. (1981). The Importance of Trip Length Changes
in Commercial Vehicle Forecasting. Paper presented at the
PTRC Summer Annual Meeting, Warwick.

Hutchinson, B.G., O'Brien, W.B., Dawson, I.N. (1975).
A National Commodity Flow Model, Canadian Journal of Civil
Engineering 2, pp. 292-304.

Hutchinson, B.G., Wilkinson, K. (1982). Techniques for
Estimating Highway Truck Volumes. Research Report, University of Waterloo, Canada.

Kessel, P., Konanz, W., Hautzinger, H., Gresser, K.,
Tritschler, P. (1981). Güterfernverkehrsmodell (Long Distance
Goods Transport Model). Interim Report, research project
no. 98 037/80 of the Ministry of Transport, Bonn, W-Germany.

Kessel, P., Konanz, W., Hautzinger, H., Gresser, K. (1982).
Güterfernverkehrsmodell (4 volumes). Final Report, research
project no. 98 037/80 of the Ministry of Transport, Bonn,
W-Germany.

Noortman, H.J., van Es, J. (1978). Traffic Model. Manuscript
on the Dutch freight transport model.

Southworth, F. (1982). An Urban Goods Movement Model: Framework and some Results. Papers of the Regional Science Association, Vol. 50, pp. 165-184.

Ninth International Symposium on
Transportation and Traffic Theory
© 1984 VNU Science Press, pp. 397–411

ESTIMATING TIME-DEPENDENT TRIP MATRICES FROM TRAFFIC COUNTS

L. G. WILLUMSEN

Transport Studies Group, University College London, London, U.K.

Abstract

The analysis of dynamic congestion problems in urban areas often requires
the use of trip matrices estimated for different consecutive time periods,
typically 15-20 minutes. The paper describes an extension of an entropy
maximising model to estimate trip matrices from traffic counts to cover
this requirement. The new model was implemented and tested in combination
with CONTRAM, a traffic management simulation model developed at the
Transport and Road Research Laboratory. The results from these tests were
very encouraging as the matrices obtained were more accurate than those
resulting from using simpler approaches. A further extension to the basic
model to incorporate variable accuracy of traffic counts is also discussed.

1.   INTRODUCTION

There is considerable interest in the development of low cost techniques
for the estimation of trip matrices. One of the most promising approaches
seeks to estimate trip matrices from traffic counts and a prior trip
matrix, if available. Most methods only try to estimate a single trip
matrix from a set of counts and they often assume that congestion does
not play an important role in determining route choice. For a number of
applications these restrictions are of minor importance.

There are, however, cases in which congestion is likely to play a major
role in route choice. Nguyen (1981), Fisk and Boyce (1982) and Willumsen
(1982), among others, have investigated this problem. Furthermore, it is
often important in congested urban areas to study the variations of demand
over time to identify temporarily overloaded links, changes in delay and
route choice, and to design traffic management measures to alleviate these
problems.

One way of considering the variation of trip matrices over time is to
estimate different trip matrices for different periods of the day, or
'time slices'. A trip is allocated to a time slice if it enters the study
area during that particular time period. It is interesting to explore
then whether traffic counts, which are simple to obtain for almost any
size of time slice, can be used to estimate time-dependent trip matrices.

This paper is organised as follows. Section 2 presents the basic entropy
maximising model (ME2) proposed by the author (Willumsen, 1982) to
estimate trip matrices from traffic counts and Section 3 discusses the
main interpretations of the solution to this model. Section 4 discusses
the problem of estimating time-dependent trip matrices and Section 5
presents a modification to ME2 which tackles this problem and
provides details of its implementation in combination
with the route choice model CONTRAM (Leonard et al, 1978). The results
of validation tests are discussed in Section 6. Finally, Section 7
presents an extension of the model that allows attaching different levels
of confidence to traffic counts and to the prior trip matrix.

2.   THE ENTROPY-MAXIMISING APPROACH

Willumsen (1978, 1981a) has put forward and tested a model for estimating trip matrices from traffic counts and a prior trip matrix using an entropy maximising framework. The model is the result of the following mathematical programme:

$$\text{Maximise } S_1 = - \sum_{ij} T_{ij}(\log T_{ij}/t_{ij} - 1) \tag{1}$$

subject to

$$\sum_{ij} T_{ij}p_{ij}^a - V_a = 0 \tag{2}$$

$$T_{ij} \geq 0 \tag{3}$$

where

T_{ij} is the number of trips from i to j
t_{ij} is a prior estimate of these trips
V_a is the observed flow on link a
p_{ij}^a is the proportion of trips between i and j using link a ,
      in general $0 \leq p_{ij}^a \leq 1$.

The proportions p_{ij}^a is obviously related to the choice of route between i and j and may be estimated for each link and O-D pair using an appropriate assignment model. Wherever route choice may be considered to be independent of the O-D flows, the set of p_{ij}^a may be calculated beforehand using a suitable proportional assignment model (see Robillard, 1975) such as all-or-nothing or a stochastic assignment technique. In this case, Equations (2) will form a system of <u>linear equations</u>, one for each traffic count available.

The objective function (1) is concave and provided there is a feasible solution in the constraints (2) and (3) the solution to the mathematical programme can be found to be

$$T_{ij} = t_{ij} \, \pi_a \, X_a^{p_{ij}^a} \qquad\qquad \text{for each i,j} \tag{4}$$

where X_a is a variable associated with the Lagrange multiplier

corresponding to the Equations (2) (for a more detailed description see
Willumsen, 1981b).

An efficient solution algorithm for this problem has been developed based
on Bregman's balancing method (Lamond and Stewart, 1981) for bi-proport-
ional problems. The X_a factors play a role analogous to the balancing
factors in a doubly contrained gravity model and the exponents p_{ij}^a are
simply weights attached to each link/O-D pair set. For all-or-nothing
route choice proportions p_{ij}^a take only the values 0 or 1; an O-D pair
is only modified by an X_a factor if it uses that particular link a .
As shown by Lamond and Stewart (1981), the solution method can also be
extended to inequality constraints.

The solution to this mathematical program requires the existence of a
feasible solution to Equations (2) in non-negative T_{ij} . This condition
may be presented in matrix form by considering the following elements:

> $\underline{V}$, a column vector containing the L_c observed link flows
> $\underline{T}$, a column vector containing the Z^2 unknown T_{ij} 's
> $\underline{P}$, a matrix of coefficients p_{ij}^a with L_c rows and Z^2 columns
> $\underline{Pv}$, an extended matrix obtaining by adding the column $\underline{v}$ to $\underline{P}$ and
>   has L_c rows and $Z^2 + 1$ columns

The system of linear equations is now represented by

$$\underline{PT} = \underline{V} \tag{2a}$$

and the solution depends on the ranks of the matrices $\underline{P}$ and $\underline{Pv}$.

Three situations are possible.

(a) If $\underline{P}$ is an $Z^2 \cdot L_c$ matrix of rank M^2 the fundamental equations
     $\underline{PT}{=}\underline{V}$ are soluble for all constants V_a and this solution is unique.
     This condition is unlikely to occur in practice for this type of
     problem.
(b) If $\text{rank}(\underline{P}) \leqslant Z^2$, a solution only exists if $\text{rank}(\underline{P}){=}\text{rank}(\underline{Pv})$ and in
     this case infinitely many distinct solutions will exist. The rank of
     the matrix $(\underline{P})$ may be interpreted as the maximum number of linearly

independent equations in (2a). If $\text{rank}(\underline{P}) < L_c$ (number of links with counts) some rows in $\underline{P}$ may be generated as linear combination of other rows. The fact that $\text{rank}(\underline{P}) = \text{rank}(\underline{Pv})$ implies that these linear combinations will also be applicable to the extended matrix $(\underline{Pv})$ and the equations will be consistent and there will be feasible solutions.

(c) Rank$(\underline{P}) < M^2$ and $\text{rank}(\underline{P}) \neq \text{rank}(\underline{Pv})$. In this case the system of equations is inconsistent and has no feasible solution. A row in $\underline{P}$ may be expressed as a linear combination of other rows but these operations do not apply equally to the extended matrix. One of the sources of this inconsistency is the violation of link flow continuity conditions. (Total flow into a node should equal total flow out of it.) A second source resides in the relationship between $[p_{ij}^a]$ and $[v_a]$, when the observed counts and the implied traffic assignment model are incompatible. A simple case of this would be a link with non-zero observed flow but where all $p_{ij}^a = 0$ because the implied assignment model does not load any trips onto it. In practice though, the inconsistencies may be of a more subtle nature which can only be detected by matrix operations. As this second source of inconsistency is related to the interplay of flows on assignment it will be referred to in the future as the 'path flow' continuity conditions in contrast with the first and more obvious source of 'link flow' continuity equations.

If there are many feasible solutions to the problem (case b) one may define the degrees of freedom (DF) of these solutions as $DF = M^2 - \text{rank}(\underline{P})$, that is the number of unknowns less the number of independent equations (counts). The greater de degrees of freedom the greater the underspecificationof the problem. The degrees of freedom of the solution can be reduced by increasing the number of independent counts or by reducing the number of unknowns.

If the conditions discussed under (b) and (c) above are met, the ME2 model will estimate a trip matrix which when loaded onto the network reproduces the observed link flows to any reasonable degree of accuracy. While this is a desirable property for a model of this kind it does not constitute a validation of the model. This requires a data set which includes both traffic counts and a directly observed trip matrix.

Van Vliet and Willumsen (1981) reported tests on the model using data for
Reading (UK). Willumsen (1982) has also extended the model to situations
where congestion influences route choice. After testing different
approaches he found that by first identifying feasible routes and then
estimating path flows (origin-destination-route flows) a more accurate
trip matrix could be obtained. Data from Reading were again used to
validate his approach.

This extended model can be written:

$$T_{ij\ell} = \frac{t_{ij}}{D} \prod_a X_a^{\delta_{ijd}^a} \tag{5}$$

where the extra subscript d represents one of the D feasible routes.
The trip matrix is then obtained as:

$$T_{ij} = \sum_d T_{ijd} \tag{6}$$

and the effective route choice proportions as:

$$p_{ij}^a = \frac{1}{T_{ij}} \sum_d T_{ijd} \delta_{ijd}^a \tag{7}$$

3.   MODEL INTERPRETATION

The entropy-maximising framework used to derive the ME2 model as a source
of well rehearsed interpretations to the estimated matrix, see for example
Wilson (1970), Webber (1981), Fisk (1983) and in the context of this model
Willumsen (1981b). In this section a less well known pragmatic interpretation will be highlighted following the work of Murchland (1980).

Consider the function

$$S_{ij}' = T_{ij} \log T_{ij}/t_{ij} - T_{ij} + t_{ij} \tag{8}$$

the Taylor's series for $\log x$ for $x \geq \frac{1}{2}$ is

$$\log x = \left(\frac{x-1}{x}\right) + \frac{1}{2}\left(\frac{x-1}{x}\right)^2 + \frac{1}{3}\left(\frac{x-1}{x}\right)^3 + \dots$$

and for x close to 1, $\log x$ can be approximated by the first two terms.

By assuming that T_{ij} is sufficiently close to t_{ij} so that $x = T_{ij}/t_{ij}$ is ~1 for all ij, S'_{ij} can be approximated by

$$S'_{ij} \approx T_{ij}\left(\frac{T_{ij}/t_{ij}-1}{T_{ij}/t_{ij}}\right) + \frac{T_{ij}}{2}\left(\frac{T_{ij}/t_{ij}-1}{T_{ij}/t_{ij}}\right)^2 - T_{ij} + t_{ij}$$

$$\approx T_{ij}\left(\frac{t_{ij}}{T_{ij}}\frac{(T_{ij}-t_{ij})}{t_{ij}}\right) + \frac{T_{ij}}{2}\left(\frac{t_{ij}}{T_{ij}}\frac{(T_{ij}-t_{ij})}{t_{ij}}\right)^2 - T_{ij} + t_{ij}$$

$$\approx \frac{1}{2}\frac{(T_{ij}-t_{ij})^2}{T_{ij}} \tag{9}$$

In this equation the approximated objective function has the form of an error-like measure between prior and estimated trip matrices.

As $\sum_{ij} t_{ij}$ is a constant the maximisation of S_1 and the minimisation of $S_2 = \sum_{ij} S'_{ij}$ subject to the constraints (2) and (3) are equivalent problems. The estimated matrix $\underline{T}$ is then the matrix closest to the prior $\underline{t}$ (in the sense defined by the measure of distance (8)) and that reproduces the observed traffic counts.

If the prior matrix is considered to be not very close to the current one, it may be useful to introduce a scaling factor θ to ensure that current flow levels are reproduced (on average). An estimate of this scaling factor can be obtained from:

$$\theta = \frac{1}{L_c}\sum_a^{L_c}(V_a/\sum_{ij} t_{ij}p_{ij}^a) \tag{10}$$

This scaling factor is explicitly included in a variation of the ME2 model proposed by Bell (1983) and solved using a Newton method. This variation is more demanding in computer resources but has additional advantages in terms of speed and an estimation of confidence intervals for the resulting matrix.

An error function analogous to (8) will be used later to account for errors in the traffic counts.

4. TIME-DEPENDENT TRIP MATRICES

It is recognised that travel demand, in particular in urban areas, suffer variations over time. Many of the most difficult congestion problems are

associated to the build up and decay of queues and delay at key junctions
and links in response to these fluctuations in demand. When these
variations in demand are associated to well defined regular journeys (e.g.
to/from work) they may follow regular patterns on similar days.

A thorough analysis of these dynamic congestion problems is assisted by
the use of suitable traffic simulation models, for example CONTRAM
(Leonard et al, 1978) and SATURN (Hall et al, 1980) in the UK. These
models require, of course, estimations of demand for different periods of
the day. In general terms it has been found difficult to treat time as a
continuous variable in these cases and most applications consider suitable,
discrete time intervals. In this way the time period of interest (say
morning peak) is subdivided into S time slices (of say, 15 minutes each)
and trip matrices are required for each one of these.

In some cases, the trip matrix for one time slice may be considered to
affect flows on that time slice only (although queues may be passed on to
the next time interval). But if the time slices are not much longer than
the time it takes to drive through the study area the assumption above is
not very realistic. A more strict definition requires including in the
trip matrix of slice s only trips beginning in that time period (i.e.
appearing into the study area). Trips that started in time period s may,
however, affect flows and delays in one or more subsequent time slices.

The estimation of time-dependent trip matrices from traffic counts requires
incorporating these effects consistently with the estimation process. This
has been attempted with ME2 along the following lines.

The variable p_{ijs}^{ar} is used to indicate the proportion of trips between
i and j starting in time slice s which travel over link a during
time slice r . It should be noted that this variable is playing two roles.
On one hand it represents time-dependent route choice between origin and
destination and in this effect could be modelled by a number of good
route choice models. On the other hand, it also accounts for the travel
time of a vehicle since its entrance to the network up to its passage
over a particular link and identifies the time slice during which this
takes place. We can now write the equation for the flow on a link a at
time slice r as:

$$V_{ar} = \sum_{ijs} T_{ijs} p_{ijs}^{ar} \tag{11}$$

where T_{ijs} is the number of trips from i to j starting at time s.

Within this framework ME2 can easily be extended to estimate matrices for different time intervals by maximising a modified objective function

$$\text{Maximise } S_3 = - \sum_{ijs} [T_{ijs} \log (T_{ijs}/t_{ijs}) - T_{ijs}] \tag{12}$$

subject to (11) and $T_{ijs} \geq 0$.

If multiple paths between i and j are allowed for this means introducing an additional index d as in Equations (5) to (7). This will be omitted for simplicity here.

The same algorithm originally developed for ME2 could be used to solve this extended model. However, the arrays p_{ijs}^{ar} and T_{ijs} are now huge and impose important requirements in computer capacity. A heuristic approach based on the route choice model of CONTRAM was then adopted as described below.

5. MODEL IMPLEMENTATION

The extended model requires a route choice model capable of identifying the contribution of a trip starting in time r over the flow on a link in time $r+\Delta r = s$. The new version of ME2 has now been implemented in combination with CONTRAM, a traffic management simulation program developed by the Transport and Road Research Laboratory (Leonard et al, 1978). CONTRAM provides an ideal route choice model to this end.

During the development of CONTRAM particular attention was given to the modelling of variation of travel demand over time, including periods of oversaturation, because it is in such periods that traffic conditions are likely to have a large effect on driver's route choice. The movement of traffic is simulated in terms of 'packets' of vehicles, typically 1 to 10 vehicles depending on the size of the network and computer capacity. The model provides estimates of the variation of link travel times over time, and packets are assigned to quickest routes through the network on the

basis of estimates of travel times that they would encounter on each link
at the time at which they would traverse it.

As part of its normal operation CONTRAM produces the set of routes used
by packets of vehicles and the resulting modelled flows in each time
interval. It was decided to use basically this information to implement a
solution algorithm as follows:

(1) Make $N=0$ and $n=0$ and make the current modelled matrix $\underline{T}^{Nn}$ equal
to the prior matrix $\underline{T}^{Nn}=\underline{t}$.

(2) Run CONTRAM with $\underline{T}^{N0}$ and obtain a set of routes. Make $n=0$ and
$N=N+1$.

(3) Make $n=n+1$, load $\underline{T}^{Nn}$ using CONTRAM paths and obtain modelled
flows v_a . Compare them with the observed flows and obtain the
ratios $y_{as}=V_{as}/v_{as}$ where V_{as} is the observed flow and v_{as} the
modelled flow on time interval s .

(4) If all y_{as} are reasonably small, proceed to step (7), otherwise

(5) Trace the paths followed by each group of vehicles and calculate the
accumulated factor

$$Y_{ijs} = \left(\prod_a y_{au} \right)^{1/m_{ij}}$$

where u is the time interval during which T_{ijs} travel over link
a , and m_{ij} is the number of observed links in the path.

(6) Make $T^n_{ijs} = T^{n-1}_{ijs} \cdot Y_{ijs}$, and proceed to step (3).

(7) If $\underline{T}^{N,n}$ and $\underline{T}^{N-1,0}$ are sufficiently similar, stop, otherwise
make $\underline{T}^{N,0}=\underline{T}^{N,n}$ and proceed to step (2).

The inner (n) and outer (N) loops are common to earlier versions of
the model (Willumsen, 1982). The power $1/m_{ij}$ in the calculation of the
accumulated balancing factor is required because in this algorithm all
O-D pairs are updated at the same time. In the previous algorithm after

updating one O-D pair all flows were also updated, this is no longer practical in this case.

CONTRAM allows up to 3 vehicle types in the network and our implementation of ME2-CONTRAM catered for these. It is now even possible to have a mixture of classified and unclassified traffic counts; in the latter case flows for each vehicle type are added up and the corresponding balancing factor for that influences the 3 modes.

6. VALIDATION

The same data base used to validate ME2 was used to test this extension. These were data collected by the Transport and Road Research Laboratory and their main characteristics as well as some initial validation results were reported in the 8th International Symposium (Van Vliet and Willumsen, 1981). Further validation of ME2 was reported in Willumsen (1982) including capacity restrained route choice.

The following indicators were used to compared estimated and observed matrices.

(a) Mean Absolute Difference, defined as

$$MAE = \sum_{ij} |T_{ij}^a - T_{ij}^b|/n$$

where T_{ij}^a and T_{ij}^b are the two matrices and n is the number of cells.

(b) Relative MAE

$$\%MAE = 100 \sum_{ij} |T_{ij}^a - T_{ij}^b| / \sum_{ij} T_{ij}^a \quad .$$

(c) Root Mean Square Error

$$RMSE = \sqrt{\frac{\sum_{ij} (T_{ij}^a - T_{ij}^b)^2}{n}} \quad .$$

(d) Relative RMSE

$$\%RMSE = 100 \, RMSE / (\sum_{ij} T_{ij}^a /n) \quad .$$

In our case these comparisons could be done for each time slice or for an aggregation of all trips over the whole observation period (2 hours). The time slices used were of 15 minutes each.

The following table summarises comparable results from the two models with Reading data.

Table 1: Tests of different versions of the ME2 model
with Reading data and a full set of counts.

Indicator	ME2 all-or-nothing	ME2 capacity restrained	ME2 CONTRAM	Average day-to-day variation
MAE	1.2	1.0	45.6	0.9
%MAE	101	83	44	85
RMSE	1.8	1.8	68.1	1.8
%RMSE	149	148	66	150

Time intervals affect the average size of the O-D flows hence making comparable only the relative values of the indicators (%MAE and %RMSE). It can be seen that ME2-CONTRAM produces a marked improvement in performance over the earlier versions of the model. In other words the improved route choice model used by CONTRAM has a very positive effect on the accuracy of the general ME2 approach.

The tests were carried out in different computers thus making CPU times not directly comparable. However, it is estimated that the use of CONTRAM requires between 10 and 20 times the CPU times of a conventional route choice model with capacity restraint.

7.　FURTHER EXTENSIONS

It should be recalled that one of the conditions for the existence of a feasible solution to the entropy maximising problem was that the observed link volumes should satisfy, where relevant, normal flow continuity (flow 'into' a node should equal flow 'out of' it). Errors in the traffic counts and the use of flows obtained on different days are likely to violate this condition and prevent convergence of the solution algorithm.

A computer program has been written to test this condition and to modify the observed link flows so that they do satisfy it using a maximum likelihood method (Van Zuylen and Willumsen, 1980). However, one would like to incorporate the possibility of errors in the counts directly into the estimation process.

The implementation of ME2 in combination with SATURN (Hall et al, 1980) adopted the pragmatic approach of constraining the values taken by the balancing factors to limits imposed by the user, typically 0.2 to 5.0. In other words, traffic counts are not permitted to modify the prior matrix beyond certain limits.

However, one would like a more consistent method to trade-off modifications to the prior trip matrix against error levels in the counts; in qualitative terms to obtain a matrix which is not too different from the prior matrix at the cost of not quite reproducing the traffic counts.

One possibility is to replace each link flow equation by two inequalities recognising the expected range of the 'true' value of the traffic count. An alternative approach, which seems easier to integrate to the ME2 model, may be outlined as follows.

Consider the entropy function as a measure of separation or error between the prior and the 'true' or current value of a variable. In the case of link flows the true values is represented by V^* and the observed value by V. One could then try to obtain a better estimate of the true values for $\underline{V}$ by minimising

$$\sum_a V^*_a (\log V^*_a/V_a - 1) \tag{13}$$

subject to constraints reflecting observations on the flows. But it is also possible to combine these two problems (matrix and flow estimation) into a single mathematical programme:

$$\text{Minimise } S_4 = \sum_{ij} T_{ij} (\log T_{ij}/t_{ij} - 1) + \alpha \sum_a V^*_a (\log V^*_a/V_a - 1) \tag{14}$$

subject to

$$\sum_{ij} T_{ij} p^a_{ij} - V^*_a = 0 \tag{2}$$

and

$$T_{ij} \geq 0$$

where α represents the relative weight attached to errors in the counts compared to changes to the prior matrix. The solution to this problem becomes

$$T_{ij} = t_{ij} \, \pi \, x_a^{p_{ij}^a} \tag{4}$$

and

$$V_a^* = V_a \, x_a^{-1/\alpha} \tag{15}$$

This now looks like a problem in which the balancing factors X_a are used to modify the prior trip matrix and the observed flows and the relative magnitude of these corrections is governed by the weight α . For a very large α only minor modifications to V_a are allowed and the ME2 model can be seen as one in which $\alpha=\infty$. One may also envisage an option in which different weights are attached to different counts or to different parts of the prior trip matrix, for example according to their reliability or age.

The structure of the problem in Equations (2), (4) and (15) suggests that only minor modifications to Bregman's algorithm would be required. This model has not been yet tested but it seems to show some promise in combining both problems in a consistent and efficient manner.

8. CONCLUSIONS

The ME2 model under different implementations, has been tested using data for Reading and applied on several occasions. There is a version written for use in 8 and 16 bit microcomputers and it has been used to estimate and update matrices in several cities and local areas.

It is clear that the route choice model used to implement ME2 plays a key role in determining its accuracy. In particular, problems where dynamic changes in travel demand are very important (as they seem to be during the afternoon peak in Reading) a model like CONTRAM seems to be well justified. If congestion is important but time variation of demand is less so or if the study area is such that it can be transversed in a short time compared with relevant time slices a model like SATURN may be more efficient. Less demanding situations may be satisfied with simpler and faster route choice models.

Acknowledgements

The original version of the model was developed as part of a research
project financed by the Science and Engineering Research Council. The
research on the implementation of time-dependent trip matrices with the
CONTRAM model was carried out under a contract placed with University
College London by the Urban Networks Division of the Transport and Road
Research Laboratory. Any views expressed are not necessarily those of
the Transport and Road Research Laboratory nor the Department of Transport.

REFERENCES

Bell, M.G.H. (1983) The estimation of an origin destination matrix from
traffic counts. Transportation Science, 17(2).
Fisk, C. (1983) Entropy and information theory: are we missing something?
Department of Civil Engineering, University of Illinois at Urbana-
Champaign. Publication Number 14, Urbana, Illinois.
Fisk, C., Boyce, D. (1982) A note on trip matrix estimation from link
traffic count data. Department of Civil Engineering, University of Urbana-
Champaign. Publication Number 5. Urbana, Illinois.
Hall, M., Van Vliet, D., Willumsen, L.G. (1980) SATURN a simulation
assignment model for the evaluation of traffic management schemes. Traffic
Engineering and Control, 21(4).
Lamond, B., Stewart, N.F. (1981) Bregman's balancing method. Transport-
ation Research, 15B(4).
Leonard, D.R., Tough, J.B., Baguley, P.C. (1978) CONTRAM: a traffic
assignment model for predicting flows and queues during peak periods.
Transport and Road Research Laboratory Report LR 841. Crowthorne, UK.
Nguyen, S. (1981) Modèles de distribution spatiale tenant compte des
itinéraries. Publication No. 225, Centre de Recherche sur les Transports,
Université de Monréal, Montréal.
Van Vliet, D., Willumsen, L.G. (1981) Validation of the ME2 model for
estimating trip matrices from traffic counts. 8th International Symposium
on Transportation and Traffic Theory. Toronto University, June 1981.
Webber, M.J. (1981) Information Theory and Urban stated Structure. Croom
Helm, London.
Willumsen, L.G. (1977) O-D matrices from network data: a comparison of
alternative methods for their estimations. Proceedings of PTRC Summer
Annual Meeting, University of Warwick, July 1978. PTRC Education and
Research Services, Ltd., London.
Willumsen, L.G. (1981a) Simplified transport models based on traffic counts.
Transportation, 10, 257-278.
Willumsen, L.G. (1981b) An entropy maximising model for estimating trip
matrices from traffic counts. Transport Studies Group, University College
London (unpublished).
Willumsen, L.G. (1982) Estimation of trip matrices from volume counts:
validation of a model under congested conditions. PTRC Summer Annual
Meeting, University of Warwick, July 1982. PTRC Education and Research
Services Ltd., London.
Wilson, A.G. (1970) Entropy in Urban and Regional Modelling. Pion Press,
London.

Ninth International Symposium on
Transportation and Traffic Theory
© 1984 VNU Science Press, pp. 413–430

MATRIX ENTRY ESTIMATION ERRORS

CHRIS HENDRICKSON[1] and SUE McNEIL[2]
[1] *Department of Civil Engineering, Carnegie-Mellon University, Pittsburgh, PA, U.S.A.*
[2] *Garmen Associates, Whippany, NJ, U.S.A.*

Abstract

Estimation of trip tables or other matrices is often a first step in transportation systems analysis, but such estimates are typically uncertain. We discuss the various errors which can arise in estimation using the framework provided by constrained, generalized least squares regression. An advantage of this approach is that error estimates can be calculated. Incremental sampling strategies and the use of different types of information are introduced as means to reduce estimate uncertainty. Illustrative applications are made to *ab initio* estimation of inter-zonal work trips in a metropolitan area and to expansion of a work trip matrix to a full trip table for a transit service.

1 INTRODUCTION

Matrices of various types form a basic input for transportation planning and management analysis.   Typical uses include representation of origin/destination flows, direct and indirect economic impacts (via economic input/output models), and characterization of existing situations (such as vehicle miles of travel by vehicle class and type of roadway).   We can distinguish four approaches to estimate the current entries of such matrices: (1) surveys, including household or on-board questionnaires, (2) updating of past matrices to reflect changes over time, (3) expanding incomplete data, and (4) *ab initio* estimation from time series or distribution models.   Even with surveys, the problem of reconciling conflicting data sources often arises, so the estimation problem is not trivial.   In each of these approaches, estimation and measurement errors are likely to occur.

Consideration of the errors arising in such estimations is important for several reasons. First, errors may lead to incorrect investment or management decisions.   Since the costs and benefits of alternatives generally depend upon travel volumes, errors in volume forecasts can seriously hamper investment and operation planning (see, for example, Wohl and Hendrickson, 1984).   Second, effectively planning sampling or surveying strategies requires some knowledge of the magnitude and distribution of such errors. This latter point is now of increasing importance since general travel surveys of the population are becoming increasingly expensive.   In many areas in the United States, analysts now rely on journey-to-work data obtained from the decennial population census coupled with volume counts at particular locations.   Although technological improvements have permitted cost effective continuous monitoring of volumes in particular locations, general travel surveys are now much less frequent and extensive than in previous decades.

This paper considers the magnitude and interpretation of matrix entry estimation errors. Some implications for appropriate sampling strategies are drawn using the uncertainty associated with matrix entry estimates.   The results derived in the paper are illustrated by applications using travel data from Portland, Oregon (USA).

As a framework for the paper, constrained, generalized least squares (CGLS) regression is used.   This technique enables the calculation of estimates of the error of matrix entries.   As the actual error in an estimate is usually unknown, the estimated variance can be used as a measure of the uncertainty and hence the error associated with a matrix entry estimate.   With an assumption of normal distributions, confidence intervals for matrix entries can also be calculated.   In this technique, constraints on estimated values may be explicitly introduced.   Such constraints may arise for physical reasons (e.g. proportions must sum to one) or from measurements (e.g. volume counts).

Numerous other methods exist for matrix estimation, of course. Common methods use maximum likelihood estimation with underlying models that can be described as maximum entropy (Wilson, 1970) or information minimization (Uribe et al., 1965). For example, Landau et al. (1983) assume that a cordon sample is multinomial distributed and their maximum likelihood estimation procedure results in estimates which are identical to those obtained from information minimization. More recently, attention has been devoted to origin/destination matrix estimation with the use of link volume counts and underlying models of both trip distribution and assignment (eg. VanZuylen, 1980 or Fisk and Boyce, 1983).

Consideration of estimation accuracy and sampling properties have also appeared. For example, Sikdar and Hutchinson (1981) examined the accuracy of gravity models to estimate work trips in Edmonton, Alberta. They concluded that gravity models estimated with only row and column totals are likely to be substantially in error. Indeed, the gravity model trip tables they estimated had goodness of fit characteristics equivalent to the actual trip table with an added random error of about 75%.

The next section of the paper describes the CGLS technique. Following this, the sources and characterization of estimation errors are discussed. Section 4 presents an *ab initio* estimation example using a gravity model. Section 5 describes alternative sampling strategies for improving matrix entry estimates. A matrix estimation and incremental sampling example involving the expansion of a origin/destination matrix of transit work trips to all transit trips is described in Section 6.

2 CONSTRAINED, GENERALIZED LEAST SQUARES ESTIMATION

Our objective is to obtain an estimate $\hat{Q}$ of an actual but unknown matrix Q. We choose our entry estimates in $\hat{Q}$ such that deviation or generalized "distance" between $\hat{Q}$ and some hypothesized or modeled values Y is minimized. Entries in the matrix Y might be the result of a model with known or unknown parameters or might be base entries obtained from a past survey (McNeil, 1983). For constrained, generalized least squares regression, our generalized "distance" measure is the sum of squared errors. In addition, we can introduce a set of (linear) constraints on the estimates. Thus, the regression problem is:

$$P1: \quad \underset{q, a}{\text{Min}} \quad \{(q - Xa)' \, V^{-1} (q - Xa) \mid Rq = r\}$$

where q is a vector of entries in Q,
           X is a matrix of attributes,
           a is a vector of model parameters,

V is a weighting matrix described below,
R is an incidence matrix, and
$\underline{r}$ is a vector of constraint totals.

With known parameters $\underline{a}$, hypothesized or modeled values of $\underline{y}$ can be calculated and the regression problem is simplified to:

$$\text{P2: } \underset{\underline{q}}{\text{Min}} \ \{(\underline{q}-\underline{y})' \ V^{-1} \ (\underline{q}-\underline{y}) \ | \ R\underline{q}=\underline{r}\}$$

where $\underline{y}$ is the vector of hypothesized values. In this model, we assume that redundant linear constraints are eliminated so that the rank of the incidence matrix R is equal to the number of rows in R. Elimination of such redundant constraints permits the use of matrix inversion approaches for the solution of P1 and P2.

The underlying regression model for P2 is:

$$\underline{y} = I\underline{q} + \underline{\epsilon} \tag{1}$$

where $\underline{\epsilon}$ is a vector of residual terms. In this equation, $\underline{q}$ can be interpreted as a set of regression parameters. If the set of constraint totals $\underline{r}$ were uncertain, then the constraints $R\underline{q} = \underline{r}$ could also be entered as observations:

$$\begin{bmatrix} \underline{y} \\ \underline{r} \end{bmatrix} = \begin{bmatrix} I \\ R \end{bmatrix} \underline{q} + \begin{bmatrix} \underline{\epsilon} \\ \underline{\xi} \end{bmatrix} \tag{2}$$

in which $\underline{\xi}$ is a vector of constraint total errors. The model (2) is an example of generalized least squares (Theil, 1971).

As an example, suppose that average transit vehicle on/off counts were available to compare with user survey–derived totals. If these on/off counts were thought to be highly accurate, then the totals might be included as constraints on the survey/based trip table observations. Alternatively, the counts could enter as observations on sums of matrix entries with an error term $\underline{\xi}$ reflecting the counts' uncertainty relative to the user survey's uncertainty.

In CGLS, the vector of residual terms $\underline{\epsilon}$ is assumed to have zero mean, $E[\underline{\epsilon}] = \underline{0}$, and covariance proportional to the matrix V, $V[\underline{\epsilon}] = \sigma^2 V$. Moreover, we commonly assume that $\underline{\epsilon}$ is multivariate normally distributed in order to calculate confidence intervals and conduct hypothesis testing on $\underline{q}$.

As an example of error assumptions, the variance of residual terms ϵ_j might be assumed to be proportional to hypothesized values y_i and all covariances assumed to be zero. In this case, the least squares "distance" measure in P2 is similar in form to a chi-square test statistic:

P3: $\text{Min}_q \ \{\Sigma_i(q_i-y_i)^2/y_i \ | \ Rq = r \}$

Note that this assumption implies that matrix entries with small hypothesized values are likely to be relatively closer to their hypothesized values; we believe that this assumption is often reasonable and use it below. Kirby and Lesse (1978) suggest that the diagonal elements of V should be proportional to y_i and inversely proportional to the fraction of travelers sampled.

Solution to the CGLS estimation problem P2 is developed in several econometrics texts (such as Theil, 1971), as:

$$\tilde{q} = y - VR'(RVR')^{-1}(r-Ry) \tag{3}$$

The covariance matrix of q is:

$$\text{cov}(\tilde{q}) = \sigma^2[V-VR'(RVR')^{-1}RV] \tag{4}$$

An unbiased estimate of the proportionality factor σ^2 is:

$$\tilde{\sigma}^2 = (y-\tilde{q})' \ V^{-1}(y-\tilde{q})/p \tag{5}$$

where p is the number of constraints. The diagonal terms in $\text{cov}(\tilde{q})$ (Eq. 4) represent the variance of individual matrix entry estimates; we shall use them subsequently to design incremental sampling strategies.

To form estimates for models with unknown parameters, a slightly different regression model can be used (Carey, 1982). The regression model:

$$q = Xa + u \tag{6}$$

can be substituted into the set of constraints:

$$r = Rq$$
$$= RXa + Ru \tag{7}$$
$$= X*a + \epsilon*$$

where $X* = RX$ and $\epsilon* = Ru$. This yields a generalized least squares problem. Note that observations of $\underline{q}$ can enter as constraints in Eq. 7. For this problem, the estimates of $\underline{a}$ are calculated as:

$$\hat{\underline{a}} = (X'R'(RVR')^{-1}RX)^{-1}X'R'(RVR')^{-1}\underline{r} \tag{8}$$

with

$$\mathrm{cov}[\hat{\underline{a}}] = \sigma^2(X'R'(RVR')^{-1}RX)^{-1} \tag{9}$$

and an unbiased estimate of the proportionality factor σ^2 is:

$$\tilde{\sigma}^2 = (\underline{r}-RX\hat{\underline{a}})'(RVR')^{-1}(r-RX\hat{\underline{a}})/(p-q) \tag{10}$$

where q is the number of parameters in $\hat{\underline{a}}$ to be estimated.

With estimates $\hat{\underline{a}}$, the estimated matrix entries can be calculated as $\tilde{\underline{q}} = X\tilde{\underline{a}}$. Unfortunately, the estimates $\tilde{\underline{q}}$ may not be consistent with the constraints $R\tilde{\underline{q}} = \underline{r}$. A second stage estimation may be employed in this case with the predicted entries used as hypothesized entries in P2. Other reconciliation procedures can also be used (McNeil, 1983).

As with all statistical estimation methods, we should emphasize that the accuracy of the resulting estimates and the estimated uncertainty measures depend upon the appropriateness of the underlying model formulation and on the adequacy of the data used for estimation. In particular, formulating appropriate models ($X\underline{a}$), error structures (V) and realistic constraints ($R\underline{q}=\underline{r}$) is worthy of close attention. An advantage of the CGLS formulation in this regard is that there is considerable flexibility in specifying attributes or constraints. For example, observations and constraints may be developed from cordon counts (Carey et al., 1981) or small samples (Landau et al., 1982).

Alternative matrix estimation techniques exist, of course. These alternative techniques can be characterized as alternative measures for generalized distances (Bacharach, 1970). For example, the information minimization estimation problem with hypothesized values can be formulated as:

P4: Min $\{\Sigma_i\ y_i\ \ln(y_i/q_i)\ |\ R\underline{q} = \underline{r}\}$

The biproportional or RAS method has a similar form with a transposition of q_i and y_i in the objective function:

P5: Min $\{\Sigma_i\ q_i\ \ln(q_i/y_i)\ |\ R\underline{q} = \underline{r}\}$

These are non-linear deviation or distance metrics.

While these alternative techniques may result in different estimates, the estimation problems P3 (CGLS with a chi-square error structure), P4 and P5 are similar since the objective function of P3 is a first order approximation to those of P4 and P5 (see Bacharach, 1970 or Hewings and Janson, 1980). As a result, differences in estimates from the three methods are small as long as Q is in the neighborhood of Y. We shall use the CGLS technique below to make use of the uncertainty measures $\text{cov}(\tilde{q})$.

As a final note, it is also possible to add inequality constraints to the estimation problems P1-P3. For example, such inequality constraints might be imposed to insure that all matrix entries are non-negative. The CGLS problem P3 would then become:

P6: Min $\{(q-y)'V^{-1}(q-y) \mid Rg=r, Sg \geq s \}$

where S is a matrix of inequality constraint incidence and s is a vector of inequality totals. Solution to P6 can be obtained with any general purpose quadratic programming package (Carey et al., 1981). While not strictly a CGLS model, P6 does represent a statistical estimation procedure; McNeil (1983) gives a Bayesian derivation of this model.

3 SOURCES AND CHARACTERIZATION OF ESTIMATION ERROR

The derivation of the constrained, generalized least squares (CGLS) model explicitly assumes an additive error structure such as Eq. (1). The error term in the regression equation is an unobservable quantity that accounts for discrepancies between the observed variables y and the actual but unknown matrix entries q. Such discrepancies may arise from any combination of three possible sources: specification error, errors in measuring the dependent variable and other random variations. Specification error includes omitted variables, incorrect functional form and the incorrect inclusion or exclusion of variables. Unfortunately, each of these various errors are likely to arise in practice, so that the absolute and relative values of ϵ are likely to be large.

As an example of the possible structure of these errors, suppose that there is an error in sampling elements of estimated entries y such that $\tilde{y} = y + \omega$. With known constraints on matrix entries, we can formulate a CGLS regression problem identical to P2 with the regression equation:

$$y = Iq + \epsilon* \tag{11}$$

where y are true values and $\epsilon*$ is an error term to account for omitted variables and specification errors. We assume that $E[\epsilon*] = 0$, the expected value of $\epsilon*$ is zero and that $Var[\epsilon*] = T$. With the measurement error in y, we are actually estimating:

$$\tilde{y} = Iq + \underline{e}* + \underline{\omega} \tag{12}$$

If we assume $\tilde{y}$ is an unbiased estimate of y (that is, $E[\underline{\omega}] = 0$), then the best linear unbiased estimate of q is

$$\tilde{q}* = \tilde{y} + V*R'(RV*R')^{-1}(\underline{r}-R\tilde{y}) \tag{13}$$

where $V* = (T + 1/\sigma^2\ W)$ and

W is the variance–covariance matrix of $\underline{\omega}$.

It can be shown that $\tilde{q}*$ is the best linear unbiased estimate of q following a proof parallel to the one presented on page 286 of Theil (1971). As in the mixed estimation problem, σ^2 is unknown but a consistent estimate of σ^2 may be found by assuming that there is no error in the vector y. The problem is usually avoided as it is common to assume that W is proportional to T (or vice versa) so that $V* = V$. Often the structure of W and T is defined by the sampling procedure or the method used to obtain y. Furthermore, if W is assumed to be proportional to T when it actually has some other structure the estimates are statistically inefficient but remain unbiased provided the expected values of each of the components of the error terms are zero. If the estimation problem is one of reconciliation, then T is small or zero and may be neglected. Specification error occurs when there is change in the structure of the matrix entries. The amount of such change relative to the amount of measurement error in y determines the relative magnitudes of T and W. Provided the V matrix is correctly specified the parameter σ accounts for the absolute magnitude of these errors.

As a practical matter, we have found that the use of entries in V proportional to the values of y gives the best general results. This assumption is illustrated in the estimation problem P4 above.

Whether or not the errors represented by these different estimation procedures are too large depends upon their use and the analyst's judgement. In this *estimation* problem (as contrasted with a *forecasting* problem), estimates could be improved by gathering additional data. We show below that relatively inexpensive data (such as volume counts or roadside interviews) can be used effectively in the estimation process.

4 EXAMPLE OF AN AB INITIO ESTIMATION

As a simple example of the extent of matrix entry estimation errors, suppose that we wish to estimate an origin/destination matrix of metropolitan work trips. For this purpose, we shall use 48 zones in the Portland, Oregon (USA) metropolitan area so that 48 x 48 = 2304 trip elements are to be estimated. For estimation data and comparison of sampled and estimated trip volumes, we shall use results from the 1976 Annual Survey of Housing conducted by the US Census Bureau.

The proposed model has one unknown parameter and is of a gravity model type:

$$q_{ij} = a \, W_i E_j \, / \, d_{ij}^2 + \epsilon_{ij} \qquad (14)$$

where q_{ij} is the entry to be estimated,

a is a parameter to be estimated,

W_i is the number of workers resident in zone i,

E_j is the number of workers employed in zone j,

d_{ij} is the inter-centroid peak period travel time from i to j; and

ϵ_{ij} is an error term.

We assume that the weighting matrix V of error terms is a diagonal matrix with elements proportional to $W_i E_j / d_{ij}^2$.

We use four sets of constraints on values of q_{ij} to achieve our estimation. These include:

1. the number of workers resident or employed in each zone. These constraints are row and column totals for Q and represent 95 independent constraints.

2. the volume crossing a north-south cordon line through the Portland metropolitan area. This cordon coincides with the path of the Willamette River for which few crossings exist. A description of the use of cordon count data appears in Carey et al. (1981).

3. the origin zone of commuters crossing the north-south cordon. This data could be obtained from simple roadside or mailback surveys. It provides an additional 46 independent observations.

4. the origin and destination zones of commuters crossing the north-south cordon. Again, this data could be obtained relatively cheaply and provides an additional 942 independent observations.

These various observations are used as constraints on the estimation problem.

Table 1 reports estimated values of the parameter "a" along with estimated t–statistics. McNeil (1983) describes the computational details. For comparison, results from a simple regression on all 2,304 measured matrix values are also reported. Rather surprisingly, each estimation resulted in the same estimate of the parameter. However, the uncertainty in the estimate of the parameters declined with additional information.

Table 1:   Parameter Estimates for Alternative Formulations
            of the Ab Initio Estimation Problem

Problem Formulation	Number of Constraints	Parameter Estimate (x 10^{-4})	t–statistic
(1)	95	3.80	8.15
(2)	96	3.80	7.79
(3)	142	3.80	8.76
(4)	1084	3.80	22.31
Regression	2,304 obs.	3.80	18.80

While the parameter estimates reported in Table 1 are consistent, it is not surprising that estimates of actual travel volumes have substantial differences from the surveyed totals. As shown by the various aggregate measures in Table 2, average errors are still fairly large. Additional data did reduce the error, as we might expect. Estimates from the unconstrained regression model would be worse than the results reported in Table 2 since constraint information is not used and the parameter estimates are all identical. For formulation 2 (with 96 constraints), 363 or 6% of estimates were negative; constraining these entries to zero and re–estimating reduced the aggregate error measures.

5 INCREMENTAL SAMPLING STRATEGIES

When obtaining estimates of matrix entries, a typical objective is to minimize the error in the estimates. As the actual error is unknown and cannot be obtained without knowing the entry values, we might attempt to minimize the *uncertainty* associated with the estimates using the available data. For example, the constrained generalized least squares estimator derived in Section 2 is a best linear estimator in the sense that the variance–covariance of any other linear unbiased estimate found using the same data exceeds it by a positive definite matrix, so that the expected sum of the squared errors from any other (linear) estimator will exceed the CGLS sum of squared errors.

If the uncertainty associated with some of the entry estimates obtained using the

Table 2:   Aggregate Error Measures for the Ab Initio Estimation Example

Problem Formulation	Average Absolute Error	Average Relative Error *	Average Aggregate Error **
(1)	330	1.2	0.90
(2)	340	1.4	0.92
(with negatives set to zero)	320	1.3	0.87
(3)	310	1.2	0.84
(4) (for unconstrained elements only)	270	1.5	0.73

* Calculated as the average of errors divided by surveyed entries omitting zero entry values.

** Calculated as the average error divided by the average surveyed entry value.

available information is above some acceptable level, we can reduce the uncertainty by undertaking some additional sampling. To derive sampling strategies, we need some criteria or rules to determine

- what to sample or survey, and

- how large our survey should be.

There are three types of surveys that might be undertaken:

- Surveys to obtain better estimates of the initial values. These may be better measurements of the base values y or improvements in the estimates provided by a distribution function. Improvement in the distribution function may include a better specification to account for omitted variables, improved estimates of the attributes in the distribution function or more data to re-estimate the parameters of the function.

- Surveys to add more constraints, including constraints on individual elements, cutsets or any linear combination of entries.

- Surveys to improve the estimates of the constraint totals.

The derivation of *optimal* sampling strategies is complex and problem specific. The impact on the uncertainty of each type of survey alone varies with the criteria used to measure the reduction in the uncertainty and the resources available. Although the three different types of surveys can be used simultaneously, the number of possible strategies using a combination of the different types of surveys is usually large and even more difficult to evaluate. Further, we are using *estimates* of the uncertainty associated with each entry to suggest possible improvements. Therefore, no optimal sampling strategies are derived in this section but several approaches to deriving reasonable sampling strategies are suggested.

There are several possible formulations of our objective to reduce the uncertainty associated with the estimates. In general, the objective is to minimize a function of the estimates uncertainty

$$\text{Minimize } f(\hat{\text{Var}}[\tilde{q}_{new}]) \qquad\qquad (15)$$

where $\tilde{q}_{new}$ is the estimate after the sampling has been completed and the additional information added, and $f(\hat{\text{Var}}[\tilde{q}_{new}])$ is some error function. The objective function may be subject to constraints such as budget constraints related to the cost of sampling, the maximum or minimum number of surveys that may be undertaken or the maximum or minimum number of constraints that may be added.

Daganzo (1980) suggests three forms for the objective function $f(\hat{\text{Var}}[\tilde{q}_{new}])$.

- minimize the trace of $\hat{\text{Var}}[\tilde{q}_{new}]$
- minimize the value $\underline{z}' \hat{\text{Var}}[\tilde{q}_{new}]\underline{z}$ for a given vector $\underline{z}$ of unit length, and
- minimize the largest eigenvalue of $\hat{\text{Var}}[\tilde{q}_{new}]$ or, equivalently, maximize the smallest eigenvalue.

The first objective simply minimizes the sum of the variances. The second objective is to optimize an amount which is proportional to the expected sum of the squared error (when $\hat{\text{Var}}[\tilde{q}_{new}]$ is not large). The final objective is equivalent to minimizing the maximum value of the expected sum of squared errors, as from Daganzo (1980):

$$\underline{z}' \hat{\text{Var}}[\tilde{q}_{new}] \ \underline{z} \le \text{largest eigenvalue of } \hat{\text{Var}}[\tilde{q}_{new}] \qquad\qquad (16)$$

$$\text{when } \underline{z}'\underline{z} = 1$$

Other functions of the variance–covariance may be used as the objective function. For example, we might wish to minimize some function of the coefficient of variation if we are interested in minimizing the relative uncertainty compared with the magnitude of the matrix entries. Another example is to minimize some function of the variance of the estimates expressed as a proportion of each row or column total when it is the estimated proportion that will be used.

Often the objective for defining additional sampling strategies is related to the definition of the weighting matrix V in formulation P1. For example, if formulation P1 is equivalent to minimizing the percentage change, we may want to obtain sampling strategies that reduce the coefficients of variation of the elements. If the estimation and sampling design objectives are not the same, then we are being inconsistent. Although the objective function in P1 refers to the change from the *base* value to the *estimated* value, the objective function used to derive the sampling strategies relates the difference between the *actual* and *estimated* values. It can be shown that the two objectives are indeed equivalent.

As an example, suppose that we can undertake special surveys to improve our estimates of base values y_i. At the extreme, we could sample thoroughly enough to impose a constraint on the estimated entry $r_i = q_i$. If we are limited to sample no more than n entries, a simple strategy would be to rank the matrix entries with respect to their estimated uncertainty ($\text{Var}[\tilde{q}_i]$) and select the top n for special sampling. An example appears below. More complicated strategies could be developed for different budget constraints or different data gathering strategies (Cochrane, 1977). For example, we could compare alternative cordon counts with respect to their impact on entries' uncertainty, as illustrated below.

6 AN EXAMPLE OF INCREMENTAL SAMPLING

As an example, consider the case of a transit operator who would like to estimate a trip matrix of all transit trips without conducting a complete on-board survey. We assume that three types of data are available for estimation: a matrix of work trips (which can come from census surveys), route volume counts, the number of transfers issued on each route, and the possibility of special on-board surveys for a limited number of routes. Thus, our example assumes a variety of aggregate, typically available data. Our application will be to the Portland, OR (USA) tri-met system which has 71 routes. For comparison purposes, we shall use the results of a 12% on-board survey conducted by Tri-Met in 1980.

For simplicity, we will assume that rather than desiring a complete OD matrix, the transit operator is interested in the number of trips to the central business district and the number of transfer trips on each route. This gives four possible destinations: j=1 if the destination is outside the CBD for a non-transfer trip, j=2 if the destination is inside the CBD for a non-transfer trip, j=3 for a transfer trip destined outside the CBD, and j=4 for a transfer trip destined inside the CBD. This gives 71 x 4 = 284 matrix entries to estimate.

To formulate the problem, we factor up the work trip OD matrix entries to base estimates of all trips by multiplying by the ratio of all trips to all transit trips $y = a\underline{w}$. Row and column totals derived from route counts provide 74 independent constraints. If the total number of transfers is also available, an additional 70 constraints are available. For the quadratic or CGLS estimations, we assume that entry variances are proportional to hypothesized values as in P3.

Table 3 reports several measures of errors for estimations with differing numbers of constraints. Comparisons are made to estimates derived from an on-board survey. Since there is no one best measure of the difference between matrices, other measures could also be calculated (McNeil, 1983). For comparison, the errors of estimates obtained using the factored up work trip O/D matrix (referred to as an ad hoc method in Table 3) and from the biproportional method (P5) are also reported. As can be seen, the quadratic and biproportional methods give similar results. As expected, additional constraints reduced the estimation errors. The magnitude of error is substantial, however.

Table 3:   Average Absolute and Relative Errors for Expanding the Portland Work Trip Transit Matrix

Method	Average Absolute Error	Average Relative Error	Average Aggregate Error
No Constraints			
ad hoc	97.0	0.240	0.166
With Row and Column Constraints			
Quadratic	49.7	0.184	0.085
Biproportional	48.4	0.179	0.083
With Row and Column Constraints and Constraints on the Number of Transfers			
Quadratic	35.0	0.155	0.060
Biproportional	34.4	0.151	0.059

To attempt to improve our estimates, we could do a limited survey of several transit routes. How should we select these routes? As discussed earlier, such incremental sampling should attempt to substantially reduce errors. Here, we consider three different strategies:

1. Survey the five routes that only operate during peak hours. This strategy

was chosen as the errors in estimating the entries associated with these routes are expected to be (and are) large.

2. Survey the five routes with the largest cumulative variances for the entries associated with each route. This strategy is an approximation to a strategy that would minimize the trace of the estimated variance–covariance matrix after the survey was complete and the problem re–estimated.

3. Survey the five routes with the highest coefficient of variations. The coefficient of variation for the route is defined to be the square root of the sum of the variances for each destination on the route divided by the total number of passengers using the route.

The two latter survey strategies are based on the assumption that a high variance in an entry's estimate is strongly correlated with its error. Table 4 shows that this is the case. As we might expect, the standard deviation and average absolute error are positively correlated, while the average relative error and the coefficient of variation are positively correlated. Thus, strategy 2 is likely to influence the average *absolute* error more whereas strategy 3 is likely to influence the average *relative* error more.

Table 4: Correlation Coefficients Between Errors and Uncertainty of the Estimates for the Estimated Entries in the Transit Origin–Destination Matrix

Method	Uncertainty	Average Absolute Error	Average Relative Error
With Row and Column Constraints	Standard Deviation	0.509	−0.542
	Coefficient of Variation	−0.389	0.542
With Constraints on Transfers, Rows, and Columns	Standard Deviation	0.475	−0.187
	Coefficient of Variation	−0.204	0.453

Table 5 indicates some resulting aggregate errors due to incremental sampling. The additional sampling did improve the estimates. Sampling the five routes with maximum variance (representing a 7% sample of routes) resulted in about a 15% reduction in average absolute errors. Still, about a 17% error exists after estimation even with 20 known matrix entries and 74 constraints. Figure 1 illustrates the histogram of errors before and after sampling. As can be seen, the histogram shifts to the left, indicating lower errors.

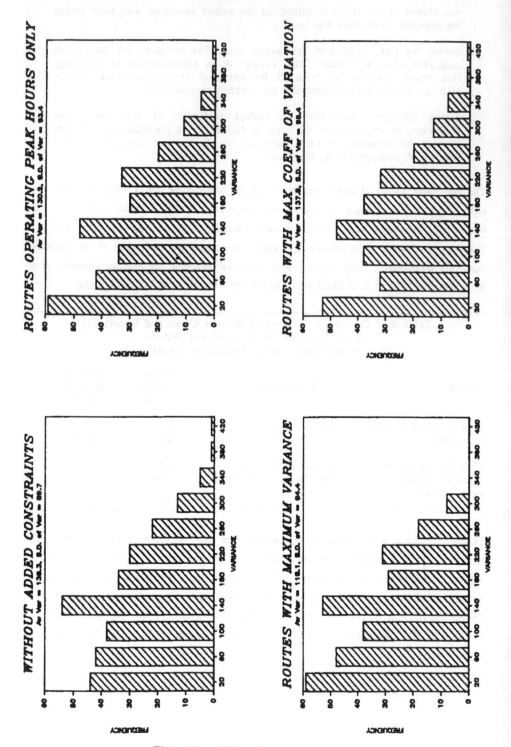

Figure 1:   Histogram of Errors

Table 5: Error Measures for the Portland Transit Origin-Destination Matrix with Incremental Sampling

	Trace of the Variance–Covariance Matrix	Average Absolute Error	Average Relative Error	Average Aggregate Error
With Row and Column Constraints Alone				
Quadratic	7,710,000	50	0.18	0.086
Biproportional	–	48	0.18	0.082
With an Additional Five Routes Surveyed				
1. With Peak Service Only				
Quadratic	7,280,000	48	0.17	0.082
Biproportional	–	47	0.17	0.081
2. With Maximum Variance				
Quadratic	5,980,000	42	0.17	0.072
Biproportional	–	41	0.17	0.070
3. With Maximum Coefficient of Variation				
Quadratic	7,970,000	49	0.17	0.084
Biproportional	–	48	0.16	0.081

7 CONCLUSIONS

We have discussed the various types of errors which arise in matrix estimation within the framework of the constrained, generalized least squared (CGLS) regression method. Both a review of past studies and the example applications described here suggest that the uncertainty in matrix entries is likely to be substantial.

An advantage of the CGLS method is that error estimates can often be calculated. We outlined the use of such estimates to develop incremental sampling strategies to improve estimates. In particular, we discussed the use of relatively inexpensive special surveys such as volume counts, roadside interviews or on-board surveys. With such additional information incorporated in the estimation process, matrix entry estimates generally improved.

Despite such improvements, travel volume estimates and forecasts are likely to remain highly uncertain. Implications of such uncertainty for the planning, design and management of transportation systems is an interesting research topic.

Acknowledgements

The authors are grateful for research support from the American Association of University Women and the U.S. National Science Foundation (Grant CEE-8209755).

REFERENCES

Bacharach, M. (1970). Biproportional Matrices and Input–Output Change, Cambridge University Press.

Carey, M., C. Hendrickson and K. Siddharthan (1981). A Method for Estimation of Origin/Destination Trip Matrices, Transportation Science 15(1), 32–49.

Carey, M. (1982). Constrained Estimation of Direct Demand Functions and Trip Matrices, Working paper, Carnegie–Mellon University.

Cochran, W.G. (1977). Sampling Techniques, John Wiley and Sons, New York, 1977.

Daganzo, C. (1980). Optimal Sampling Strategies for Statistical Models with Discrete Dependent Variables, Transportation Science 14(4), p. 324–345, November.

Fisk, C.S. and D.E. Boyce (1983). A Note on Trip Matrix Estimation from Link Traffic Count Data, Transportation Research 17B(3), p. 245–250, June.

Hewings, G.J.D. and B.N. Janson (1980). Exchanging Regional Input–Output Coefficients: A Reply and Further Comments, Environment and Planning A 10, p. 843–854.

Kirby, H.R. and M.N. Lesse, Trip Distribution Calculations and Sampling Error: Some Theoretical Aspects, Environment and Planning A 10, p. 837–851.

Landau, U., E. Hauer and I. Geva (1982). Estimation of Cross–Cordon Origin–Destination Flows From Cordon Studies, Transportation Research Record 891, p. 5–9.

McNeil, Sue (1983). Quadratic Matrix Entry Estimation, PhD. Thesis, Carnegie–Mellon University.

Sikdar, P.K. and B.G. Hutchinson (1981). Empirical Studies of Work Trip Distribution Models, Transportation Research 15A, p. 233–243.

Theil, H. (1971). Principles of Econometrics, John Wiley and Sons, Inc., New York, NY.

Uribe, P., C.G. deLeuw and H. Theil (1965). The Information Approach to the Prediction of Interregional Trade Flows, Tech. Report 6507, Econometric Institute of the Netherlands School of Economics.

Van Zuylen, H.J. and L.G. Willumsen (1980). The Most Likely Trip Matrix Estimated from Traffic Counts, Transportation Research 14B(3), p. 281–294.

Wilson, A.G. (1970). Entropy in Urban and Regional Modelling, Pion, London.

Wohl, M. and C. Hendrickson (1984). Principles of Transportation Pricing and Investment, John Wiley.

PAPER 21

Ninth International Symposium on
Transportation and Traffic Theory
© 1984 VNU Science Press, pp. 431–450

A SYSTEMS DYNAMICS APPROACH TO THE ESTIMATION OF ENTRY AND EXIT O–D FLOWS

MICHAEL CREMER[1] and HARTMUT KELLER[2]
[1]*Universität Hamburg-Harburg, Hamburg, F.R.G.*
[2]*Technische Universität München, München, F.R.G.*

ABSTRACT

After a review of recent approaches to determine traffic flows
from traffic counts at complex traffic facilities a new method
for the estimation of the origin-destination matrix is presented.
For this purpose traffic flow in the facility is treated as a
dynamic process. The model uses the information contained in
the time sequences of the traffic volumes at the entries and
exits at a complex intersection and solves the problem with a
parameter optimisation technique satisfying the given constraints exactly.
The comparison with real and synthetic data proved the new
method to be superior to previously published approaches to
the problem. Because of the efficiency of the algorithm, the
procedure could be implemented on a microcomputer at the con-
trolsite giving online the data basis for traffic responsive
control.

1.   INTRODUCTION

The knowledge of the flows from any entry to each exit is an
important information for the design and control of a complex
intersection or a network. However, while the volumes passing
a cross-section can easily be collected by conventional traffic
counters, the flows from any entry (or origin) to the different
exits (or destinations) can be measured only with a rather
expensive instrumentation or even not at all. For this reason
considerable effort has been invested in the past to develop
methods which provide satisfactory estimates for the unknown
origin-destination flows (for a survey see Willis and May (1981)
and Bell (1983)).

Since measurements of the accumulated traffic counts at the
entries and the exits of an intersection or of a network provide only a highly underdetermined set of equations for the
unknown OD flows, the main interest of the methods proposed
in recent years was directed toward the problem of finding
additional information and assumptions for the origin-destination pattern, to obtain reliable estimates for the unknown
flows.

For example, in Beil (1979) a solution is proposed which minimizes a certain norm of the O-D-matrix. Van Zuylen and Willumsen (1980) try to find a solution which is most likely with
respect to some measure of likelihood (see also Sammer et al
(1982)). Other estimation methods involve certain apriori
assumptions of the structure of traffic demand (Carey et al,
1981) or try and use observations about volumes on the system
in order to make inferences about the prevailing trip pattern
embodied in an O-D-matrix (Hauer, Tom Shin, 1981). Most of
these and other related methods are unsatisfactory from several
points of view. First, they still yield only approximate solutions and don't give any statements to what extent the additional assumptions meet the real situation. Second, they often

need considerable computational effort which might be a serious
impediment for an online implementation. Third, they are designed to estimate more the long time means of the unknown flows
rather than the short time variations which might be needed for
traffic adaptive control. However, as has been found by Van Vliet
and Willumsen (1981) using a comprehensive data base the origin-
destination pattern might often change from hour to hour as well
as from day to day.

In two recent approaches Cremer and Keller (1981) and Cremer
(1983) have made an attempt to get the missing information from
the measurements itself treating the generation of OD flows
within an intersection or within a small network as a dynamic,
causal process. While the first approach uses a recursive,
dynamic adaptation formula the second approach is built up on
cross correlation techniques. These approaches have the advantage that the whole information to be processed is taken from
the real system without any assumption. By that way the methods
are able to track even slow variations of the O-D-matrix making
the methods especially useful for an online application at the
site of data collection for traffic intersection control.

It is the purpose of this paper to present a third new approach
in which the traffic flow through an intersection again is
treated as a dynamic, causal process. However, here the problem
of identification of the OD flows is formulated as a parameter
optimization problem for a reference model which is tuned to
give best coincidence with the real system. In contrast to the
methods mentioned above, this approach can directly and exactly
satisfy constraints which are given in the form of certain
equality and inequality constraints. Moreover, it can be shown
that the expectation of the estimates of the OD flows converges
to the real long term average flows when the observation period
is increased. The method has been tested with real as well as
with synthetic data with good results. Comparative studies with
the two previous methods are included.

2.   PROBLEM STATEMENT

To derive a mathematical model of the process of OD flows generation we consider a complex intersection with m entries and n exits as shown in Fig. 1.

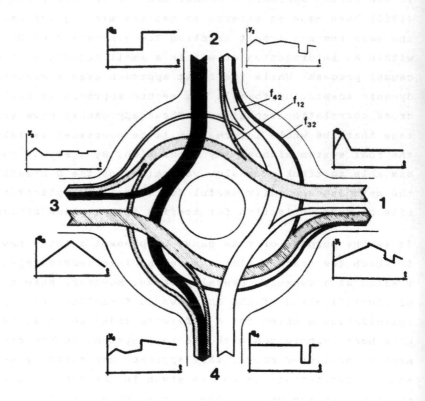

Fig. 1. Topology of a complex intersection with 4 entries and 4 exits.

To describe the time-dependent process of traffic flows through the intersection, the following variables are introduced:

$q_i(k)$:   volume which enters entrance i during time inter-
              val $(k-1)$ $T \leq t < k \cdot T$
              where $i = 1, \ldots m$

$y_j(k)$:   volume which leaves exit j during time interval
              $(k-1)$ $T + \tau \leq t < kT + \tau$
              where $j = 1, \ldots n$

$f_{ij}(k)$:   that part of volume $q_i(k)$ which leaves the inter-
              section through exit j during time interval
              $(k-1)$ $T + \tau \leq t < kT + \tau$

Here T is a sampling interval defining short measurement periods. For an intersection this interval should be chosen
to be only a few minutes in order that the time variations of the volumes are expressed representatively. The parameter τ denotes the average travel time a vehicle needs to
pass from an entry to an exit. In the case that a network is
considered, it might become necessary to choose a longer sampling interval T and to introduce individual travel times τ_{ij}
between entry i and exit j.

The basic idea of the presented approach might be explained
best by regarding the volumes as functions over time as depicted in Fig. 1. Suppose the arriving rates q_i at the entries
of the intersection have the particular profiles shown in Fig. 1
Assuming further that the splitting pattern into the OD flows
is strictly constant over time, this leads to the shown shapes
of departing rates over time, simply by putting together the
corresponding parts of the different entry flows. Now, under
these idealistic assumptions it can be deduced by inspection
from the time responses of the departing volumes what part of
the different entry flows is contained in each exit flow. In
reality, however, the arrival rate curves are now here near so
distinct, but look more like a random sequence and the splitting of the entry flows into the OD flows varies from time
interval to time interval randomly about a certain mean.

We now take the balance of the departing and entering vehicles which gives for each time interval

$$y_j(k) = \sum_{i=1}^{m} f_{ij}(k) . \tag{1}$$

Since each OD flow $f_{ij}(k)$ is a certain portion of the i-th entering flow, we may introduce split parameters $b_{ij}(k)$

$$f_{ij}(k) = q_i(k) \cdot b_{ij}(k) \tag{2}$$

where the parameters b_{ij} are obviously bounded by the inequality condition

$$0 \leq b_{ij}(k) \leq 1 \quad \text{for all } i,j,k \tag{3}$$

and must satisfy by the law of conservation of vehicles

$$\sum_{j=1}^{n} b_{ij}(k) = 1 \quad \text{for all } i,k . \tag{4}$$

Inserting Eq. (2) into Eq. (1) yields

$$y_j(k) = \sum_{i=1}^{m} q_i(k) \, b_{ij}(k) \quad \text{for all } j,k \tag{5}$$

Introducing the 1 x n row vector $\underline{y}'(k)$ with elements $y_j(k)$, the 1 x m row vector $\underline{q}'(k)$ with elements $q_i(k)$ and forming a m x n matrix $\underline{B}(k)$ from the elements $b_{ij}(k)$, Eq. (5) may be written for all indices j in compressed form

$$\underline{y}'(k) = \underline{q}'(k) \cdot \underline{B}(k) \tag{6}$$

Here ' denotes a row vector or transposition of a matrix, vectors without the prime are column vectors. Next we consider a period of K sampling periods and introduce the following mean values

$$\bar{q} = \frac{1}{K} \sum_{k=1}^{K} q(k)$$

$$\overline{\underline{y}} = \frac{1}{K} \sum_{k=1}^{K} y(k) \tag{7}$$

$$\overline{\underline{B}} = \frac{1}{K} \sum_{k=1}^{K} \underline{B}(k)$$

Then we may write $\underline{q}(k)$, $\underline{y}(k)$ and $\underline{B}(k)$ as the sum of their mean and a (random) deviation

$$\underline{q}(k) = \overline{\underline{q}} + \Delta \underline{q}(k)$$
$$\underline{y}(k) = \overline{\underline{y}} + \Delta \underline{y}(k) \tag{8}$$
$$\underline{B}(k) = \overline{\underline{B}} + \Delta \underline{B}(k)$$

Inserting this into Eq. (6) gives the model equation of an intersection

$$\underline{y}'(k) = \underline{q}'(k) \, (\overline{\underline{B}} + \Delta \underline{B}(k)) \tag{9}$$

or

$$\overline{\underline{y}}' + \Delta \underline{y}'(k) = (\overline{\underline{q}}' + \Delta \underline{q}'(k)) \cdot (\overline{\underline{B}} + \Delta \underline{B}(k))$$

It can be easily shown by the definition of $\overline{\underline{B}}$ (Eq. (7)) that the mean split parameters $\overline{b}_{ij}$ must meet the restrictions of Eq. (3) and Eq. (4) too.

Once the split parameters $b_{ij}(k)$ are known, the OD flows are easily computed from Eq. (2). Unfortunately for any interval k, Eq. (4) and Eq. (6) give only $m + n - 1$ linearly independent relations for the unknown $m \cdot n$ split parameters when the measurements of $q_i(k)$, $y_j(k)$ are known. That means that this set of equations can be satisfied by a large multiplicity of solutions for the unknown split parameters. On the other hand data from several time intervals can be used together, and indeed for the purpose of traffic signal control as well as for design purposes the knowledge of the mean split parameters $\overline{\underline{B}}$ is much more important than the knowledge of the matrix $\underline{B}(k)$ of a single sampling interval. Thus, the problem may be stated now in the following way: given measurements of sequences of entering and departing flow samples, $q_i(k)$, $y_i(k)$ over a period $k=1,...K$, find the mean split parameters $\overline{\underline{B}}$ of this intersection.

Taking the mean of Eq. (9) over the period of K intervals
yields

$$\bar{\underline{y}}' = \bar{\underline{q}}' \cdot \bar{\underline{B}} + \frac{1}{K} \sum_{k=1}^{K} \Delta q'(k) \, \Delta \underline{B}(k) \tag{10}$$

Under the realistic assumption that the deviations $\Delta q_i(k)$
and $\Delta b_{ij}(k)$ are statistically independent, the second term
on the right hand side of Eq. (10) becomes negligible for K
not too small a number, which gives

$$\bar{\underline{y}}' = \bar{\underline{q}}' \cdot \bar{\underline{B}} \tag{11}$$

This equation might be regarded as an additional condition
which the unknown split parameters $\bar{b}_{ij}$ must satisfy.

Subtracting Eq. (11) from Eq. (9) results in

$$\Delta \underline{y}'(k) = \underbrace{\Delta \underline{q}'(k)\bar{\underline{B}}}_{\substack{\text{deterministic} \\ \text{information}}} + \underbrace{\bar{\underline{q}}'\Delta \underline{B}(k) + \Delta \underline{q}'(k)\Delta \underline{B}(k) - \frac{1}{K} \sum_{k=1}^{K} \Delta \underline{q}'(k)\Delta \underline{B}(k)}_{\text{random noise term}} \tag{12}$$

This equation represents with its first part the additional
deterministic information we obtain by recording the entering
and leaving volumes as time sequences. The last three terms
on the right hand side show that this information is corrupted
by random noise terms.

The question which arises at this point is how can the additional information be used optimally to estimate the unknown
split matrix $\bar{\underline{B}}$ and to what extend the result is detracted by
the noise terms.

As mentioned above, two different methods have been proposed
recently to estimate the split parameters. The first method
(Cremer and Keller, 1981) proposes a dynamic recursion formula
for each split parameter

$$\hat{b}_{ij}(k) = \hat{b}_{ij}(k-1) + \gamma \cdot \Delta q_i(k) \cdot [\Delta y_j(k) - \Delta \hat{y}_j(k)] \qquad (13)$$

where $\Delta \hat{y}_j(k)$ is the deviation of the exit volume j as predicted applying the model Eq. (6) with the estimates $\underline{B}(k-1)$ from the last sampling period. γ is a weighting factor which has to be chosen small enough to guarantee stability of the recursion formula.

In the second approach (Cremer, 1983), Eq. (9) was premultiplied by the column vector $\underline{q}(k)$ and then the mean of the resulting expression was taken of the period of K samplings. This results in

$$\Phi_{qy}(K) = \underline{\Phi}_{qq}(K) \cdot \overline{\underline{B}} + \frac{1}{K} \sum_{k=1}^{K} \underline{q}(k) \cdot q'(k) \, \Delta\underline{B}(k) \qquad (14)$$

where

$$\Phi_{qy}(K) = \frac{1}{K} \sum_{k=1}^{K} \underline{q}(k) \cdot \underline{y}'(k) \qquad (15)$$

$$\Phi_{qq}(K) = \frac{1}{K} \sum_{k=1}^{K} \underline{q}(k) \cdot \underline{q}'(k) \qquad (16)$$

are the finite interval cross-correlation matrices of the time sequences $\underline{q}(k)$, $\underline{y}(k)$ and $\underline{q}(k)$, $\underline{q}(k)$, respectively. Neglecting the second term on the right hand side, which becomes smaller as K increases, Eq. (14) can uniquely solved for $\overline{\underline{B}}$ as long as the input sequences $q_i(k)$ are not totally correlated.

A new approach is now presented which determines estimates for the split matrix $\overline{\underline{B}}$ by solving a parameter optimization problem. This approach has the advantage that it is capable to include fully the equality and inequality constraints of Eqn's (3), (4) and (11). Following up are the results of this method when compared with the results which are obtained from the former two approaches.

3. ESTIMATION PROCEDURE

Let us assume that the real process of partial flow generation is represented sufficiently well by the stochastic model
Eq. (9). We then formulate a similar deterministic model of
the intersection introducing an estimate $\hat{\underline{B}}$ for the unknown
mean split matrix $\bar{\underline{B}}$:

$$\hat{\underline{y}}'(k) \;=\; \underline{q}'(k) \cdot \hat{\underline{B}} \qquad\qquad (17)$$

This model is driven by the measured entering flow samples
q(k) and generates on the basis of $\hat{\underline{B}}$ estimates $\underline{y}(k)$ of the
exit volumes.

Then we reformulate the problem in the following manner:

Given the sequences $q_i(k)$, $y_j(k)$ of flow samples at the
entries and exits of an intersection over a period of K
sampling intervals.

Find a matrix $\hat{\underline{B}}$ of estimates for the split parameters
which minimizes the sum of the squared error between
the exit flows of the model and the measured exit flows

$$J \;=\; \frac{1}{K} \sum_{k=1}^{K} \; ||\; \underline{y}(k) - \hat{\underline{y}}(k)\;||_2 \;\rightarrow\; \underset{\hat{\underline{B}}}{\text{Min}} \qquad (18)$$

where the estimates $\hat{b}_{ij}$ have to fulfill conditions of Eq.
(4) and Eq. (11) and inequality constraints of Eq. (3).

By this formulation the problem of estimating the split parameters $\bar{b}_{ij}$ and by that way the partial flows $\bar{f}_{ij}$ is posed as
a parameter optimization problem with constraints.

Inserting the model Eq. (17) into Eq. (18) gives

$$(19)$$
$$J = \frac{1}{K} \sum_{k=1}^{K} \; [\underline{y}'(k)\underline{y}(k) - 2\,\underline{q}'(k) \cdot \hat{\underline{B}}\,\underline{y}(k) + \underline{q}'(k)\,\hat{\underline{B}}\hat{\underline{B}}'\underline{q}(k)]$$

Since $\underline{y}(k)$ and $\underline{q}(k)$ are measured counts and by that way known numbers, Eq. (19) is a quadratic expression for the unknown split parameters $\hat{b}_{ij}$. That means that J is an unimodal function of the split parameters which has only one minimum which is the global minimum of that function. Assuming at this moment that the minimum of J lies inside the space of feasible parameter values defined by the constraints of Eqn's (3), (4) and (11), then the optimal solution for $\hat{\underline{B}}$ can be determined by

$$\frac{\delta J}{\delta \underline{B}} \overset{!}{=} \underline{0} = -\frac{2}{K} \sum_{k=1}^{K} \underline{q}(k)y'(k) + \left(\frac{2}{K} \sum_{k=1}^{K} \underline{q}(k)q'(k) \right) \hat{\underline{B}}_{opt} \qquad (20)$$

With the definition of the finite interval cross-correlation matrices of Eq. (15) and Eq. (16) this equation can be solved for $\hat{\underline{B}}_{opt}$ yielding

$$\hat{\underline{B}}_{opt} = \phi_{qq}^{-1} \cdot \phi_{qy} \qquad (21)$$

which is the solution we would obtain from Eq. (14) neglecting the random term on the right hand side. This shows, that the solution obtained from the cross-correlation matrices approach is an optimal solution in the sense stated above as long as this solution meets the given constraints. Unfortunately, as case studies have shown, this does not hold in many cases. In these cases the optimal solution lies on the boundary of the admissible space and has to be determined by numerical optimization methods.

Furthermore, it is convenient to introduce the additional condition

$$\hat{b}_{ii}(k) = 0 \quad \text{for } i = 1,\ldots\min(m,n) \qquad (22)$$

which expresses the realistic assumption that no car leaves the intersection through the same port he entered. From the total of $n \cdot m$ elements of the split matrix $\hat{\underline{B}}$ m elements are bound by condition of Eq. (4), n - 1 elements are bound by

the condition of Eq. (11) for the balance of the accumulated
flows (one equation of Eq. (11) is already contained within
the n - 1 other equations together with condition of Eq. (4))
and min (m,n) elements are fixed by Eq. (22). Thus, only

$$n_p = n \cdot m + 1 - n - m - \min (m,n) \tag{23}$$

elements are left as free parameters p_i for the optimization
procedure while the other elements are linearly dependent on
them. For the parameter optimization algorithm applied here a
simple routine was written by which the vector of free parameters p was assigned to a certain subset of split parameters
while the remaining set was determined using the equality
conditions in a consistent manner.

To investigate the quality of the solution $\hat{B}$ obtained by minimizing the performance index J, we substitute the measured
exit flow vectors $\underline{y}(k)$ using the process Eq. (9). After some
simple algebraic manipulations we get

$$J = \frac{1}{K} \sum_{k=1}^{K} \left[\underline{q}'(k) \; (\bar{\underline{B}} - \hat{\underline{B}}) \cdot (\bar{\underline{B}}' - \hat{\underline{B}}') \underline{q}(k) + 2 \; \underline{q}'(k)(\bar{\underline{B}} - \hat{\underline{B}}) \cdot \Delta \underline{B}'(k)\underline{q}(k) \right.$$

$$\left. + \underline{q}'(k) \; \Delta \underline{B}(k) \cdot \Delta \underline{B}'(k) \underline{q}(k) \right] \tag{24}$$

The first term within the brackets is a positiv semidefinit
expression which obviously is minimal if $\hat{\underline{B}}$ is equal to $\bar{\underline{B}}$.
However, the second term which depends on $\hat{\underline{B}}$ too is an indefi-
nit expression and might become negativ for some matrices $\hat{\underline{B}}$.
The last term is a positiv definit term and does not depend
on $\hat{\underline{B}}$. From this it is seen that for a finite interval K the
matrix $\hat{\underline{B}}_{opt}$ which minimizes (24) might (or mostly will) be
a matrix $\hat{\underline{B}} \neq \bar{\underline{B}}$ which yields a negativ value of the second
term compensating the positiv definite terms to some extent.
However, due to the quadratic form of the first term the op-
timal estimate $\hat{\underline{B}}_{opt}$ will be near to the real matrix $\bar{\underline{B}}$. Fig. 2
shows this situation where only one free element $\hat{b}_{ij}$ is va-
ried while all other elements are kept at their optimal values.

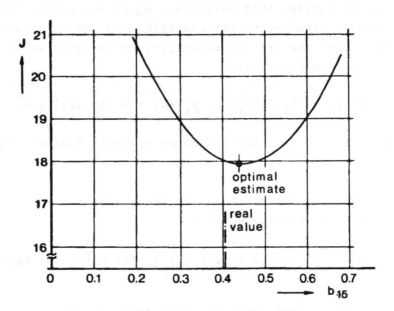

Fig. 2. Performance index J as function of element $\hat{b}_{15}$ for an intersection with m = 5 entries and n = 5 exits.

For a deeper investigation of the estimation error $\overline{\underline{B}} - \hat{\underline{B}}_{opt}$ we assume again that the optimal split parameter set lies in the interior of the admissible subspace of the parameter space. Inserting then Eq. (9) into Eq. (20) gives for the estimation error

$$\hat{\underline{B}}_{opt} - \overline{\underline{B}} = \underline{\Phi}_{qq}^{-1}(K) \cdot \frac{1}{K} \sum_{k=1}^{K} \underline{q}(k)\underline{q}'(k) \cdot \Delta\underline{B}(k) \tag{25}$$

Introducing the following notation

$$\begin{aligned} \underline{\psi}_i'(k) \quad &: \quad \text{i-th row of}(\underline{\Phi}_{qq}^{-1} \cdot \underline{q}(k) \cdot \underline{q}'(k)) \\ \Delta\underline{b}_j(k) \quad &: \quad \text{j-th column of } \Delta\underline{B}(k) \end{aligned} \tag{26}$$

We can write for the estimation error of the i,j-th element

$$\hat{b}_{ij} - \overline{b}_{ij} = \frac{1}{K} \sum_{k=1}^{K} \underline{\psi}_i'(k) \cdot \Delta\underline{b}_j(k) \tag{27}$$

Taking now on both sides the statistical expectation under the assumption that deviations $\Delta \underline{q}(k)$ are statistically independent from the variations $\Delta \underline{B}(k)$ and taking into consideration that the arrival rates $q(k)$ and by that way $\underline{\psi}_i(k)$ are known numbers we get

$$E \{\hat{b}_{ij} \dot{-} \overline{b}_{ij}\} = \frac{1}{K} \sum_{k=1}^{K} \underline{\psi}_i{}'(k) \cdot E \{\Delta \underline{b}_j(k)\} = 0 \qquad (28)$$

which shows that the estimated split parameters are bias free.

Considering the variances of the estimation error we obtain under the same assumptions

$$\qquad\qquad\qquad\qquad\qquad\qquad\qquad\qquad\qquad (29)$$
$$\mathrm{Var} \{\hat{b}_{ij} - \overline{b}_{ij}\} = \frac{1}{K^2} \sum_{k=1}^{K} \sum_{l=1}^{K} \underline{\psi}_i{}'(k) \cdot E \{\Delta \underline{b}_j(k) \cdot \Delta \underline{b}_j{}'(l)\} \cdot \underline{\psi}_i(l)$$

Assuming further that the variations $\Delta b_{ij}(k)$ are random white noise with

$$E \{\Delta \underline{b}_j(k) \cdot \Delta b_j{}'(l)\} = \begin{cases} O & \text{for } k \neq l \\ \underline{R}_j & \text{for } k = l \end{cases} \qquad (30)$$

Then we have

$$\mathrm{Var} \{\hat{b}_{ij} - \overline{b}_{ij}\} = \frac{1}{K^2} \sum_{k=1}^{K} \underline{\psi}_i{}'(k) \cdot \underline{R}_j \cdot \underline{\psi}_i(k) \qquad (31)$$

Since the sum on the right hand side increases approximately linearly with the observation period K, the premultiplication by K^{-2} effects that the variances of the estimation errors decay with K^{-1} and converge to zero as K grows to infinity.

4. RESULTS

The proposed method was tested with a number of synthetic data sets as well as with real data which were collected from an intersection in the city of Munich during a two hour interval.

In both cases the results were compared with the results obtained by the approaches of Cremer and Keller (1981) and Cremer (1983) mentioned above. For the parameter optimization procedure we have applied the Complex Algorithm of Box (1965) which is capable to handle inequality constraints.

First, traffic flow was simulated through an intersection with five entries and five exits yielding a total number of twenty OD flows to be identified. Both the variances of the random variations of the entering volumes $\Delta q_i(k)$ as well as the variances of the deviations of the split parameters $\Delta b_{ij}(k)$ were chosen according to real observations. In Fig. 3 the representative time responses are shown for two split parameters b_{14} and b_{15}. Though the random variations of the individual sampling intervals are rather high, it is seen that the average value of the split parameters is estimated fairly good. The quality of the estimates improves with the length K of the optimization period.

To test the capability of the estimates to track even variations of the short time means of the split parameters, the two parameters b_{13} and b_{14} were altered according to a cosine half wave. This capability might be highly important in cases when the procedure is to be applied for traffic responsive control by traffic lights. As Fig. 4 shows, the tracking capability of the estimates is fairly good.

For the purpose of comparison, the maximum estimation error e_{max} and the average estimation error $\bar{e}$ were computed from the total set of errors $|\hat{b}_{ij} - \bar{b}_{ij}|$ again for an intersection with five entries and five exits. The same case study was carried out for the three dynamic estimation procedures: the procedure of parameter optimization with constraints proposed here, the method of inversion of the cross-correlation

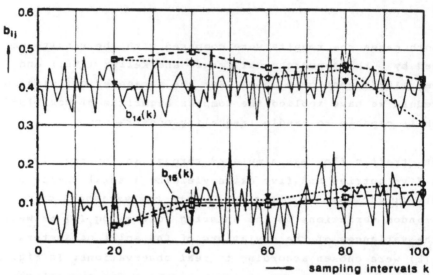

Fig. 3. Real split parameters $b_{14}(k)$ and $b_{15}(k)$ and estimates
  ▼ real mean values over a period of K = 20
  ◉ estimates from an optimization period of K = 20
  ▢ estimates from optimization periods of K = 20,40,60,80,100

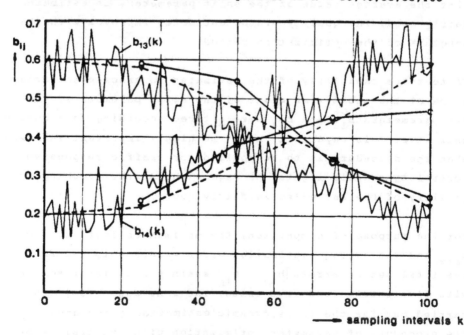

Fig. 4. Tracking behaviour of the estimates
    short time mean values of $b_{13}(k)$ and $b_{14}(k)$ (K = 25)
    estimates for $b_{13}(k)$ and $b_{14}(k)$ (K = 25)

matrix of Cremer (1983) (which was shown above to be equivalent to an unconstrained optimization) and the procedure of
Cremer and Keller (1981) using a recursive adaptation formula
of Eq. (13). The results are listed in Table 1. As expected,
it is seen that the new method presented here is superior to the
previous ones because it is capable to incorporate all real
constraints.

Method	Duration of the estimation period (number of sampling intervals)					
	K = 20		K = 40		K = 80	
	e_{max}	$\bar{e}$	e_{max}	$\bar{e}$	e_{max}	$\bar{e}$
I Constrained minimization (Eq. (18))	0.151	0.059	0.138	0.063	0.075	0.036
II Finite interval cross-correlation (Eq. (14), Eq. (21))	0.191	0.089	0.176	0.074	0.094	0.040
III Dynamic recursive adaption (Eq. (13))	0.186	0.065	0.188	0.071	0.139	0.047

Table 1. Estimation errors for the presented procedure as
compared to the methods of Cremer and Keller (1981) and Cremer
(1983)

With regard to the computational effort, the method presented here needs the most computation time due to the iterative
character of the optimization routine. The main computational
work of method II is due to the inversion of a m x m - matrix.
Though this might mean less effort than required for method
I, from the view point of error propagation it may be even
more critical. Finally method III which yields slightly worse
results needs the less computational effort. Nevertheless,
for intersections with up to ten entries and ten exits each me-

thod can be implemented on a standard microcomputer for an online application at the site of an intersection.

Finally, we have tested the procedure using real data which were collected from an intersection in the city of Munich. The intersection is the crossing of a one way road with a two-way street. In terms of our model Eq. (6) this gives a three by four matrix $\underline{B}$ with a zero column and zero diagonal elements. Though this is a simple intersection, taking into account all equality conditions there are still two parameters completely free according to Eq. (23). Table 2 gives the matrices of the real and the estimated OD flows f_{ij} for three subsequent periods, each of K = 20 sampling intervals. The results demonstrate that the method works quite well too when applied to real data.

	Period I				Period II				Period III			
real flows	0	0	197	31	0	0	227	22	0	0	238	42
	64	0	38	521	60	0	50	470	67	0	41	462
	153	0	0	25	151	0	0	29	146	0	0	17
estim. flows	0	0	196	32	0	0	216	33	0	0	242	38
	77	0	39	507	67	0	61	452	82	0	37	451
	140	0	0	38	144	0	0	36	131	0	0	32

Table 2. Estimation results (veh.) for real data matrix of OD flows f_{ij} for successive periods.

5. CONCLUSIONS

A new method is presented for estimating the matrix of origin
and destination flows from the entries to the exits of a
traffic facility on the basis of traffic counts at these ports.
Traffic flow in the facility is treated as a dynamic process,
considering the traffic volumes at and between the counting
stations in consecutive time intervals.

The model is formulated for complex intersections making use
of the whole information contained in these time sequences,
without the need of arbitrary assumptions on driver behaviour. Based on the information from the time sequences of vehicle arrivals and departures at the intersection a parameter optimization method is developed which satisfies given
constraints exactly.

The performance of the method has been tested with real and
synthetic data. It was shown that it gives good estimates of
the desired OD flows and proved to be suited for the identification of time variant origin and destination structures,
too. The comparison with the results of previously published
methods based on similar problem formulations showed that the
new method is superior to these. Because of the efficiency
of the algorithm the procedure could be implemented on microcomputers to give at the site of control online the data basis
for traffic responsive control.

REFERENCES

Beil, D. (1979)."Matrixmodelle zur Ermittlung von Teilströmen
aus Querschnittsmessungen." - Ph. D.-Thesis, Universität
Karlsruhe.
Bell, M.G.H. (1983). "The estimation of an origin-destination
matrix from traffic counts." - Transp. Science, 17, pp. 198-
217.
Box, M.J. (1965). "A new method of constrained optimization
and a comparison with other methods." - Comput. Journal, 8,
pp. 42-52.
Carey, M., Hendrichson, C. and Siddhartan, K. (1981). "A method for direct estimation of origin-destination trip matrices." - Transp. Science, 15, pp. 32-49.
Cremer, M. and Keller, H. (1981). "Dynamic identification of
flows from traffic counts at complex intersections." - Prepr.
of the 8th Int. Symp. on Transp. and Traffic Theory, Toronto,
Canada, pp. 199-209.
Cremer, M. (1983). "Determining the time dependent trip distribution in a complex intersection for traffic responsive
control." - Prepr. of the 4th Int. IFAC/IFIP/IFORS-Conference
on Control in Transportation Systems, Baden-Baden, Germany,
pp. 145-151.
Hauer, E., Tom Shin B-Y. (1981). "Origin-Destination matrices
from Traffic counts: application and validation on simple
systems. Traffic Engineering and Control, p. 118-121.
Sammer, G., Zelle, K., Schechtner, O. (1982). "Fortschreibung
einer Matrix der Verkehrsbeziehungen mittels Querschnittszäh-
lungen." - Straßenverkehrstechnik, 26, Heft 1.
Van Vliet, D. and Willumsen, K.G. (1981). "Validation of the
ME2 Model for estimation trip matrices from traffic counts."
- Prepr. of the 8th Int. Symp. on Transp. and Traffic Theory,
Toronto, Canada, pp. 191-198.
Van Zuylen, H.J. and Willumsen, L.G. (1980). "The most likely
trip matrix estimated from traffic counts." - Transp. Res.,
14B, pp. 281-293.
Willis, A.E. and May, A.D. (1981). "Deriving origin-destina-
tion Information from routinely collected traffic counts." -
Research Report UCB-ITS-RR-81-8, Institute of Transportation
Studies, University of California.

PAPER 22

Ninth International Symposium on
Transportation and Traffic Theory
© 1984 VNU Science Press, pp. 451–469

LOG-LINEAR MODELS FOR THE ESTIMATION OF ORIGIN–DESTINATION MATRICES FROM TRAFFIC COUNTS: AN APPROXIMATION

MICHAEL G. H. BELL

Institut für Verkehrswesen, Universität (TH) Karlsruhe, Karlsruhe, F.R.G.

ABSTRACT

Origin–destination matrices play a fundamental role in transport planning and analysis as well as in traffic management and control. However, their estimation by direct means such as interviews, licence plate surveys and aerial photography is both time consuming and expensive, as well as being unsuitable for on-line applications where the matrix is to be frequently updated. Therefore indirect procedures based on traffic counts and various forms of route assignment information have received much attention recently. These procedures are reviewed in the paper. In particular, a family of log-linear models has proved popular although iterative calibration of the models is required. This paper suggests a non-iterative procedure for obtaining an approximate log-linear model solution. It is suggested that this approximation is particularly suitable for on-line applications. Certain statistical properties of the approximate solution are discussed, and a relationship to Bayesian estimation is shown. The closeness of the approximation is examined empirically and is found to depend strongly on the quality of the prior information.

1. BACKGROUND

The origin-destination (OD) matrix constitutes part of the basic
information about transport demand required both for transport planning
and for traffic control. However, the collection of OD information
directly by interview (roadside, household or workplace), by licence
plate surveys or by aerial photography is expensive and often not
appropriate. On the one hand, the structure of OD flows can vary
substantially over relatively short periods of time (see Van Vliet and
Willumsen, 1981). On the other, on-line traffic monitoring and control
procedures are currently being developed which require frequently updated
estimates of OD flows.

In contrast, the collection of link flow data by electronic vehicle
presence detectors or other types of traffic counter is inexpensive, can
be carried out at a large number of sites simultaneously and does not
necessitate the disruption of existing traffic patterns. Moreover, in
many cases traffic measuring equipment is already in situ. Thus there
is a continuing interest in procedures for the inference of OD information
from link flow data.

The problem has two commonly encountered forms. In the first and
essentially simpler, the OD matrix relates to vehicle turning volumes
within a junction which may be complex. Turning volumes that are not
directly measurable are estimated from observations of the junction
inflows and outflows, which correspondend respectively to the OD matrix
row and column sums. Route assignment information is not required. In
the second form, the estimated OD matrix relates to networks of more
than one junction. In this case, the OD matrix row and column sums are
not measured directly. Unless the network is of a single path type,
route assignment information is required.

2. PREVIOUS WORK

Most of the procedures developed expressly for the single intersection resemble growth factor methods. Jeffreys and Norman (1977) and Norman, Hoffman and Harding (1979) have suggested heuristic, non-iterative procedures. Mekky (1979) and Van Zuylen (1979) derived models by the consideration of constrained optimisation problems. Perhaps the most attractive of these is the biproportional model. Ways of obtaining prior turning probabilities for the biproportional model have been investigated by Hauer, Pagitsas and Shin (1981) and by Mountain and Westwell (1983). The performance of alternative growth factor methods for the estimation of turning volumes has been assessed by Buehler (1983).

A rather different procedure, suitable for the on-line estimation of turning volumes, has been suggested by Cremer and Keller (1981). Essentially, turning volumes are estimated by correlating variations in measurements of junction inflows and outflows, allowing for the average time spent by vehicles in the junction.

In the case of the network of more than one junction, a wide range of procedures have been developed. These are reviewed in Willumsen (1978), in Chan et al. (1980) and more recently in Nguyen (1983).

When proportional or all-or-nothing assignment is assumed, the values of the elements of the OD matrix are related to the link volumes by a set of linear equations, defined in Section 4. It may be that the equations have no feasible solution for the OD matrix. Infeasibility might come from two sources, namely the observed flows or the route assignment proportions. Inconsistency in the observed flows, resulting from the violation of the conservation of flow at some nodes, can be remedied by the procedures suggested by Van Zuylen and Willumsen (1980) and by Van Zuylen and Branston (1982).

In general, these equations do not uniquely define the OD matrix because the number of OD flows to be estimated exeeds the number of linearly independent equations. Beil (1979) sought a solution by the calculation of a generalised inverse for the matrix of route assignment proportions

(see Section 6). Some have sought to reduce the number of variables to
be estimated by introducing a gravity or direct demand distribution
model. In this way, Robillard (1975) and Högberg (1976) used non-linear
regression, and Carey et al. (1981) constrained optimisation, in order
to obtain the estimated OD matrix. Others have introduced a convex
objective function which is minimised with respect to the estimated OD
flows subject to the constraints imposed by the assignment equations.
Generally, the objective functions have the form of a distance measure
and therefore require a target trip matrix. Through the target trip
matrix, prior information about the OD matrix may be incorporated.
Objective function formulations have been derived from various
considerations, such as information (Van Zuylen, 1979, and Van Zuylen
and Willumsen, 1980) and entropy (Willumsen, 1978, as well as Van Zuylen
and Willumsen, 1980). The resulting constrained optimisation problems
have closely related log-linear models as explicit solutions.

In practice, measurements of link volumes are sampled quantities and
therefore random variables. Maher (1983) allows for this within a
Bayesian statistical framework. Cascetta (1983) suggests a weighted
least squares approach. Bell (1983a,b) has described a procedure for
the estimation of the variances and covariances of log-linear model
fitted values from variances and covariances for the link volumes.
This procedure is outlined in Section 5.

It has been argued that the assumption of constant route assignment
proportions (as defined in Section 4) is under certain circumstances
unreasonable. Moreover, a method which combines OD matrix estimation
with the determination of route assignment would obviate the need to
determine these proportions externally. In this context, equilibrium
route assignment has some attractions. By making strong assumptions
about the form of the link cost functions, which imply the absence of
interactions at junctions, equilibrium assignment emerges as the solution
to a tractable optimisation problem (see, for example, Nguyen, 1983).
Nguyen (1977) has shown how an OD matrix can be found which, under
equilibrium assignment, reproduces the observed link volumes. The route
assignment proportions are not explicity considered, and indeed under
equilibrium assignment are usually not uniquely defined. Furthermore,

as in the case of constant route assignment proportions, the OD matrix
is generally not uniquely defined. Hence, the introduction of various
forms of secondary objective function has been considered. Turnquist
and Gur (1979) suggested a heuristic adjustment process and Le Blanc
and Farhangian (1982) a quadratic difference function, in both cases
centered on a target trip matrix. Jörnsten and Nguyen (1979) and Fisk
and Boyce (1983) consider the introduction of an entropy function.

3. SCOPE OF THE PAPER

Subsequent sections of the paper relate to the log-linear models as
derived by entropy or information optimisation subject to linear
constraints. Such models are applicable to the single junction, to the
single path network as well as to the multiple path network when constant
route assignment proportions are assumed. In the case of the single
junction, the biproportional model is referred to, and in the case of
the multiple path network with all-or-nothing assignment, the
multiproportional model is referred to.

The calibration of such log-linear models (namely the determination of
parameter values so that given link volumes are reproduced) requires an
iterative numerical method. Such methods include Bregman's balancing
method (of which the multiproportional balancing procedure is a special
case), the Multiplicative Algebraic Reconstruction Technique (MART) and
the Newton method. If a solution for the parameter values exists, then
the Bregman and the MART procedures will converge (see Lamond and
Stewart, 1981), although perhaps not rapidly. Newton's method is
quadratically convergent in a neighbourhood of the solution, but is not
in general globally convergent.

Despite the need for iterative calibration, the log-linear models have
proved popular in practice because of their relative simplicity and
robustness. Furthermore, in many multiple path network applications,
estimates of route assignment proportions are available. In this paper,
a non-iterative procedure yielding an approximation to the log-linear
model solution is suggested. Certain statistical properties of the
approximation as well as a relationship to Bayesian estimation are shown.

The suggested procedure is most suited to situations where the OD matrix
is frequently updated, where a solution that accurately satisfies the
constraints is required and where confidence intervals (or variances)
for the fitted values are to be repeatedly calculated. It is therefore
envisaged that the procedure might have applications in on-line traffic
monitoring and control systems.

The approximate solution generated by the procedure satisfies the
constraint equations exactly. The closeness of the correspondence
between the approximate and exact solutions depends on the closeness of
target trip matrix to the exact solution. When the correspondence is
sufficiently poor, the procedure may yield negative fitted values. The
closeness of the approximation and the problem of negative fitted values
is examined empirically in Section 7.

4. SOME NOTATION

Let

t_j : = the flow between the jth pair of zones (note that the OD
matrix is thus treated as a vector).

v_i : = the volume of traffic on the ith link or inflow/outflow.

p_{ij} : = the proportion of the flow between the jth pair of zones
that takes the ith link or inflow/outflow.

t_j^o : = a target value for t_j.

μ_i : = the ith parameter of a log-linear model.

The following column vector and matrix representations are used

$$\underline{t} = \{t_1, t_2, \ldots, t_q\}^T$$

$$\ln\underline{t} = \{\ln t_1, \ln t_2, \ldots, \ln t_q\}^T$$

(similar definitions apply for $\underline{t}^o$ and $\ln\underline{t}^o$)

$$\underline{v} = \{v_1, v_2, \ldots, v_n\}^T$$

$$\underline{\mu} = \{\mu_1, \mu_2, \ldots, \mu_m\}^T$$

$$P = \begin{Bmatrix} P_{11}, P_{12}, \ldots, P_{1q} \\ P_{21}, P_{22}, \ldots, P_{2q} \\ \vdots \quad \vdots \qquad \vdots \\ P_{n1}, P_{n2}, \ldots, P_{nq} \end{Bmatrix}$$

Superscript T denotes a matrix or vector transposition, vectors are underlined and matrices are represented by capital letters.

Thus P represents the matrix of route assignment proportions introduced in Section 2. It is assumed that the elements of P are determined externally. In the case of the single junction, P is an <u>incidence</u> <u>matrix</u> with elements 0 or 1. In single path networks or under all-or-nothing route assignment, the elements of P are also 0 or 1. Otherwise, their values lie between 0 and 1 inclusively.

The OD matrix is related to the link volumes by the relationship

$$\underline{v} = P\underline{t} \tag{1}$$

In subsequent sections, it is required that only the linearly independent constraints be included. Thus, linearly dependent rows of P should be identified and removed. This can be done by a numerical technique such as <u>Gaussian</u> <u>elimination</u> (see Bell, 1983a). The corresponding elements of vector $\underline{v}$ should also be removed. The rank of matrix P is therefore equal to the number of remaining rows.

When t_j^o is equal to zero, t_j is also set equal to zero. Let A be the set of indices for the t_j terms not set, or constrained by (1), to be equal to zero. Components t_j for which j is not an element of set A should be removed from vectors $\underline{t}$, $\underline{t}^o$, $\ln\underline{t}$ and $\ln\underline{t}^o$. The corresponding columns of matrix P should also be removed.

5. THE LOG-LINEAR MODELS

The entropy and information based log-linear models have the general
form

$$\ln \underline{t} = \ln \underline{t}^{o} + S^{T} \underline{\mu} \tag{2}$$

In the case of the model suggested by Willumsen (in Van Zuylen and
Willumsen, 1980, and in Van Vliet and Willumsen, 1981) matrix S^{T} is
simply equal to P^{T}. For the model suggested by Van Zuylen (1981) and
Bell (1983a,b), matrix S^{T} is equal to P^{T} angmented by a vector all of
whose elements are one, namely $\{\underline{1}, P^{T}\}$.

Given $\underline{v}$, $\underline{t}^{o}$ and P (and therefore S), model parameters $\underline{\mu}$ are calculated
by an iterative numerical method, such as one of those mentioned in
Section 4, so that

$$\underline{v} = P\underline{t} = P\exp(\ln \underline{t}^{o} + S^{T} \underline{\mu}) \tag{3}$$

where the exp operator, like the ln operator, applies to each element
of the subsequent vector. In the case of the Van Zuylen and Bell model,
there is the additional requirement that

$$\mu_{1} = \ln \sum_{j \in A} t_{j} - \ln \sum_{j \in A} t_{j}^{o} \tag{4}$$

Let $V_{(\underline{v})}$ be the covariance matrix for the elements of $\underline{v}$, and $V_{(\ln \underline{t})}$ the
corresponding covariance matrix for $\ln \underline{t}$. Furthermore, let D be a diagonal
matrix with the elements $\{t_{1}, \ldots, t_{q}\}$ on the principal diagonal. It can
be shown (see Bell, 1983a) that

$$V_{(\ln \underline{t})} \simeq S^{T} J V_{(\underline{v})} J^{T} S \tag{5}$$

where, in the case of Willumsen's model

$$J = (PDP^{T})^{-1} \tag{6}$$

and in the case of Van Zuylen and Bell's model

$$J = \frac{\alpha \underline{v}^T M}{M - M \underline{v} \alpha \underline{v}^T M} \tag{7}$$

where

$$M = (PDP^T)^{-1} \tag{8}$$

and

$$\alpha = (\underline{v}^T M \underline{v})^{-1} \tag{9}$$

A proof of the non-singularity of PDP^T is to be found in Bell (1983a).

6. THE LINEAR APPROXIMATION

Let $f(\underline{t})$ be a twice differentiable entropy or information based function from which a log-linear model is derived. When this is expanded as a Taylor series about the target trip matrix, $\underline{t}^o$, and when the first three terms only are considered, we obtain the approximation

$$f(\underline{t}) \simeq f(\underline{t}^o) + (\underline{t} - \underline{t}^o)^T \underline{\delta} + \frac{1}{2}(\underline{t} - \underline{t}^o)^T N (\underline{t} - \underline{t}^o) \tag{10}$$

where

$$\underline{\delta} = \frac{\partial f(\underline{t})}{\partial \underline{t}} \quad \text{evaluated at } \underline{t} = \underline{t}^o \tag{11}$$

and

$$N = \frac{\partial^2 f(\underline{t})}{\partial \underline{t} \partial \underline{t}} \quad \text{evaluated at } \underline{t} = \underline{t}^o \tag{12}$$

Since $f(\underline{t})$ is a measure of distance from the target trip matrix, $\underline{t}^o$, $\underline{\delta}$ has elements all equal to zero. Hence the second term of the expansion vanishes. The values of the elements of matrix N depend on the form of $f(\underline{t})$ and the values of $\underline{t}^o$.

Dropping the constant term, we obtain

$$g(\underline{t}) = \frac{1}{2} (\underline{t} - \underline{t}^o)^T N (\underline{t} - \underline{t}^o) \tag{13}$$

When (13) replaces the original objective function, and when the non-negative conditions are omitted, the following constrained minimisation problem is obtained

$$\underset{\underline{t}}{\text{Min}}\ g(\underline{t}) \text{ subject to } \underline{v} = P\underline{t}$$

The Lagrange equation for the optimisation problem is

$$L(\underline{t},\underline{\lambda}) = \frac{1}{2} (\underline{t} - \underline{t}^o)^T N (\underline{t} - \underline{t}^o) + \underline{\lambda}^T (\underline{v} - P\underline{t}) \tag{14}$$

where $\underline{\lambda}$ is a vector of Lagrange multipliers. The necessary conditions for a minimum are

$$\frac{\partial L}{\partial \underline{t}} = N(\underline{t} - \underline{t}^o) - P^T\underline{\lambda} = 0 \tag{15}$$

Additionally

$$\frac{\partial^2 L}{\partial \underline{t} \partial \underline{t}} = N \tag{16}$$

Since $f(\underline{t})$ is in all cases convex, matrix N is positive definite. Therefore the necessary conditions (15) define a minimum for $g(\underline{t})$ subject to the constraints. Furthermore, since positive definite matrices are non-singular, N^{-1} exists.

From (15) we obtain

$$\underline{t} = \underline{t}^o + N^{-1}P^T\underline{\lambda} \tag{17}$$

Substituting into (1) we obtain

$$\underline{v} = P\underline{t} + PN^{-1}P^T\underline{\lambda} \tag{18}$$

Since P is of full row rank (see Section 4) and N is positive definite, the matrix

$$X = (PN^{-1}P^T)^{-1} \tag{19}$$

is non-singular. Thus from (18) we obtain

$$\underline{\lambda} = X\underline{v} - XP\underline{t}^o \tag{20}$$

Substituting (20) into (17) and rearranging we obtain

$$\underline{t} = N^{-1}P^T X\underline{v} + (I - N^{-1}P^T XP)\underline{t}^o \tag{21}$$

where I is an identity matrix of appropriate dimension. Let

$$B = N^{-1}P^T X = N^{-1}P^T(PN^{-1}P^T)^{-1} \tag{22}$$

then (21) may be rewritten as

$$\underline{t} = B\underline{v} + (I - BP)\underline{t}^o \tag{23}$$

Thus when the form of $f(\underline{t})$ and the values of $\underline{t}^o$ are known, N may be obtained. Given P, matrix B can be calculated. Finally, given link volumes, $\underline{v}$, (23) yields an estimated OD matrix, $\underline{t}$.

As (22) shows, premultiplication of B by P yields an identity matrix. Therefore B is a generalised inverse of P (in the sense that PBP = P) and the first part of (23), namely $B\underline{v}$, corresponds to a model suggested by Beil (1979). Premultiplication of (23) by P demonstrates that the estimate for $\underline{t}$ satisfies the constraint equations (1).

In on-line applications where P and $\underline{t}^o$ remain unaltered, only the matrix-vector product $B\underline{v}$ needs to be recalculated in order to update the fitted values. In relation to iterative calibration techniques, this procedure could therefore result in a considerable computational saving. Since the target trip matrix, $\underline{t}^o$, remains unaltered, a covariance matrix for the fitted values, $V_{(\underline{t})}$, is yielded by

$$V_{(\underline{t})} = B V_{(\underline{v})} B^T \tag{24}$$

where $V_{(\underline{v})}$ is a given covariance matrix for the link volumes. When matrix B has already been evaluated, it is computationally much simpler to solve (24) than it is to evaluate approximation (5).

Alternatively, it is possible to regard (23) as an updating equation. Let $\underline{v}(k)$, $\underline{t}(k)$, B(k) and P(k) be the values of $\underline{v}$, $\underline{t}$, B and P evaluated at period k. Route assignment proportions are allowed to respond to changing link flows. The updating equation therefore becomes

$$\begin{aligned}\underline{t}(k) &= B(k)\underline{v}(k) + (I - B(k)P(k))\underline{t}(k-1) \\ &= \underline{t}(k-1) + B(k)(\underline{v}(k) - P(k)\underline{t}(k-1))\end{aligned} \tag{25}$$

where

$$B(k) = N^{-1}(k-1)P^T(k)(P(k)N^{-1}(k-1)P^T(k))^{-1} \tag{26}$$

Let $V_{(\underline{v}(k))}$ be the covariance matrix for the traffic counts in period k and $V_{(\underline{t}(k))}$ the covariance matrix for the fitted values in period k. Furthermore, let us assume that $\underline{v}(k)$ and $\underline{t}(k-1)$ are independent sets of random variables. Then the following equation

$$V_{(\underline{t}(k))} = B(k)V_{(\underline{v}(k))}B^T(k) + (I - B(k)P(k))V_{(\underline{t}(k-1))}(I - B(k)P(k))^T \tag{27}$$

updates the covariance matrix for the fitted values.

7. SOME PROPERTIES OF THE APPROXIMATION

In this section an expression is derived for B(k) such that (25) yields a minimum variance estimator for $\underline{t}(k)$. By this means, conditions under which the proposed algorithm would yield the minimum variance estimator are shown. A relationship to Bayesian estimation is also pointed out.

It will be assumed that the covariance matrices $V_{(\underline{v}(k))}$ and $V_{(\underline{t}(k-1))}$ are positive definite. This in turn implies that all the variances should be greater than zero and that the correlations should not be perfect.

Let w_{ii} be the $(ii)^{th}$ element of matrix $V_{(\underline{t}(k))}$, $\underline{b}_i^T$ the ith row of $B(k)$ and $\underline{e}_i^T$ the ith row of the identity matrix I. From (27) we obtain

$$w_{ii} = \underline{b}_i^T V_{(\underline{v}(k))} \underline{b}_i + (\underline{e}_i^T - \underline{b}_i^T P(k)) V_{(\underline{t}(k-1))} (\underline{e}_i - P^T(k)\underline{b}_i)$$

$$i\epsilon A \qquad (28)$$

The necessary condition for a minimum is

$$\frac{\partial w_{ii}}{\partial \underline{b}_i} = 2\underline{b}_i^T V_{(\underline{v}(k))} - 2(\underline{e}_i^T - \underline{b}_i^T P(k)) V_{(\underline{t}(k-1))} P^T(k) = 0$$

$$i\epsilon A \qquad (29)$$

or alternatively

$$B(k) V_{(\underline{v}(k))} = (I - B(k)P(k)) V_{(\underline{t}(k-1))} P^T(k) \qquad (30)$$

Differentiating (29) again, we obtain

$$\frac{\partial w_{ii}}{\partial \underline{b}_i \partial \underline{b}_i} = 2V_{(\underline{v}(k))} + 2P(k) V_{(\underline{t}(k-1))} P^T(k) \qquad i\epsilon A \qquad (31)$$

This matrix is positive definite since $V_{(\underline{v}(k))}$ and $V_{(\underline{t}(k-1))}$ are positive definite (it is assumed that matrix $P(k)$ has full row rank). Thus the sufficient condition for a minimum is fulfilled. Rearranging (30) we get

$$B(k) (V_{(\underline{v}(k))} + P(k) V_{(\underline{t}(k-1))} P^T(k)) = V_{(\underline{t}(k-1))} P^T(k) \qquad (32)$$

Since positive definite matrices are also non-singular we can therefore write

$$B(k) = V_{(\underline{t}(k-1))} P^T(k) (V_{(\underline{v}(k))} + P(k) V_{(\underline{t}(k-1))} P^T(k))^{-1} \qquad (33)$$

When $B(k)$ is as defined in (33) rather than as in (26), updating equation (25) yields the minimum variance estimate for the OD matrix. Furthermore, updating equations (25) and (27) correspond to Bayesian estimation under the assumption that both the traffic counts, $\underline{v}(k)$, and the prior estimates, $\underline{t}(k-1)$, are multivariate normally distributed variates

(Maher, 1983, gives the relevant Bayesian updating equations).
Equations (33) and (26) would be equivalent if $N^{-1}(k-1)$ were equal to
$V_{(\underline{t}(k-1))}$ and if $V_{(\underline{v}(k))}$ were equal to zero. The latter condition
implies that the traffic counts contain no random error.

8. AN EXAMPLE

The performance of the suggested non-iterative procedure is here
assessed in relation to Willumsen's model. The entropy based function
suggested in Van Zuylen and Willumsen (1980) and in Van Vliet and
Willumsen (1981) has the following form

$$f(\underline{t}) = \sum_{j \in A} t_j (\ln(t_j/t_j^o) - 1) \tag{34}$$

It follows that

$$\frac{\partial^2 f(\underline{t})}{\partial t_i \partial t_j} = \begin{cases} 1/t_j & \text{when } i = j \\ 0 & \text{otherwise} \end{cases} \tag{35}$$

Therefore from (10) we obtain

$$N = \begin{Bmatrix} 1/t_1^o & & & \\ & 1/t_2^o & & \\ & & \ddots & \\ & & & 1/t_q^o \end{Bmatrix} \tag{36}$$

Both the function $f(\underline{t})$ and its approximation are convex and separable
with a minimum at $f(\underline{t}^o)$. Furthermore, at point $\underline{t}^o$ both have the same
curvature.

Data relating to a complex intersection in Stuttgart (the Pragsattel)
is used. There are 36 turning movements and a total of 13 inflows and
outflows. In this case, Willumsen's model reduces to a biproportional
model, namely the product of an inflow factor, an outflow factor and a

prior estimate (or target value).

In Fig. 1, the turning volumes yielded by (22) and (23) are compared
with the exact solution and with the actual values. Part (a) shows the
results when no prior information is introduced. In this case, the
elements of the target trip matrix were derived as follows

$$t_j^o = \sum_i v_i \; / \sum_{ik} p_{ik} \qquad\qquad \text{all } j\varepsilon A \qquad\qquad (37)$$

Both the exact and approximate solutions to Willumsen's model fit the
data poorly (the values of R^2 are 0.49 for the exact and 0.37 for the
approximate solutions). As one would expect, there is a tendency for
the larger turning volumes to be underestimated and the smaller to be
overestimated.

Part (b) of Fig. 1 shows the marked effect that even limited prior
information can have. In order to obtain values for the elements of the
target trip matrix, the actual turning volumes were assigned to 4 classes
according to their size. The group of largest flows were assigned a
weight of 4, the group of second largest flows 3, the group of third
largest flows 2 and finally the group of smallest flows 1. Let w_j be
the weight assigned to the jth flow. Then the elements of the target
trip matrix were obtained as follows

$$t_j^o = w_j (\sum_i v_i \; / \sum_{ik} p_{ik} w_k) \qquad\qquad \text{all } j\varepsilon A \qquad\qquad (38)$$

In relation to Part (a), the goodness of fit of both the exact and
approximate solutions is considerably better (the R^2 values are now
0.86 for both solutions). The correspondence between the two solutions
is also much closer. Furthermore, the absolute size of the negative
estimates yielded by (22) and (23) are markedly smaller.

The better the quality of the prior information, the better the goodness of
fit of the two solutions, the closer the correspondence between the two
solutions and the smaller the absolute sizes of the negative estimates
yielded by the approximation. These tendencies have been confirmed by
increasing the number of classes to which the actual turning volumes

Part (a)

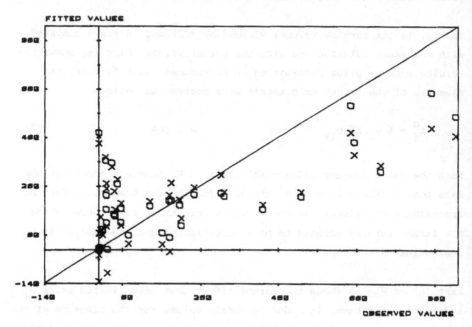

Part (b)

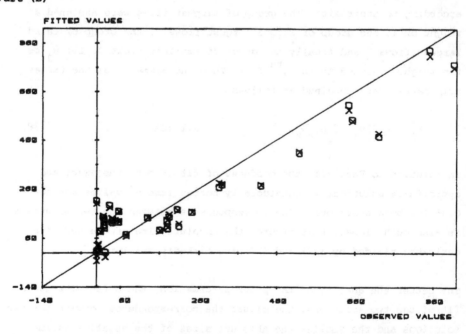

Fig. 1. Exact and approximate solutions for Willumsen's model.
Part (a): Without prior information; Part (b): With prior information.
X denotes the approximate solution and □ the exact solution.

are assigned, thus increasing the level of discrimination in the weighting used to derive values for the target trip matrix.

9. CONCLUSIONS

Log-linear models have proved popular for the estimation of OD matrices from traffic counts. A non-iterative procedure for the derivation of an approximation to the log-linear model solution is presented in this paper. In relation to iterative calibration procedures, the suggested non-iterative method may be significantly advantageous in on-line applications where an OD matrix has to be frequently updated in the light of fresh traffic counts. The empirical results presented in the previous section indicate that when a reasonable amount of prior information is available, a good correspondence can be expected between the exact solution for the log-linear model and the approximate solution yielded by the proposed method.

Acknowledgements

The author wishes to express appreciation to Professor Leutzbach for encouragement and to the Alexander von Humboldt-Stiftung for the post-doctoral research fellowship which facilitated this research.

REFERENCES

Beil, D. (1979). Matrixmodelle zur Ermittlung von Teilströmen aus Quer-schnittsmessungen. Thesis for Dr.-Ing., Fakultät für Bauingenieur- und Vermessungswesen der Universität Karlsruhe.

Bell*, M.G.H. (1983a). The estimation of an origin-destination matrix from traffic counts. Transportation Science 17, pp. 198-218.

Bell, M.G.H. (1983b). The estimation of origin-destination flows and their confidence intervals from measurements of link volumes: a computer program. Traffic Engineering and Control 24, pp. 202-205.

Buehler, M.G. (1983). Forecasting intersection traffic volumes. Journal of Transport Engineering 109, pp. 519-533.

Carey, M., Hendrickson, C., Siddharthan, K. (1981). A method for direct estimation of origin-destination trip matrices. Transportation Science 15, pp. 32-49.

Cascetta, E. (1983). An estimator of the origin-destination matrix that combines the result of surveys and/or models with traffic flows. Proceedings of the Associazione Italiana di Ricerca Operativa, Napoli, September.

Chan, V.K., Dowling, R.G., Willis, A.E. (1980). Deriving origin-destination information from routinely collected traffic counts. Phase I: Review of current research and applications. Working Paper UCB-ITS-WP-80-1. Institute of Transportation Studies, University of California, Berkeley.

Cremer, M., Keller, H. (1981). Dynamic identification of flows from traffic counts at complex intersections. Proceedings of the 8th International Symposium on Transportation and Traffic Theory, Toronto University, June.

Fisk, C.S., Boyce, D.E. (1983). A note on trip matrix estimation from link traffic count data. Transportation Research 17B, pp. 245-250.

Hauer, E., Pagitsas, E., Shin, B.T. (1981). Estimation of turning flows from automatic counts. Transportation Research Record 795, pp. 1-7.

Högberg, P. (1976). Estimation of parameters in models for traffic prediction: a non-linear regression approach. Transportation Research 10, pp. 263-265.

Jeffreys, M., Norman, M. (1977). On finding realistic turning flows at road junctions. Traffic Engineering and Control 18, pp. 19-21, 25.

Jörnsten, K., Nguyen, S. (1979). On the estimation of a trip matrix from network data. Publication No. 153, Centre de recherche sur les transports, Université de Montreal.

Lamond, B., Stewart, N.F. (1981). Bregman's balancing method. Transportation Research 15B, pp. 239-248.

Le Blanc, L.J., Farhangian, K. (1982). Selection of a trip table which reproduces observed link flows. Transportation Research 16B, pp. 83-88.

Maher, M.J. (1983). Inference on trip matrices from observations on link volumes: a Bayesian statistical approach. Transportation Research 17B, pp. 435-447.

Mekky, A. (1979). On estimating turning flows at road junctions. Traffic Engineering and Control 20, pp. 486-487.

Mountain, L.J., Westwell, P.M. (1983). The accuracy of estimation of turning flows from automatic counts. Traffic Engineering and Control 24, pp. 3-7.

Nguyen, S. (1977). Estimating an OD matrix from network data: A network equilibrium approach. Publication No. 60, Centre de recherche sur les transports, Université de Montreal.

Nguyen, S. (1983). Inferring origin-destination demand from network data. Proceedings of the Associazione Italiana die Ricerca Operativa, Napoli, September.

Norman, M., Hoffman, N., Harding, F. (1979). Non-iterative methods for generating a realistic turning flow matrix for a junction. Traffic Engineering and Control 20, pp. 587-589.

Robillard, P. (1975). Estimating the OD matrix from observed link volumes. Transportation Research 9, pp. 123-128.

Turnquist, M.A., Gur, Y.J. (1979). Estimation of trip tables from observed link volumes. Transportation Research Record 730, pp. 1-6.

Van Vliet, D., Willumsen, L.G. (1981). Validation of the ME2 Model for estimating trip matrices from traffic counts. Proceedings of the 8th International Symposium on Transportation and Traffic Theory, Toronto University, June.

Van Zuylen, H.J. (1979). The estimation of turning flows on a junction. Traffic Engineering and Control 20, pp. 539-541.

Van Zuylen, H.J. (1981). Some improvements in the estimation of an OD matrix from traffic counts. Proceedings of the 8th International Symposium on Transportation and Traffic Theory, Toronto University, June.

Van Zuylen, H.J., Branston, D.M. (1982). Consistent link flow estimation from counts. Transportation Research 16B, pp. 473-476.

Van Zuylen, H.J., Willumsen, L.G. (1980). The most likely trip matrix estimated from traffic counts. Transportation Research 14B, pp. 281-293.

Willumsen, L.G. (1978). Estimation of an OD matrix from traffic counts: A review. Working Paper No. 99, Institute of Transport Studies, Leeds University.

Ninth International Symposium on
Transportation and Traffic Theory
© 1984 VNU Science Press, pp. 471–491

TRANSFERABILITY OF DISAGGREGATE TRIP GENERATION MODELS

GEOFFREY ROSE and FRANK S. KOPPELMAN[1]

Department of Civil Engineering and [1] Transportation Center,
Northwestern University, Evanston, IL, U.S.A.

ABSTRACT

This paper investigates the transferability of travel demand models describing the frequency of travel undertaken by members of a household. Travel activity is represented in terms of the number of tours and number of non-home stops made by members of a household instead of the conventional representation in terms of the number of trips. Two disaggregate data bases, collected in Minneapolis and Baltimore, are used to empirically investigate intra- and inter-regional transferability of tours- and stops-generation models. A set of measures is used to evaluate the effectiveness of model transfer. The high variability in transfer effectiveness obtained from the empirical analyses suggests that model transfer should be employed with care to assure that the transferred model is useful in the application context. The results indicate that adjustment of the transfer model, by updating the intercept constant of the regression model using data from the application context, produces an important increase in transfer effectiveness.

1.   INTRODUCTION

The concept of transferability has received considerable attention in the
literature (Atherton and Ben Akiva, 1976; Lerman, 1981; Koppelman and
Wilmot, 1982, 1983).   We define model transfer as the application of a
model formulated and estimated in one context to a new context. Model
transferability implies that the transferred models can provide useful
information about the behavior or phenomenon of interest in the
application context.

We further identify two types of model transfer: temporal and spatial.
When a model developed at one point in time is used to provide predictions
for another point in time, temporal transferability is implicitly
evoked.   Spatial transferability is invoked when a model developed in one
region is used to predict behavior in another region.   There is
considerable interest in spatial transferability because it provides an
analysis base when time or resources are not available for the development
of a model in the application context.

The research reported here is part of an ongoing study of model transferability, and complements research undertaken by Koppelman and Wilmot
(1982, 1983), Koppelman and Pas (1983) and Koppelman et al. (1983) into
the transferability of discrete choice models. This report investigates
the transferability of disaggregate, linear regression, trip generation
models.

The trip generation models to be used in this study are formulated and
some fundamental concepts of model transfer are discussed in Section
Two.   In Section Three, we describe a set of measures for evaluating model
transfer.   In Section Four, we discuss the results of an empirical study
which uses this set of measures to evaluate both intra-regional and inter-
regional model transfer.   Finally, in Section Five, we summarize the
important issues raised in this paper and use these as a basis to draw
conclusions and suggest directions for future research.

2. TRANSFERRING TRIP GENERATION MODELS: ISSUES AND CONCEPTS

In this section, we establish the foundations on which this study of model transferability is based. We begin by formulating the trip generation models which will be used in this study. We then discuss some concepts relevant to the issue of model transfer. These include the distinction between model transfer and transferability and methods of updating to improve model transferability.

2.1 Model formulation

Conventional trip generation models focus on single destination trips, commonly classified as home-based work, home-based other and non-home based trips. Unfortunately, this approach cannot adequately represent multidestination travel (Horowitz, 1978) where for example, travellers may link two destinations into a single tour and eliminate two home based trips. The usefulness of this approach for analysing urban travel demand is limited since recent studies demonstrate the significance of multidestination travel (Peskin et al., 1975; Hanson, 1977; Adler and Ben-Akiva, 1979; Oster, 1978).

More flexible representations of urban travel activity have been developed. Hensher (1976), Daly and vanZwam (1981), and Horowitz (1976, 1978) represent travel activity in terms of stops and/or tours. Their representation is adopted in this study. A stop, or sojourn, is a visit to a non-home place. A tour is a movement which begins and ends at home and includes one or more stops. In this framework we define the conventional unit of analysis, the trip, as a movement between two stops on a tour. These three measures of travel activity are intimately related (Hautzinger, 1981) since the number of tours plus the number of stops is to equal the number of trips.

In the models developed in this study we use the number of tours (TOURS) and the number of stops (STOPS) as separate dependent variables. We use linear least squares regression rather than non-linear regression because the non-linearities tested were not significant and the linear functional form has found most application in practice. Poisson regression (Ruijgrok and vanEssen, 1980) and negative binomial regression (Litinas et al., 1981) models have been proposed for modelling trip generation.   One advantage of these models is that they predict only positive integer valves for the dependent variable whereas regression models are unrestrained in this respect. Additionally, it has been argued that the error distributions implied by these advanced models are more appropriate than the normal distribution commonly used in regression estimation. However, the consistancy property of all these models suggests that they will obtain approximately equal parameter estimates in moderate size samples.   This result has been confirmed empirically by Timmermans (1982). Therefore, the use of linear regression models in this study is not only consistent with common practice but is also justified with respect to estimation theory.

2.2 Transfer concepts

Although the issue of model transfer has been widely discussed in the literature, there appears to he some confusion between the transfer of a model and the appropriateness of that transfer. We define model transfer as the application of a model formulated in one context to another context. According to this definition, model transfer does not necessarily imply that the application is appropriate or useful. We define model transferability, or effective model transfer, as a measure of the usefulness of the information about behavior in the application context which is provided by the transferred model.

It is appropriate to consider the degree of transferability of a model from one context to another (Lerman, 1981) rather than to describe transferability as a dichotomous property. It is generally recognised that models are not perfectly transferable between contexts. Further, different components of models are likely to be more or less transferable.   For discrete choice models, the alternative specific constants are less transferable than either the scale or relative parameter values (McFadden, 1978; Westin and Manski, 1979). Whenever it is possible, the alternative specific constants and the scale of disaggregate choice models should be adjusted. Similarly we expect that the intercept constant of regression models are least transferable and should be adjusted when possible.

We refer to the process of adjusting the alternative specific constants and scale of disaggregate choice models, or the intercept constant of regression models, as model updating. For choice models, updating procedures are usually distinguished on the basis of whether they use aggregate or disaggregate local data (Koppelman and Rose, 1983a). This distinction is not necessary in the context of linear regression models, since updating with aggregate data produces precisely the same result as updating with disaggregate data.

Consider a model estimated in context A which is to provide predictions in context B. We refer to this as the full transfer model

$$\hat{Y}_B = \hat{\alpha}_A + X_B \hat{\beta}_A \qquad (1)$$

where

$\hat{Y}_B$ is a vector of predicted values for the dependent variable in context B using the full transfer model.

X_B is the matrix of explanatory variables in context B.

$\hat{\alpha}_A, \hat{\beta}_A$ are parameters obtained from model calibration in context A.

The performance of this model in context B can be improved by adjusting, or updating the intercept constant using data from context B. The model with the updated constant is given by

$$Y_B = \alpha_B^U + X_B \hat{\beta}_A + \varepsilon \qquad (2)$$

where

Y_B is a vector of observed values for the dependent variable in context B, and

α_B^U is the updated intercept constant in context B.

The estimation of the updated intercept constant from the normal equations is

$$\hat{\alpha}_B^U = \frac{1}{N_B} \left\{ i'Y_B - i'X_B \hat{\beta}_A \right\}$$

$$= \overline{Y}_B - \overline{X}_B \hat{\beta}_A \tag{3}$$

where

$\hat{\alpha}_B^U$ is the estimator of the updated constant in context B given $\hat{\beta}_A$,

N_B is the number of disaggregate observations in context B,

i' is a row vector of ones of length N_B,

$\overline{Y}_B$ is the average of observed values in context B, and

$\overline{X}_B$ is the vector of average explanatory variables in context B.

The updated constant, along with the transferred coefficients of the explanatory variables, can be used to provide improved predictions in the application context (context B). We refer to this as the partial transfer model:

$$\tilde{Y}_B = \hat{\alpha}_B^U + X_B \hat{\beta}_A \tag{4}$$

where $\tilde{Y}_B$ is a vector of predicted values for the dependent variable in context B using the partial transfer model.

It is apparent from equation 3, that the estimated updated constant obtained by use of disaggregate data is identical to that which would be obtained if aggregate data $(\overline{Y}_B, \overline{X}_B)$ across the entire context were used instead. Thus, in the case of linear regression models, updating with aggregate data does not imply any loss of predictive accuracy relative to updating with disaggregate data.

3.   MEASURES OF MODEL TRANSFERABILITY

In this section, we describe a set of measures which can be used to evaluate the transferability of regression models. Analogous measures can be defined to evaluate the transferability of discrete choice models (Koppelman and Rose, 1983a). These measures are organized into two different classification systems. The first classification is by the level of analysis, disaggregate or aggregate, at which they evaluate

transfer effectiveness. The second classification is by type of measure. We employ two types of measures in this study. First, we use
absolute measures of the ability of the transferred model to replicate
observed behavior in an application context. Second, we measure this
replication ability relative to the replication ability of a similarly
specified model estimated on the observed data in the application
environment. These classes can be crossed so that we identify measures of
each type at each level of analysis. There are clear analytic relations
among the measures at each level of analysis; however, use of both types
of measures provides more insight into model transferability than either
single measure. The relationships between aggregate and disaggregate
measures are less easily defined. However, aggregate and disaggregate
measures generally give consistent results although they are not strictly
monotonic.

All the tests described here evaluate transferability for an entire model
estimated in one environment and applied to another. Each of these
measures can be reformulated to test the tranferability of portions of the
model by modifying the transferred model so that selected portions are re-
estimated to best fit the application context. These partial transfer
measures (Koppelman and Wilmot, 1982; Koppelman and Rose, 1983a) measure
the transferability of the balance, the non-updated portion, of the
model. The tests are formulated and interpreted identically to the full
transfer tests except that the statistical test must be adjusted for
differences in the degrees of freedom to represent the number of
restrictions in the transferred model.

We describe four general measures, one for each of the joint classes.
First, we describe the disaggregate measures starting with an absolute
goodness-of-fit measure followed by a relative goodness-of-fit measure.
Next, we review aggregate measures in the same order.

3.1 Disaggregate measures of model transferability

These measures evaluate the effectiveness of the transferred model based
on the ability to replicate observed behavior at the disaggregate level.
The first of these measures is the transfer goodness-of-fit measure. It

is formulated in Table 1 and is analogous to the common R-square measure. The differences between this transfer measure and the corresponding local goodness-of-fit measure is the use of transferred model sum of squared errors in place of the corresponding value for the local model. Thus, we call this specific measure the transfer R-square for regression models. Although the transfer R-square has no lower bound we expect it to be positive.

The second measure is the **transfer index**. This measure describes the disaggregate transfer prediction accuracy or goodness-of-fit relative to the corresponding goodness-of-fit of a similarly specified model estimated in the application context. The transferability index has an upper bound of one which it attains when the transferred model is identical to the locally estimated model. Although this index has no lower bound we expect it to be positive.

3.2 Aggregate measures of model transferability

These measures evaluate the effectiveness of the transferred model based on the ability to replicate observed aggregate behavior for aggregate population groups generally defined by spatial proximity. Each of the aggregate measures is based on the error in prediction in each aggregate group. The basic error measure for each aggregate prediction of travel activity is given by

$$BEM_z = \frac{N_z^p - N_z^o}{NHH_z} \qquad (5)$$

where BEM_z = basic error measure for an aggregate group z.

   N_z^p = travel activity predicted for group z.

   N_z^o = travel activity observed for group z.

   NHH_z = number of households in aggregate group z.

The basic error measure can be interpreted as the error in the predicted aggregate trip rate.

DISAGGREGATE MEASURES	AGGREGATE MEASURES
TRANSFER GOODNESS OF FIT	ROOT MEAN SQUARE ERROR
$$R_{ij}^2 = 1 - \frac{ESS_i(\beta_j)}{TSS_i}$$	$$RMSE = \left\{ \frac{\sum_z N_z^P \, BEM_z^2}{\sum_z N_z^P} \right\}^{1/2}$$
TRANSFER INDEX	RELATIVE ROOT MEAN SQUARE ERROR
$$TI_{ij} = \frac{TSS_i - ESS_i(\beta_j)}{TSS_i - ESS_i(\beta_i)}$$ $$= R_{ij}^2 / R_{ii}^2$$	$$REL. \ RMSE = \frac{RMSE_{ij}}{RMSE_{ii}}$$

Table 1 Measures of transferability for regression models.

Key:

R_{ij}^2 = Transfer R^2 (local R^2 if i=j).

TI_{ij} = Transfer Index for transfer from context j to context i.

$ESS_i(\beta_j)$ = Error sum of squares for model estimated in context j applied in context i.

TSS_i = Total sum of squares in context i.

N_z^P = Number of tours/stops predicted in zone z.

BEM_z = Basic error measure for zone z defined by equation 5 in text.

The Root-Mean-Square-Error (Table 1) summarizes the prediction error for all aggregate predictions. This measure is an index of the average error in prediction weighted by the size of the prediction element and structured to place emphasis on large errors.

The root-mean-square-error can be used to measure the absolute error in aggregate prediction by either transferred or local models. We define the relative root mean square error as the ratio of these measures shown in Table 1. This measure indicates the degree to which aggregate prediction errors produced by the transferred model exceed those produced by a corresponding local model. The relative root square error has a value of 1.0 when the transfer model predicts as well as the local model. Higher values indicate a greater error in the transfer model predictions than in the local model predictions

4.   EMPIRICAL ANALYSIS

The empirical research conducted as part of this study investigates intra-regional and inter-regional transferability of TOURS and STOPS generation models. In this section we discuss the results of the empirical analysis. We begin by briefly describing the data bases and discussing the model specification. The results from the intra-regional transfers are then presented, followed by the inter-regional results.

4.1 Data, model specification, and model estimation results

Data sets from two distinct urban regions in the United States, Baltimore and Minneapolis-St. Paul, were used in the analysis. Intra-regional transferability is examined by transferring models between sectors within each region while inter-regional transferability is investigated by transferring models between the urban regions.

The household was chosen as the unit of analysis because individuals within the household interact to determine travel arrangements which satisfy time and vehicle availability constraints as well as satisfying household subsistence and maintenance needs. Thus, the dependent variables, TOURS and STOPS, are defined per household on a daily basis. Models of total TRIPS were not estimated because the relationship among these variables (i.e., STOPS + TOURS = TRIPS) ensures that the parameters of a TRIPS model will be the sum of corresponding parameters for identically specified STOPS and TOURS models. Therefore, the transfer effectiveness of a TRIPS model would be a weighted average of the transfer effectiveness of the corresponding STOPS and TOURS models.

The estimation results for TOURS and STOPS models in the north and south sections of both regions and for the regions as a whole are reported in Table 2. The specification employed in this analysis was selected after extensive testing of a wide range of model specifications (Koppelman and Rose, 1983b). The first specifications for both TOURS and STOPS models include three explanatory variables: persons per household, vehicles available per household and number of workers per household. The coefficients of all the explanatory variables are significant and have the expected signs.

Conventional practice is to estimate trip generation models by purpose categories. The models used in this analysis combine all trip purposes because initial investigation indicated the substantive results with respect to model transfer are not sensitive to disaggregation by purpose. However, models estimated by disaggregating TOURS and STOPS into work/non-work tours (identified by at least one workplace stop on the tour) and work/non-work stops clarified the role of the explanatory variables. Specifically, the number of workers is primarily associated with work tours and work stops. Number of persons per household and vehicles available are primarily associated with non-work tours and stops (Koppelman and Rose, 1983b).

The definitional relationship between TOURS and STOPS requires that the number of stops must be greater than or equal to the number of tours. The estimation results reported in Table 2 satisfy that constraint for every pair of estimations. Further, we expect the predicted average number of tours and stops for every household type to be positive. This expectation is violated only for tour prediction with the Minneapolis-North Sector TOURS model for a single person, non-worker, no automobile household (none of which exist in the data).

(A) TOURS

FOR TRANSFERS	ESTIMATED IN	EXPLANATORY VARIABLES				# OF HOUSE-HOLDS	R^2
		CONST.	PHH	VEHA	NWORK		
WITHIN BALTIMORE	NORTH SECTOR	0.11 (0.23)	0.71 (0.06)	0.29 (0.10)	0.56 (0.10)	393	0.45
	SOUTH SECTOR	-0.07 (0.23)	0.65 (0.06)	0.41 (0.11)	0.54 (0.13)	382	0.46
WITHIN MINNEAPOLIS	NORTH SECTOR	-0.58 (0.19)	0.54 (0.04)	0.52 (0.11)	1.04 (0.10)	878	0.37
	SOUTH SECTOR	-0.45 (0.19)	0.63 (0.05)	0.46 (0.11)	0.83 (0.10)	1,073	0.33
BETWEEN REGIONS	BALTIMORE	0.02 (0.16)	0.68 (0.04)	0.35 (0.07)	0.56 (0.08)	775	0.45
	MINNEAPOLIS	-0.50 (0.14)	0.59 (0.03)	0.49 (0.08)	0.92 (0.07)	1,951	0.34

(B) STOPS

FOR TRANSFERS	ESTIMATED IN	EXPLANATORY VARIABLES				# OF HOUSE-HOLDS	R^2
		CONST.	PHH	VEHA	NWORK		
WITHIN BALTIMORE	NORTH SECTOR	0.83 (0.51)	0.90 (0.13)	0.42 (0.22)	1.08 (0.23)	393	0.26
	SOUTH SECTOR	0.20 (0.48)	0.78 (0.12)	1.19 (0.22)	0.58 (0.26)	382	0.31
WITHIN MINNEAPOLIS	NORTH SECTOR	-0.53 (0.35)	0.69 (0.08)	0.84 (0.22)	1.39 (0.18)	878	0.24
	SOUTH SECTOR	-0.54 (0.36)	0.76 (0.09)	1.09 (0.21)	1.18 (0.18)	1,073	0.22
BETWEEN REGIONS	BALTIMORE	0.49 (0.35)	0.83 (0.09)	0.78 (0.15)	0.91 (0.17)	775	0.28
	MINNEAPOLIS	-0.53 (0.25)	0.73 (0.06)	0.99 (0.15)	1.26 (0.13)	1,951	0.23

Table 2. Estimation results. KEY: CONST. = CONSTANT
PHH = PERSONS PER HOUSEHOLD
VEHA = VEHICLES AVAILABLE PER HOUSEHOLD
NWORK = NUMBER OF WORKERS PER HOUSEHOLD
( ) = STANDARD ERROR

There are some large differences in estimated parameters between pairs of models reported in Table 2. These differences are especially large for the parameters associated with vehicles available and number of workers in the Baltimore north and south sector STOPS models. The impact of these differences on model transferability is examined in the next section.

4.2 Transfer analysis

We now report the results of intra- and inter-regional transfer of TOURS and STOPS generation models. Two forms of model transfer are reported. First, the full model is transferred to predict behavior in the application context. Second, the coefficients of the explanatory variables are transferred but the constants are updated using data from the application context. These applications are referred to as full and partial transfer, respectively.

4.2.1 Intra-regional transfer

The disaggregate measures for local fits and intra-regional transfer of the TOURS and STOPS generation models for sectors in Baltimore and Minneapolis are presented in Tables 3 and 4 respectively. The R^2 values, local and transferred, provide an indication of the degree to which variability of behavior is explained by the local or transferred model. These measures are in a satisfactory range for all models considered. The higher values in Table 3 than in Table 4 indicate a greater ability to explain differences in number of TOURS than number of STOPS. The transfer index measures the goodness of fit performance of the transferred model relative to the local model. These values are uniformally high, ranging from 0.85 to 0.98 for full transfer and 0.92 to 1.00 for partial transfer. Thus in excess of 85 percent of the information which could be gained by estimating all the parameters in the application context is provided by the full transfer model. Local adjustment of the constant (partial transfer) produces a moderate improvement in model transfer.

The aggregate transfer measures for the TOURS and STOPS generation models are presented in Tables 5 and 6, respectively. Three important issues are raised by these results. First, we note that the root mean square error

		TRANSFERRED TO SECTORS IN:			
		BALTIMORE		MINNEAPOLIS	
		North Sector	South Sector	North Sector	South Sector
TRANSFERRED FROM	North Sector	0.45 (1.00)	0.45 (0.98) / 0.46 (1.00)	0.37 (1.00)	0.32 (0.98) / 0.32 (0.98)
	South Sector	0.44 (0.98) / 0.45 (0.99)	0.46 (1.00)	0.36 (0.98) / 0.37 (0.98)	0.33 (1.00)

Table 3. Disaggregate measures for intra-regional transfer of TOURS generation models: R-square (Transfer Index)

		TRANSFERRED TO SECTORS IN:			
		BALTIMORE		MINNEAPOLIS	
		North Sector	South Sector	North Sector	South Sector
TRANSFERRED FROM	North Sector	0.26 (1.00)	0.27 (0.88) / 0.29 (0.93)	0.24 (1.00)	0.22 (0.98) / 0.22 (0.99)
	South Sector	0.22 (0.85) / 0.24 (0.92)	0.31 (1.00)	0.23 (0.97) / 0.23 (0.99)	0.22 (1.00)

Table 4. Disaggregate Measures for intra-regional transfer of STOPS generation models: R-Square (Transfer Index).

Key:
Full transfer
Partial transfer

| | TRANSFERRED TO SECTORS IN: | | | |
| | BALTIMORE | | MINNEAPOLIS | |
TRANSFERRED FROM	North Sector	South Sector	North Sector	South Sector
North Sector	0.27 (1.00)	0.33 (1.41) / 0.25 (1.07)	0.15 (1.00)	0.26 (1.03) / 0.25 (1.00)
South Sector	0.31 (1.15) / 0.23 (0.85)	0.24 (1.00)	0.16 (1.07) / 0.15 (1.01)	0.25 (1.00)

Table 5. Aggregate Measures for intra-regional transfer of TOURS generation models: Absolute and (Relative) root mean square error.

| | TRANSFERRED TO SECTORS IN: | | | |
| | BALTIMORE | | MINNEAPOLIS | |
TRANSFERRED FROM	North Sector	South Sector	North Sector	South Sector
North Sector	0.92 (1.00)	1.01 (1.52) / 0.81 (1.22)	0.33 (1.00)	0.67 (1.15) / 0.61 (1.05)
South Sector	0.97 (1.05) / 0.80 (0.87)	0.67 (1.00)	0.43 (1.30) / 0.32 (0.98)	0.59 (1.00)

Table 6. Aggregate Measures for intra-regional transfer of STOPS generation models: Absolute and (Relative) root mean square error.

Key: Full transfer / Partial transfer

for the STOPS models are larger than the TOURS models. Although some of
this difference may be attributable to differences in the magnitude of the
TOURS and STOPS variables, a substantial component is attributable to
greater difficulty in explaining variability of STOP behavior. Second,
the absolute and relative transfer errors between sectors in Baltimore and
Minneapolis are nonsymmetric.   Third, the use of partial transfer
generally produces substantial reductions in errors.

4.2.2 Inter-regional transfer

The disaggregate transfer measures for the inter-regional transfer of the
TOURS and STOPS generation models are presented in Tables 7 and 8
respectively.   The transfer R^2 and transfer index indicate a high degree
of inter-regional transferability. Full transfer models provide in excess
of 89 percent of the information provided by the local model (measured by
the transfer index) and with adjustment of the constant (partial transfer)
the transferred models provide in excess of 94 percent of the information
provided by the local model.

The aggregate transfer measures are reported in Tables 9 and 10 for the
TOURS and STOPS models respectively. The absolute and relative root mean
square errors consistently indicate the importance of using partial rather
than full transfer in inter-regional transfer applications.   While the
relative root mean square errors for full transfer range from 1.17 to 1.66
the range for partial transfer is 1.00 to 1.07. When evaluating transfer
of TOURS models into Baltimore, the relative root mean square error
indicates that the partial transfer Minneapolis TOURS model performs
almost as well as the local Baltimore model and that the partial transfer
Minneapolis STOPS model performs as well as the local Baltimore model
(relative root mean square error of 1.0).

5.   SUMMARY, CONCLUSIONS AND DIRECTIONS FOR FUTURE RESEARCH

The research reported here provides insight into the spatial transferability of disaggregate, linear regression trip generation models, both
within and between urban regions. Although the considerable variability
in measured transferability suggests the need for caution in interpreting
these results, some important general observations can be made.

		PREDICTING ON	
		BALTIMORE	MINNEAPOLIS
ESTIMATED ON	BALTIMORE	0.45 (1.00)	0.32 (0.94) / 0.33 (0.96)
	MINNEAPOLIS	0.42 (0.93) / 0.42 (0.94)	0.34 (1.00)

Table 7. Disaggregate Measures for inter-regional transfer of TOURS generation models: R-square (Transfer Index).

		PREDICTING ON	
		BALTIMORE	MINNEAPOLIS
ESTIMATED ON	BALTIMORE	0.28 (1.00)	0.20 (0.89) / 0.22 (0.97)
	MINNEAPOLIS	0.25 (0.91) / 0.27 (0.96)	0.23 (1.00)

Table 8. Disaggregate Measures for inter-regional transfer of STOPS generation models: R-square (Transfer Index).

Key: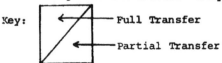

Full Transfer

Partial Transfer

		PREDICTING ON	
		BALTIMORE	MINNEAPOLIS
ESTIMATED ON	BALTIMORE	0.23 (1.00)	0.28 (1.35) 0.23 (1.07)
	MINNEAPOLIS	0.30 (1.17) 0.26 (1.02)	0.21 (1.00)

Table 9 . Aggregate Measures for inter-regional transfer of TOURS generation models: absolute and (relative) root mean square error.

		PREDICTION ON	
		BALTIMORE	MINNEAPOLIS
ESTIMATED ON	BALTIMORE	0.83 (1.00)	0.81 (1.66) 0.52 (1.07)
	MINNEAPOLIS	0.99 (1.23) 0.81 (1.00)	0.49 (1.00)

Table 10. Aggregate Measures for inter-regional transfer of STOPS generation models: absolute and (relative) root mean square error.

Key:

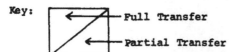

— Full Transfer

— Partial Transfer

First, the adjustment of the constant of the transfer model substantially improves its performance in the application context. The improvement is greatest in those cases in which transfer without adjustment is relatively poor.

Second, transfer effectiveness is better for intra-regional than for inter-regional transfer. This suggests that context similarity may be an important determinant of model transferability. It is not surprising that the transfer of disaggregate models is becoming common since this affords an opportunity for substantial time and cost savings in areas of data collection and analysis. However, the considerable variability of results obtained in this study suggests that model transfer should be employed with care to assure that the transferred model is useful in the application context.

Research is needed to improve the ability to understand the conditions under which model transfer is likely to be highly effective. Specifically, research should be directed at developing relationships which describe the effect of model type, specification and the effect of differences in regional contexts on the level of transfer effectiveness. Such relationships can provide a basis for a priori prediction of the expected level of effetiveness of specific model transfers and thus a basis for selecting models to be used in specific application contexts.

Acknowledgements

This paper is based on work undertaken at the Transportation Center of Northwestern University under contract with the U.S. Department of Transportation, Office of University Research. James Ryan of the Urban Mass Transportation Administration, who is contract monitor, has provided important advice, encouragement, and support to the research team. The manuscript was prepared by Merikay Smith and Marianne Thomas.

REFERENCES

Adler, T., M.E. Ben-Akiva (1979). A Theoretical and Empirical Model of Trip Chaining Behavior, Transportation Research 13B, pp. 243-257.
Atherton,T.J., M.E. Ben-Akiva (1976). Transferability and Updating of Disaggregate Travel Demand Models, Transportation Research Record 610, pp. 12-18.
Daly, A. J., H. H. P. van Zwam (1981). Travel Demand Models for the Zuidvleugel Study. In: Proc. PTRC Summer-Annual Meeting, University of Warwick, England, pp. 65-84.
Hanson, S. (1977). Urban Travel Linkages: A Review, Resource Paper, Workshop A, Third International Conference on Behavioral Travel Modelling, Tanunda, Australia.
Hautzinger, H. (1981). Combined Modelling of Activity and Trip Patterns: A New Approach to the Trip Generation Problem. In: Proc. PTRC Summer Annual Meeting, University of Warwick, England, pp. 271-283.
Hensher, D. (1976). The Structure of Journeys and Nature of Travel Patterns, Environment and Planning A 8, pp. 655-672.
Horowitz, J. (1978). Disaggregate Demand Model for Nonwork Travel, Transportation Research Record 673.
Horowitz, J. (1976). Effects of Travel Time and Cost on the Frequency and Structure of Automobile Travel, Transportation Research Record 592.
Koppelman, F.S., E.I. Pas (1983). Multidimensional Choice Model Transferability, Transportation Center Report, Northwestern University.
Koppelman, F.S., G. Rose (1983a). Geographic Transfer of Travel Choice Models: Evaluation and Procedures, Paper presented at the International Symposium on New Directions in Urban Modelling, University of Waterloo, Canada.
Koppelman, F.S., G. Rose (1983b). The Transferability of Disaggregate Linear Regression Trip Generation Models, Transportation Center Report, Northwestern University.
Koppelman, F.S., C.G. Wilmot (1982). Transferability Analyses of Disaggregate Choice Models, Transportation Research Record 895, pp. 18-24.
Koppelman, F.S., C.G. Wilmot (1983). The Effect of Model Specification Improvement on Transferability, Transportation Center Working Paper, Northwestern University.
Koppelman, F.S., C.G. Wilmot, G.K. Kuah (1983). The Impact of Scale Adjustments on Intra-city Transferability, Transportation Center Report, Northwestern University.
Lerman, S.R. (1981). Interspatial, Intraspatial and Temporal Transferability. In: New Horizons in Travel Behavior Research, edited by P.R. Stopher, A.J. Meyburg and W. Brog (D.C. Heath and Company, Lexington), pp. 628-632.
Litinas, N., S.L. Gonzales, S.R. Lerman, I. Salomon (1981). A Theory of Travel Demand Processes with Integer Outcomes, M.I.T. Center for Transportation Studies, Cambridge, Massachusetts.
McFadden, D. (1978). The Theory and Practice of Disaggregate Demand Forecasting for Various Models of Urban Transportation. In: Emerging Transportation Planning Method edited by W.F. Brown (U.S. Department of Transportation, Office of University Research).
Oster, C.V. (1978). Household Tripmaking to Multiple Destinations: The Overlooked Urban Travel Pattern, Traffic Quarterly 32, No. 4, pp. 511-529.
Peskin, R.L., J.L. Schofer, P.R. Stopher (1975). The Immediate Impact of Gasoline Shortages on Urban Travel Behavior, Federal Highway Administration.

Ruijgrok, C.J., P.G. van Essen (1980). The Development and Application of Disaggregate Poisson Model for Trip Generation, 59th Annual Meeting, Transportation Research Board, Washington, D.C.

Timmermans, G.V.F. (1982). A Comparative Analysis of Disaggregate Trip Generation Models, M.S. Thesis, Department of Civil Engineering, Northwestern University.

Westin, R. and C.F. Manski (1979). Theoretical and Conceptual Developments in Demand Modelling. In: Behavioral Travel Modelling, edited by D.A. Hensher and P.R. Stopher (Croom-Helm, London), Chapter 17.

PAPER 24

Ninth International Symposium on
Transportation and Traffic Theory
© 1984 VNU Science Press, pp. 493–512

A MODEL OF CONSTRAINED BINARY CHOICE

RYUICHI KITAMURA and TENNY N. LAM

Department of Civil Engineering, University of California, Davis, CA 95616, U.S.A.

ABSTRACT

This study presents a framework of incorporating constraints on the choice set, such as time and monetary budgets, into discrete binary travel choice models. A term called "inclusion probability" is used to represent the probability that a particular alternative is in a decision maker's choice set. This term makes possible the extension of the ordinary binary models to include various types of constraints that influence travel choice behavior. Numerical examples are presented in the paper to illustrate the general properties of the constrained choice models and to compare them to those of the conventional unconstrained models. Using data generated by Monte Carlo simulation, the feasibility and practicality of estimating the model coefficients by the maximum likelihood method are investigated.

INTRODUCTION

A trip maker does not always have unlimited choices in making his travel decisions. Many travel choices are subject to one form of constraint or another. These constraints are imposed by personal considerations and circumstances, the urban structure, transporation system or social institutions. Travel demand is the collective result of the choices made by many individual decision makers under the influence of such constraints. The existence of the constraints cannot be neglected in studying the pattern and behavior of urban travel.

Viewing travel demand as a collective behavior of individual decisions has led to the development of disaggregate travel behavior analysis. This approach has contributed to many recent significant advances in transportation research. However, the existing utilitarian models of travel behavior may have oversimplified some travel choices that are subject to constraints. A typical constraint is the travel time budget. For example, choosing to shop 60 miles away during a one hour lunch break is highly unlikely or infeasible. The limited lunch time thus precludes certain destinations from the trip maker's choice set. Another example is the limited availability of the family car to some members of the household during certain hours of the day. This situation would eliminate a particular travel mode from the choice.

Theoretically, it is possible for disaggregate analysis of travel decisions to incorporate these constraints into its framework by treating them as limitations imposed on the choice set. The structure of the existing choice models such as logit and probit, however, does not allow explicit consideration of the constraints. At the same time, the set of alternatives from which the choice was actually made by a trip maker cannot always be identified from survey observations. Consequently, the researcher must subjectively remove the infeasible choice alternatives, a process which may not always result in a correct choice set. This problem is known as the "choice set definition" problem (see e.g., 1).

Thus far, no statistical approach is available to prevent possible biases due to the omitted constraints and the resulting errors in the choice set definition. It seems to be worthwhile to extend the range of application of travel choice models by incorporating constraints, such as time budget, into the model structures. It is especially important that constraints be incorporated into choice models when they are applied to less affluent subgroups, underdeveloped regions, or developing countries. In these situations, constraints dominate the travel behavior (2). In addition, a number of recent studies also showed the importance of time budget and other constraints on activity scheduling and travel behavior (e.g., 3-7).

In the present paper, attempts are made mainly on determining the feasibility and practicality of modifying the existing discrete choice models to allow the incorporation of constraints. A basic model framework is proposed and plausible models representing the constraints are discussed. Merely proposing some model forms would not be useful, unless suitable calibration procedures can be developed to estimate the coefficients. Attempts are made to show the feasibility of statistically estimating the models with simulated data. The artificial data are generated by Monte Carlo simulation using a model specification assumed beforehand. A maximum likelihood estimation procedure is then applied to the simulated data set and the properties of the coefficient estimates thus obtained are examined. The analyses of this study are limited to the case of binary choice. For the cases examined, the result shows that the model coefficients can be estimated by the maximum likelihood method, although there is room for improvement in the estimation procedure. The numerical examples further show that the nature of travel choices may be incorrectly identified if the constraints influencing the choices are ignored.

BACKGROUND

The development and widespread use of disaggregate travel choice models has been one of the major advances in transportation research in the past ten years. Recent research efforts have provided theoretical bases for model estimation, model updating, inferences, sampling methods and use of non-random samples, and extensive examples of innovative applications (for reviews, see 8-12). Many issues associated with the application of the choice models have also been raised and challenged. Most notable are the aggregation problem, validity of the assumption of independence from irrelevant alternatives, and model transferrability (see e.g., 13-15). Some critics of travel choice models argue that forcing travel decision into the discrete choice framework is an inadequate simplification (e.g., 16). The question of choice set definition is one of such problems which has been long recognized, but relatively little effort has been devoted to this issue thus far.

Because of budget constraints, restricted availability and limited information, the actual choice set on which the individual's choice is based is not necessarily identical to the set defined and used in the analysis of the choice behavior.

The rest of this section discusses in turn the above three elements that influence choice sets for travel decision.

Budget Constraints: Unlike the classical microeconomic model of consumer behavior, discrete choice models do not explicitly incorporate budget constraints into their structure. These constraints are not directly observable from the actual choices and it is also uncertain whether or not they are binding on the choices observed. Accordingly, their effects on the choice set have been taken into account by the researchers only through insights and judgements. Identifying alternatives that are ruled out due to a budget constraint, however, is not obvious when the daily time budget is involved. Clearly, taking a 60-minute shopping trip during a lunch break is infeasible. In general, the existence of such time budget constraints and their effects on daily travel behavior are difficult to determine; yet, one cannot dispute the fact that they are one of the important factors affecting travel and activity choices (e.g., 4, 17,-19).

Availability: To exactly define a choice set for an individual within the travel environment unique to that person is a problem that arises even in such well defined choice problems as travel mode choices (e.g., 20-21). The reason is that the availability of, in this case, respective modes cannot always be objectively defined from the standard survey information.   For example, determining car availability for a given person in a household is not a trivial task when the number of drivers exceeds the number of cars owned by the household. Use of proxy variables (e.g., the number of cars per driver) that approximate the unknown availability is the usual remedy. Attempts have been made to introduce attitudinal measurements such as "perceived availability" into the analysis (e.g., 22), but clear relationships have not been found between the perceived availability and measureable variables whose future values can be forecast (23).

Limited Information: The uncertainty about the amount of information an individual has about alternatives is another factor that contributes to the difficulty involved in defining choice sets. A good example is a person's knowledge of bus services—routing, scheduling, and location of stops—and its effect on the formation of a choice set for (especially nonwork) mode choice. Destination choice is also affected by the extent of information the individual has. It is also expected that there exist variations in choice sets across individuals, because of the differences not only in the level of information but also in the way each individual perceives the environment in formulating his own "action space." Some empirical results in the transportation field (5) have given some support to this view. Little is known as to how a choice set can be defined under the heterogeneity across individuals.

These three elements, availability, information, and budget constraints, perhaps do not exhaust all the factors that affect the formation of choice sets. As noted earlier, these and other factors do influence the outcome of the choice, but no statistical methods have been developed in order to incorporate them into the analytical framework. The practice is to rely almost solely on proxy variables and intuitive judgements by the researcher. The following sections present an effort to develop a statistical model that is capable of incorporating various constraints imposed on the formation of choice sets within the prevailing framework of discrete choice analysis.

THE FRAMEWORK OF THE CONSTRAINED CHOICE MODEL

Whether or not an objectively identified alternative is in fact included in an individual's choice set is not always observable. This is especially true when one intends to analyze the effects of the time budget constraints. The difficulty of direct observation necessitates a treatment of the constraining effects in probabilistic terms. From this viewpoint, the observed choice behavior may be represented by two terms. One is the probability that a given alternative will be chosen over the others if it is in the choice set. The other is the probability that the alternative will be included in the set.

For example, consider the case of binary choice where the first alternative (alternative 1) is always available to the individual, while alternative 2 is available only with a certain probability because of the constraints. Then, for an individual who is making the choice,

Pr [alternative 2 chosen]

= Pr [alternative 2 preferred over alternative 1]
x Pr [alternative 2 included in choice set]

Pr [alternative 1 chosen]

= Pr [alternative 1 preferred over alternative 2]
x Pr [alternative 2 included in choice set]
+ Pr [alternative 2 not in choice set]

= 1 - Pr [alternative 2 chosen].

where $\Pr[. . .]$ denotes the probability that the event described in the brackets will occur. Further, let

$$Q_n \quad = \quad \Pr[\text{alternative 2 is included in the choice set of individual } n]$$

$$R_n \quad = \quad \Pr[\text{individual } n \text{ prefers alternative 2 over alternative 1 given that alternative 2 is in the choice set}]$$

$$P_n(1) \quad = \quad \Pr[\text{alternative 1 is chosen by individual } n]$$

$$P_n(2) \quad = \quad \Pr[\text{alternative 2 is chosen by individual } n]$$

where Q_n shall be called the "inclusion probability." Then,

$$
\begin{aligned}
P_n(1) &= 1 - Q_n R_n \\
P_n(2) &= Q_n R_n
\end{aligned}
\tag{1}
$$

The choice model proposed here shall be called the "constrained" choice model.

Suppose that Q_n and R_n are expressed as functions of explanatory variables. Using, for example, the binary logit structure, one may write

$$
\begin{aligned}
Q_n &= 1/(1 + \exp(\theta' Z_n)) \\
R_n &= 1/(1 + \exp(\beta' X_n))
\end{aligned}
\tag{2}
$$

where

$$\theta, \beta \quad = \quad \text{vectors of coefficients}$$

$$Z_n, X_n \quad = \quad \text{vectors of explanatory variables measured for individual } n.$$

and Z_n and X_n may share the same variables. Note that if Q_n is not a function of X_n but instead, is a constant for all n, then the model reduces to the captivity model proposed earlier (24). The coefficients of the model, θ and β, can be estimated by maximizing the following log-likelihood function,

$$
L = \sum_{n \in C_1} \ln(1 - Q_n R_n) + \sum_{n \in C_2} \ln(Q_n R_n)
\tag{3}
$$

where C_1 is the set of individuals in the sample who chose alternative 1 and C_2 is the set of individuals who chose alternative 2.

INTERPRETATION OF THE INCLUSION PROBABILITY

The inclusion probability, Q_n, can be specified to represent the many factors affecting the choice set. In the case of travel mode choice, for example, Q_n may be formulated to represent the probability that the car alternative is available to individual n, and the independent variable vector, Z_n, may contain such factors as household car ownership, number of drivers, the individual's employment status and age. Application of the proposed model will eliminate the subjective judgement of car availability that would otherwise need to be made by the researcher.

The proposed model is also capable of incorporating budget constraints into discrete choice analysis. For example, consider again the case of travel mode choice where total travel time is t_1 by mode 1 and t_2 by mode 2. Without any loss of generality, we assume that $t_1 \leq t_2$ i.e., the faster mode is called mode 1. Let T be the time budget (i.e., the maximum amount of time) that can be allocated to the trip. Then the travel time of the chosen mode must be no greater than the budget T. Recall that T is unobservable. Now suppose the data set contains those individuals who made the trip and the budget constraint is always satisfied by the faster mode, i.e., $t_1 \leq T$ for all individuals. The choice probabilities can now be written for those individuals who made the trips as

Pr[mode 2 chosen]
= Pr[mode 2 preferred over mode 1] Pr[$t_2 \leq T$]
Pr[mode 1 chosen] = 1 - Pr[mode 2 chosen]

By adding subscript n to refer to the individual and defining

$$Q_n = Pr[t_{2n} \leq T_n] \tag{4}$$

the above choice probabilities can be expressed by Eq. 1. Note that T_n, not t_{2n}, is treated as a random variable, and that Q_n now represents the unknown distribution of the time budget, T_n. Let the mean of this distribution be expressed as a linear combination of explanatory factors and its standard deviation be another parameter. If the distribution is assumed to be normal,

$$Q_n = \Phi[-(t_{2n} - \theta'Z_n)/s] \tag{5}$$

where θ and Z_n are respectively vectors of coefficients and explanatory variables as before, s is the standard deviation, and Φ is the standardized normal distribution function. The term $\theta' Z_n$ represents the mean of the time budget. Alternatively, the time budgets may assume a logistic distribution, i.e.,

$$Q_n = 1/(1 + \exp[\pi(t_{2n} - \theta' Z_n)/s\sqrt{3}]) \tag{6}$$

The parameters of these distributions can be estimated using Eq. 3. Consequently the unknown distribution of the time budgets and their binding effect on travel choice can be identified. Furthermore, noting that the ordinary binary choice model is a special case of the above model with $Q_n = 1$, statistical test of the budget effect can be carried out by examining the log-likelihood values between the proposed constrained model and the ordinary unconstrained choice model.

NUMERICAL EXAMPLES

The behavior of the constrained choice model is studied in this section using numerical examples. The examples are developed by assuming two artificial models that are based on the probabilistic structure defined by Eqs. 1 and 2. One of the models (model A) represents a general availability constraint on the choice set, while the other (model B) represents a time budget constraint. These two models, which are also used in the simulation experiments of the next section, are given in Table 1 together with the assumed parameter values.

Model A is defined with an inclusion probability,

$$Q_n = [1 + \exp(a_0 + a_1 Z_n)]^{-1} \tag{7}$$

Table 1. Simulation models A and B

Model A:	Model B:
$P_n(1) = 1 - P_n(2)$	$P_n(1) = 1 - P_n(2)$
$P_n(2) = [(1 + \exp(a_0 + a_1 Z_n))(1 + \exp(b_0 + b_1 X_n))]^{-1}$	$P_n(2) = [(1 + \exp(a_0 + a_1 t_{2n}))(1 + \exp(b_1 X_n + b_2 \Delta t_n))]^{-1}$
	$t_{2n} = \max(t', t''), \quad t_{1n} = \min(t', t''), \quad t_{1n} < t_{2n}$
	$\Delta t_n = t_{1n} - t_{2n} \ (< 0)$

Parameter Values:	Distribution of Independent Variables	Parameter Values:	Distribution of Independent Variables
$a_0 = -1.0 \quad b_0 = 2.0$	$Z_n \sim N(1,1)$*	$a_0 = 5.0 \quad b_1 = -3.0$	$X_n \sim N(1, 1.5)$
$a_1 = -2.0 \quad b_1 = -3.0$	$X_n \sim N(0,1)$	$a_1 = 1.0 \quad b_2 = -1.0$	$t', t'' \sim N(1, 1)$

* Note: N(u,v) refers to a normal distribution with mean u and variance v.

and a logit choice probability

$$R_n = [1 + \exp(b_0 + b_1 X_n)]^{-1} \qquad (8)$$

for alternative 2. The explanatory variable for the inclusion probability, Z_n, is assumed to have a normal distribution with mean and variance of 1. The explanatory variable for the binary logit choice probability is X_n, which also follows a normal distribution with a mean of zero and a unit variance.

In Fig. 1, the probability of alternative 1 being chosen is plotted against the model parameter for various values of the variable Z_n in the inclusion probability. Since the coefficient of Z_n is negative, a small or negative value of Z_n results in a small probability that alternative 2 is included in the choice set. Accordingly, alternative 1 has a high probability of being chosen by the individual as Fig. 1 indicates. For large positive values of Z_n, alternative 2 is available in virtually all decisions and the choice probabilities coincide with those of the ordinary logit model,

$$P_n(2) = [1 + \exp(b_0 + b_1 X_n)]^{-1} \qquad (9)$$

Model B incorporates the time budget constraint. The variables that are subject to the constraint are t_{1n} and t_{2n}, the travel times by mode 1 and mode 2 for individual n. Following previous discussions, mode 1 is assumed to be always

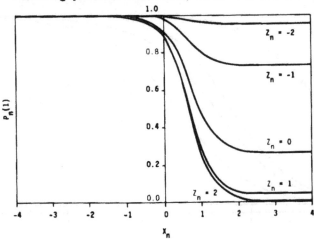

Figure 1. The probability of alternative 1 being chosen, $P_n(1)$, versus explanatory variable X_n for various values of Z_n that determines the inclusion probability. The curves are for the model A in Table 1.

faster than mode 2, i.e., $t_{1n} < t_{2n}$. It is also assumed that mode 1 is always available in the choice set. However, the availability of mode 2 depends on the unknown time budget. The inclusion probability is given by the logistic distribution,

$$Q_n = [1 + \exp(a_0 + a_1 \, t_{2n})]^{-1} \qquad (10)$$

Careful comparison of Eqs. 6 and 10 would show that a_1 is inversely proportional to the standard deviation of the time budgets, and the ratio $-a_0/a_1$ corresponds to the mean time budget $\bar{T}$. The choice probability of alternative 2, given it is included in the choice set, is a function of the travel time difference, Δt_n and an independent variable, X_n.

Model B is shown in Fig. 2. In this time budget constrained model, the probability that travel time of mode 2, t_{2n}, falls within the unobservable budget is small when t_{2n} is large. Then, the probability of mode 2 being in the choice set is small and the probability of mode 1 being chosen, $P_n(1)$, is high. The probability $P_n(1)$ also benefits from a large t_{2n} because the choice probability favors the faster mode 1. Also plotted in Fig. 2 are the corresponding logit choice probabilities, i.e., the model without the time budget constraint:

$$P_n(2) = [1 + \exp(b_1 \, X_n + b_2 \Delta t_n)]^{-1} \qquad (11)$$

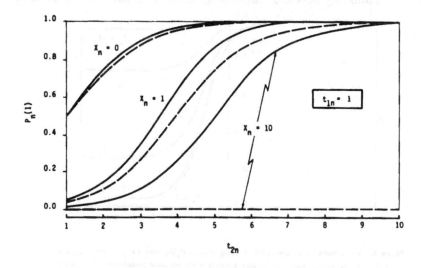

Figure 2. The probability of alternative 1 being chosen, $P_n(1)$. The solid lines are for the constrained model B in Table 1 and the dashed lines are for the unconstrained model.

The choice probability with the ordinary model is defined as a function of the travel time difference, Δt_n, and another independent variable, X_n, which represents a factor other than travel time. The coefficient of X_n is negative. Mode 1 is the favored mode when X_n is small, and there is little difference in the outcome of the choice whether or not mode 2 is considered. Accordingly, there is no discernable difference between the ordinary logit model and the time budget constrained model. As X_n increases, mode 2 becomes more favorable and the effect of the inclusion probability becomes more important. For example, when $X_n = 10$, the ordinary logit model indicates that mode 2 will almost certainly be chosen and $P_n(1)$ is virtually zero for values of t_{2n} up to 10 (t_{1n} is fixed at 1 in the figure). Because the time budget constraint affects mode 2, the constrained model on the other hand shows that mode 1 has a good chance of being chosen as well. For large values of X_n, the difference between the time constrained model and the ordinary logit model is very significant.

The examples presented here are for the purpose of illustrating the general properties of the constrained models. The numerical values shown in the figures and discussions on them do not represent real world behavior. However, the examples do indicate that the proposed models have the ability of representing the constraints placed on the choice. As shown in Figs. 1 and 2, the constrained models give an additional flexibility in modeling travel choices.

NUMERICAL ANALYSIS AND MODEL PARAMETER ESTIMATION
Several constrained choice models including the ones in the previous section are used in this section to generate artificial data sets for studying the properties of the estimated coefficients of the models. Three aspects are examined in the analysis; a) uniqueness of the coefficient estimates and convergency properties, b) sample size requirements, and c) biases in the estimates when the constraints are ignored in the choice model.

For each model considered, Monte Carlo simulation was used to generate a set of random data consisting of the explanatory variables values and outcomes of the choice. Each observation in the simulated data set was prepared as follows. Using the assumed distributions, values of the explanatory variables were first randomly generated. They were used together with the assumed values of the model coefficients to evaluate the choice probabilities of the alternatives. The choice was then simulated by generating a random number and an alternative was "chosen" to complete the record. In simulating the data using the budget constrained model (model B of the previous section), two random numbers were first obtained from a normal distribution. The larger of the two was taken to be the travel time of the slower mode (t_{2n}), while the other was taken to be t_{1n}. The value of the independent variable, X_n, was obtained again randomly, the difference Δt_n computed, the choice probability evaluated, and the choice simulated. The coefficients of the models were estimated using the data sets thus obtained. The estimated coefficient values should be in agreement with the ones used to prepare the data sets.

The maximum likelihood method was used in the analysis. The general form of the log-likelihood function for the binary case is given in Eq. 3. Given the model form and the independent variable values, the log-likelihood is a function of the model coefficients. For example, model A of the previous section has a log-likelihood function:

$$
\begin{aligned}
L &= L(a_0, a_1, b_0, b_1) \\
&= \sum_{n \in C_1} \ln[1 - 1/(1 + \exp(a_0 + a_1 Z_n))(1 + \exp(b_0 + b_1 X_n))] \\
&\quad - \sum_{n \in C_2} \ln[(1 + \exp(a_0 + a_1 Z_n))(1 + \exp(b_0 + b_1 X_n))] \quad (12)
\end{aligned}
$$

where, as before, C_1 refers to the set of records in which alternative 1 was chosen, and C_2 is the set for alternative 2. The coefficient values can be determined using the necessary condition for the maximum likelihood, that the partial derivatives of the log-likelihood function with respect to them must be all zero. If the log-likelihood is a concave function, the set of coefficient values that satisfies this condition also maximizes the likelihood and the joint probability that the observed set of choices will occur is maximized.

Unlike the case with the ordinary logit model, the constrained models entertained here create problems. While the log-likelihood function for the ordinary logit model is always concave, this is not the case for the constrained choice models. Accordingly, the extremum obtained from the first derivatives may minimize the log-likelihood function, or may be a local extremum which is neither the overall maximum or minimum. This makes the numerical search process for the desired coefficient values by far more complex than the case for the ordinary logit model.

The optimization procedure used in this analysis is the classical Newton's method, as it requires less frequent evaluation of the log-likelihood values than other algorithms that involve a unidimensional search procedure. (Evaluating the log-likelihood is quite time consuming with a large number of observations. On the other hand, evaluating the Hessian matrix and its inverse in the Newton's method requires relatively small additional computation time if the choice probability R_n is formulated in the logit form; also see reference 25.) The results of the model estimation using the simulated data are now discussed.

Uniqueness: In light of the limited knowledge we have about the log-likelihood function, the first aspect to be examined is whether the algorithm converges to the same point when initiated with different initial coefficient values. If this is the case, then we may expect that the function is unimodal at least in the region where the coefficients take on reasonable values. The result of the examination is summarized in Tables 2 and 3. Table 2 is obtained for model A, which incorporates a general constraint, and Table 3 gives the result for model B which represents a time budget constraint. The tables provide information about the convergence characteristics and compare the estimated values of the parameters (from the simulated data) with the true values of the parameters (from which the simulated data were generated).

Table 2. Convergence characteristics of model A

Estimation Result

	Initial Value					Parameter	True
Run	a_o	a_1	b_o	b_1	Convergence	Estimates	Values
1	0.0	0.0	0.0	0.0	No	$a_o = -1.503$	(-1.0)
2	0.25	0.25	0.25	0.25	No	$a_1 = -2.476$	(-2.0)
3	-1.0	-1.0	-1.0	-1.0	No	$b_o = 1.796$	(2.0)
4	-1.0	-1.0	1.0	-1.0	Yes	$b_1 = -2.421$	(-3.0)
5	1.0	-1.0	1.0	-1.0	Yes		
6	-1.0	-1.0	1.0	1.0	No		
7	-1.0	1.0	1.0	-1.0	Yes		
8	0.0	0.0	0.0	0.0	No	$a_o = -0.910$	(-1.0)
9	1.0	1.0	1.0	1.0	No	$a_1 = -1.747$	(-2.0)
10	-1.0	-1.0	-1.0	-1.0	No	$b_o = 1.591$	(2.0)
11	-1.0	-1.0	1.0	-1.0	Yes	$b_1 = -2.545$	(-3.0)
12	1.0	-1.0	1.0	-1.0	Yes		
13	-1.0	-1.0	1.0	1.0	No		
14	-1.0	1.0	1.0	-1.0	No		

Note: Runs 1 through 7 and runs 8 through 14 use different data sets.

Simulated Choice Frequencies

Alternative 1 chosen 147 times out of 200 (Runs 1 - 7)
365 times out of 500 (Runs 8 - 14)

Unfortunately, the Newton's algorithm does not guarantee convergence from all arbitrary initial coefficient vectors. This can be seen from the analysis results. Importantly, however, when it converged, it always converged at the same extremum and the coefficient estimates are reasonably close to the values used in generating the data. The result thus suggests that the log-likelihood function is unimodal in the vicinity of the true coefficient values.

Table 3. Convergence characteristics of model B

__Estimation Result__

Run	a_0	Initial Value a_1	b_1	b_2	Convergence	Parameter Estimates	True Values
1	0.0	0.0	0.0	0.0	No		
2	-0.25	0.50	-1.50	-0.50	Yes		
3	0.25	-0.50	1.50	0.50	No		
4	-7.50	1.50	-4.50	-1.50	No	$a_0 = -4.137$	(-5.0)
5	-1.0	1.0	-1.0	-1.0	Yes	$a_1 = 0.472$	(1.0)
6	1.0	1.0	-1.0	-1.0	Yes	$b_1 = -3.632$	(-3.0)
7	-1.0	-1.0	-1.0	-1.0	Yes	$b_2 = -1.012$	(-1.0)
8	-1.0	1.0	1.0	-1.0	Nò		
9	-1.0	1.0	-1.0	1.0	No		
10	0.5	0.5	0.5	0.5	No		

__Simulated Choice Frequencies__

Alternative 1 chosen 111 times out of 200.

The failure in attaining the convergence is at least partly due to the algorithm and the computer program used here which basically assumes that the log-likelihood function is concave and terminates the search process when it is not. It is expected that the problem can be corrected by adopting an alternative optimization algorithm and computer program. Development of a routine that seeks an alternative initial coefficient vector and restarts the algorithm when it fails to converge, will be helpful.

__Sample Size Requirements__: The maximum likelihood estimator is consistent, i.e., the estimates of the coefficients converge to the true coefficient values and their standard errors decrease as the sample size increases under the regularity conditions (see, e.g., _26_, p. 384). The question addressed here is how quickly the standard errors will diminish and the estimates converge to the true values for the constrained choice model proposed here. The properties of the coefficient estimates for small samples are also of important concern. To examine these aspects, a number of data sets were generated using the same parameter values in the simulation. The coefficients are then estimated for each set and the statistical variations of the estimated coefficients are evaluated. The process is repeated with sampe sizes (N) of 250, 500, and 1000, for both model A and model B. Table 4 summarizes the results.

Table 4. Consistency of coefficient estimates

Model A, initial coefficient vector $(a_0, a_1, b_0, b_1) = (-1, -1, 1, -1)$

N	No. of Runs	Convergence[a]	a_0 Mean	a_0 S.E.	a_1 Mean	a_1 S.E.	b_0 Mean	b_0 S.E.	b_1 Mean	b_1 S.E.
250	80	73.8%	-1.405	1.350	-3.281	3.99	1.966	0.333	-3.177	0.695
500	60	93.3%	-1.391	1.059	-2.887	1.878	2.002	0.344	-3.099	0.390
1000	30	100.0%	-1.196	0.492	-2.123	0.619	1.989	0.144	-3.012	0.255
True Coefficient Value			-1.0		-2.0		2.0		-3.0	

[a] Percent of runs where convergence was obtained.

Model B, initial coefficient vector $(a_0, a_1, b_0, b_1) = (-1, 1, -1, 1)$

N	No of Runs.	Convergence[a]	a_0 Mean	a_0 S.E.	a_1 Mean	a_1 S.E.	b_0 Mean	b_0 S.E.	b_1 Mean	b_1 S.E.
250	90	67.8%	-5.288	2.929	0.849	1.555	-3.171	0.565	-1.079	0.258
500	60	86.7%	-5.204	2.832	0.717	1.099	-3.134	0.338	-1.024	0.149
1000	30	100.0%	-4.712	1.344	0.822	0.624	-3.033	0.249	-0.996	0.134
True Coefficient Value			-5.0		1.0		-3.0		-1.0	

The model of nonbudgetary constraints shows clearly that the maximum likelihood estimators are consistent; the averages of the estimated coefficients approach the true coefficient values used in the simulation and their standard errors at the same time decrease as the number of observations in the data set increases. Note that the frequency of convergence also improves as the sample size increases. (In the experiment, the estimation process was started with the same fixed coefficient vector, as shown in the table, and no repeated attempts were made when the algorithm did not converge).

Examination of the result for model A indicates that, when the sample size is small, the estimates of a_0 and a_1 are on the average negatively biased away from the true coefficient values. It must be kept in mind in interpreting such results that the simulation experiment is a purely numerical exercise, and the less reliable estimates of a_0 and a_1 compared with b_0 and b_1 do not imply that the estimation of the inclusion probability involes more uncertainty than that of the choice probability, or vice versa. This is simply due to the coefficient values and independent variable distributions assumed in the simulation.

The time budget constrained model B also shows that the maximum likelihood estimators of the coefficients are consistent. The estimates of coefficient a_1, however, show rather erratic behavior. Recall that a_1 represents the reciprocal of the standard error of the (time) budgets available for the choice, and $-a_0/a_1$ is the mean of the distribution of the budget. The value of a_0 is consistently estimated with accuracy. The estimates of a_1 indicate that, at least for this particular set of true coefficient values and distributions of independent variables, estimation of the standard error of the budgets involves more uncertainty, or, requires a larger sample for the same level of accuracy.

Omission of the Constraints: The third aspect examined here is the direction and magnitude of the biases in the coefficient estimates that may arise when the model is specified without the constraints when they do in fact affect the choice. The bias is evaluated by estimating the coefficients of the ordinary choice model on the data set that was generated with the constraint. The analysis used the following models to simplify the interpretation of the results:

Model Specification with Constraints (Used to Generate Data)

$$P_n(1) = 1 - P_n(2)$$
$$P_n(2) = Q/(1 + \exp(a'X_n))$$

Alternative Specification without Constraints (Ordinary Logit Model)

$$P_n(1) = 1 - P_n(2)$$
$$P_n(2) = 1/(1 + \exp(b'X_n))$$

Note that the inclusion probability Q_n is assumed to be a constant, Q, for all observations.

Table 5 summarizes the result of the experiment in which the coefficients were estimated on 10 different data sets and the averages of the estimates of the respective coefficients are taken. With the inclusion probability Q of 0.378, alternative 2 is included in the choice set only about one-third of the time. The true model form and the distribution of the independent variables indicate that if it is included in the choice set, alternative 2 will be chosen half of the time, resulting in an aggregate choice probability for alternative 2 of 0.189 (note the symmetry that makes the derivation of the aggregate choice probability straightforward). The simulation result, where alternative 1 is chosen on average 400 times out of 500, is in agreement with this value.

Table 5. Comparison of the estimated coefficients between a constrained and an ordinary logit model. The data used were generated from the constrained model.

Constrained Model:

$P_n(1) = 1 - P_n(2)$

$P_n(2) = Q_n/(1 + \exp(a_0 + a_1 X_{1n} + a_2 X_{2n}))$

$Q_n = 1/(1 + \exp(0.5)) = 0.378$

Ordinary Model:

$P_n(1) = 1 - P_n(2)$

$P_n(2) = 1/(1 + \exp(b_0 + b_1 X_{1n} + b_2 X_{2n}))$

Parameter Values	Distribution of Independent Variables
$a_0 = 0.0$	$X_{1n} \sim N(1.0, 1.5)$
$a_1 = 2.0$	$X_{2n} \sim N(1.0, 1.5)$
$a_2 = -2.0$	

Simulated Choice Frequencies (Average of 10 runs)

Alternative 1 chosen on average 408.4 times out 500.

Averages of coefficient estimates

	Ordinary Model	True Value
$b_0 =$	1.7489	0.0
$b_1 =$	0.5734	2.0
$b_2 =$	0.4915	-2.0

Estimated Elasticity of Alternative 1 with respect to X_{1n} and X_{2n} at $(X_{1n}, X_{2n}) = (1,1)$.

	True Value	Ordinary Model
$\varepsilon_{1X_1} =$	0.232	0.0792
$\varepsilon_{1X_2} =$	-0.232	-0.0679

The ordinary logit model expresses the unbalance in the choice frequencies between the alternatives by its constant term, b_0. The result indicates that the sensitivities of the choice to the independent variables are vastly underestimated in the ordinary choice model as the sample calculation of the elasticity in the table shows. This is a result of its misrepresentation of the inclusion probability by the constant term. Accordingly, the model may underestimate the sensitivity of demand to changes in the explanatory variables when the changes are relatively small. When the model is extrapolated to assess the changes in demand due to substantial changes in the explanatory factors, the model may overestimate the change. This is illustrated in Figure 3. Of course the numerical results here do not have any relevance to the real world phenomena. They nevertheless demonstrate clearly that the incorrect representation of the constraints may lead to serious errors in the analysis and prediction of travel choice behavior.

510

Transportation and Traffic Theory

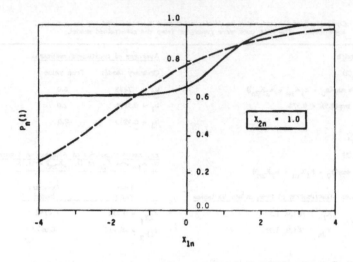

Figure 3. A comparison of the probability of choosing alternative 1 given by the input model of the simulation (solid line) and the estimated ordinary logit model with the constraint omitted (dashed line).

SUMMARY

The effect of constraints placed on the choice set in travel decisions has long been recognized. A number of recent studies have shown the importance of time budget on activity scheduling and travel behavior. During the energy crises, studies have found significant increases in trip chaining behavior under the limited and more expensive gasoline supply. The many constraints that affect the choice set of individuals in making travel decisions include time and monetary budgets, availability of transportation modes, and limitation of knowledge on alternatives.

Choice set definition has not been incorporated into most discrete travel choice models. In disaggregate demand analysis, constraints are recognized by the researcher through subjective judgements which may introduce biases and errors. In this paper, a new formulation of the binary choice model is proposed. By introducing a probability called "inclusion probability" into the traditional discrete choice models, most constraints may be appropriately incorporated into the choice models.

A number of examples were examined in the paper. The numerical properties of these examples show the flexibility that these models can offer in describing various characteristics of constrained choices. Moreover, the new formulation offers an additional dimension for modeling travel behavior and interpreting survey data. The examples show that if the choices are constrained, there are significant differences between the constrained model results and ordinary logit model results.

This paper represents a beginning of an effort to incorporate various contraints into discrete choice analysis of travel behavior, and many issues remain to be examined in the future. These include examination of the mathematical properties of the log-likelihood function, development of a computational procedure for better convergence characteristics with small data sets, extension to multinomial cases and examination of model properties such as elasticities.

REFERENCES

1. Burnett, K.P., and D.C.L. Prestwood, "The Development of Disaggregate Trip Distribution Models," in P.R. Stopher and A.H. Meyburg (eds.), Behavioral Travel-Demand Models, D.C. Heath, Lexington, MA, 1975, pp. 109-124.
2. Swait, J.D., V.J. Kozel, R.C. Barros, and M.E. Ben-Akiva, "A Model System of Individual Travel Behavior for a Brazilian City," paper presented at the World Conference on Transport Research, Hamburg, Germany, 1983.
3. Clarke, M.I., M.C. Dix, P.M. Jones, and I.G. Heggie, "Some Recent Developments in Activity-Travel Analysis and Modeling," Transportation Research Record 794, 1981, pp. 1-8.
4. Gunn, H.F., "Travel Budgets - A Review of Evidence and Modeling Implications," Transportation Research, Vol. 15A, 1981, pp. 7-23.
5. Kitamura, R., L.P. Kostyniuk, and M.J. Uyeno, "Basic Properties of Urban Time-Space Paths: Empirical Tests," Transportation Research Record 794, 1981, pp. 8-19.
6. Kitamura, R. and T.N. Lam, "A Time Dependent Markov Renewal Model of Trip Chaining," in V.F. Hurdle, E. Hauer, and G.N. Steuart (eds.), Proceedings of the Eighth International Symposium on Transportation and Traffic Theory, University of Toronto Press, Toronto, 1983, pp. 376-402.
7. Brog, W., and E. Erl, "Application of a Model of Individual Behavior (Situational Approach) to Explain Household Activity Patterns in an Urban Area and to Forecast Behavioral Changes," in S. Carpenter, and P. Jones (eds.), Recent Advances in Travel Demand Analysis, Gower Publishing Co., Aldershot, England, 1983, pp. 350-370.
8. Horowitz, J., "Random Utility Models of Urban Nonwork Travel Demand: A Review," The Regional Science Association Papers, Vol. 45, 1980, pp. 125-137.
9. Tye, W.B., L. Sherman, M. Kinnucan, D. Nelson, and T. Tardiff, Application of Disaggregate Travel Demand Models, NCHRP Report 253, Transportation Research Board, Washington, D.C., 1982.
10. McFadden, D., Quantitative Methods for Analyzing Travel Behavior of Individuals: Some Recent Developments, Working Paper No. 7704, Institute of Transportation Studies, University of California, Berkeley, CA, 1977.
11. Daganzo, C., Multinomial Probit: The Theory and Its Application to Demand Forecasting, Academic Press, New York, 1979.

12.   Lerman, S.R., and C.F. Manski, "Sample Design for Discrete Choice Analysis of Travel Behavior: The State of the Art," Transportation Research, Vol. 13A, 1979, pp. 29-44.
13.   Koppelman, F.S., "Guidelines for Aggregate Travel Prediction Using Disaggregate Choice Models," Transportation Research Record 610, 1976, pp. 19-24.
14.   McFadden, D., K. Train, and W.B. Tye, "An Application of Diagnostic Tests for the Independence From Irrelevant Alternatives Property of the Multinomial Logit Model," Transportation Research Record 637, 1977, pp. 39-46.
15.   Koppelman, F.S., and C.G. Wilmot, "Transferability Analysis of Disaggregate Choice Models," Transportation Research Record 895, 1982, pp. 18-24.
16.   Burnett, P., and S. Hanson, "Rationale for an Alternative Mathematical Approach to Movement as Complex Human Behavior," Transportation Research Record 723, 1979, pp. 11-24.
17.   Goodwin, P.B., "The Usefulness of Travel Budgets," Transportation Research, Vol. 15A, 1981, pp. 97-106.
18.   Tanner, J.C., "Expenditure of Time and Money on Travel," Transportation Research, Vol. 15A, 1981, pp. 25-38.
19.   Zahavi, Y., and J.M. Ryan, "Stability of Travel Components Over Time," Transportation Research Record 750, 1980, pp. 19-26.
20.   Charles River Associates, Inc., A Disaggregated Behavioral Model of Urban Travel Demand, PB-210-515, NTIS, Springfield, VA, 1972.
21.   Stopher, P.R., "Captivity and Choice in Travel-Behavior Models," Transportation Engineering Journal of ASCE, Vol. 106, No. TE4, 1980, pp. 427-435.
22.   Golob, T.F., and W.W. Recker, "Mode Choice Prediction Using Attitudinal Data: A Procedure and Some Results," Transportation, Vol. 6, 1977, pp. 265-286.
23.   Tardiff, T.J., "Casual Inferences Involving Transportation Attitudes and Behavior," Transportation Research, Vol. 11, 1977, pp. 397-404.
24.   Gaudry, M.J.I., and M.G. Dagenais, "The Dogit Model," Transportation Research, Vol. 13B, 1979, pp. 105-111.
25.   Sheffi, Y., R. Hall, and C.F. Daganzo, "On the Estimation of the Multinomial Probit Model," Transportation Research, Vol. 16A, 1982, pp. 447-456.
26.   Theil, H., Principles of Econometrics, John Wiley & Sons, New York, 1971.

Ninth International Symposium on
Transportation and Traffic Theory
© 1984 VNU Science Press, pp. 513–531

ESTIMATION OF DISAGGREGATE REGRESSION MODELS OF PERSON TRIP GENERATION WITH MULTIDAY DATA

FRANK S. KOPPELMAN[1] and ERIC I. PAS[2]
[1] *Department of Civil Engineering, Northwestern University, Evanston, IL, U.S.A.*
[2] *Department of Civil and Environmental Engineering, Duke University, Durham, NC, U.S.A.*

ABSTRACT

Theoretical discussions of travel behavior recognize the day to day variability of such behavior and provide a rationale for such variability. However, this variability is commonly ignored in the process of data collection and model estimation. This paper examines the estimation of disaggregate least squares regression models of person trip generation with multiday data.

The estimation properties of a variety of models are formulated, including multiday models which take account of correlated errors within individuals, averaged multiday models, and single day models. All three types of models yield unbiased estimators. However, the estimators of both types of multiday models are more efficient than those of the single day model. These analytic results are verified empirically and the magnitude of the efficiency gain is estimated using a five day data set collected in Reading, England.

1. INTRODUCTION

Trip generation models are central to the conventional four-step urban travel forecasting model sequence. Trip generation prediction establishes the level of demand for travel while the trip distribution, mode shares, and network assignment models distribute the total travel over different destinations, modes, and routes. Thus, the prediction of traffic flows in a transportation network is critically dependent on the predictions obtained from the trip generation model. In this paper, we examine the estimation of such models using repeated observations for each sample unit.

Earliest trip generation models were estimated using data aggregated over small geographical areas; however, the use of household or individual data for estimating models of residential trip generation has become accepted practice (FHWA, 1975; Dickey, 1983). Two closely related approaches are used for estimating these models; namely, least squares linear regression and category analysis. There has been considerable debate regarding the differences between these methods (Winsten, 1967; Douglas, 1973), although category analysis can be shown to be equivalent to least squares linear regression with dummy variables (Dobson, 1976). The mathematical formulations and empirical results reported in this paper deal directly with the least squares linear regression case but the results apply equally well to the category analysis model.

The methodology for estimation of household or individual trip generation models is well developed. However, these models generally are estimated using a single day's data for each household or individual even though trip generation behavior varies from day to day. Typically, the data used to estimate the models is obtained by asking each sample household or person to report the number of trips made on a single, randomly selected, day of the week. Intuitively, it seems reasonable to consider the use of data collected over a number of days to study a phenomenon which varies from day to day.

This paper explores the effects of using multiday versus single day data to estimate trip generation models. The work has two separate but related objectives. The first is to review and compare the alternative ways in which multiday data can be employed in estimating a disaggregate least squares regression trip generation model and the effect, if any, on estimation results. The second is to examine the effect of the day-to-day variability in person or household trip generation rates on the estimation properties of trip generation models estimated with single day data. We approach these objectives both analytically and empirically.

The remainder of this paper is organized as follows. In section 2, we discuss the concept of variability in daily travel behavior and we differentiate interpersonal and intrapersonal variability in this behavior. In section 3, we present analytical results concerning alternative formulations of the multiday data model and the single day data model. We examine and compare the parameter estimates and the variance-covariance matrix of the parameter estimates for these models. In section 4, we report the results of empirical analyses undertaken using a one week data set collected in Reading, England. Finally, in section 5 we summarize the paper and draw our conclusions.

2. VARIABILITY IN DAILY TRAVEL BEHAVIOR

The purpose of this section is to develop a conceptual framework for the identification of sources of variability in daily trip generation behavior. In particular, we distinguish between interpersonal and intrapersonal variability in daily travel and discuss the reasons for each type of variability.

Travel is a derived demand undertaken to satisfy the needs or desires of individuals/households to participate in various types of activity (Mitchell and Rapkin, 1954; Oi and Shuldiner, 1962). These activities can be classified into categories such as subsistence, maintenance, and leisure (Reichman, 1976) which have different characteristics with respect to flexibility of schedule or location, regularity, periodicity, duration, and importance. The participation of individuals in different types of activities is related to their underlying motivational structure (Chapin, 1968 and 1974). Characteristics of the individual and his/her household such as life cycle, role, or life style are important determinants of travel behavior (Pas, 1980; Fried et al., 1977; Koppelman et al., 1978; and others). Thus, we expect to observe differences in both travel and activity behavior between individuals. These differences are referred to as interpersonal variability.

The actions of individuals to satisfy activity needs or desires has a temporal structure which is related to the types of activities undertaken, and their periodicity and regularity. Hagerstrand's (1970 and 1973) time-geography framework provides a structure within which to examine the temporal pattern of travel and activities. Chapin (1968) suggests the usefulness of adjusting the time scale of analysis to obtain different perspectives on social behavior. Reichman (1976) argues that different types of activities have different periodicity of repetition. Thus, we expect to observe differences in travel and activity behavior for a given individual over time. These differences are referred to as intrapersonal variability.

Trip generation models commonly use variables describing the characteristics of the individual and his/her household to explain variability in travel behavior between individuals. Such variables are generally small in number and include readily observable characteristics such as household size, income, and automobile ownership rather than variables directly describing life cycle, role, and life style. These variables are useful in explaining a portion of interpersonal variability but can not explain intrapersonal variability.

A typical data set for travel behavior analysis includes a single day observation of travel behavior, description of travel service characteristics, and descriptions of the individual and his/her household. The latter two data elements are not specifically related to the date of observation but describe characteristics which are relatively invariant over short periods of time. The observed travel behavior in such data includes variability explained by or associated with descriptors of travel service and/or descriptors of individual and his/her household characteristics, unexplained variability between individuals and unexplained variability for each individual over time. Analysis of all three components of variability requires the use of a structural model of multiday data observations. If the data includes only a single daily observation for each individual, it will not be possible to separate the unexplained interpersonal variability from the intrapersonal temporal variability.

3. TRIP GENERATION MODEL FORMULATION AND ESTIMATION PROPERTIES

We formulate the linear trip generation model for daily travel as:

$$Y_{jt} = X_{jt}\beta + \epsilon_{jt} \qquad j = 1,2,\ldots,J \qquad t = 1,2,\ldots,T \qquad (1)$$

where

Y_{jt} is the number of trips by individual "j" on day "t",

X_{jt} is a column vector of variables describing individual "j" and his/her environment on day "t",

β is a column vector of parameters, and

ϵ_{jt} is an error term associated with individual "j" on day "t".

In this section, we discuss the formulation of the model for the case of single day observations and multiday observations and derive the properties of estimators under different estimation procedures.

3.1. Single day sample case.

In the single day sample case, we can write this model in matrix notation as

$$Y = X\beta + \epsilon , \qquad (2)$$

where the elements of Y and the rows of X are single day observations for individual "j" on day "t". Using ordinary least squares (OLS) estimation under the common assumption that the error terms are independently and identically distributed, $N(0, \sigma^2)$, we obtain

$$\hat{\beta} = (X'X)^{-1}X'Y \tag{3a}$$

$$\Sigma_{\hat{\beta}} = \sigma^2 (X'X)^{-1} \tag{3b}$$

The estimator, $\hat{\beta}$, is the best linear unbiased estimator in this case.

3.2. Multiday sample case.

In the multiday sampling case, we can represent the trip generation relationship in matrix form by

$$Y_* = X_* \beta + \varepsilon \tag{4}$$

where each T elements in Y_* and each T rows in X_* are the repeated daily observations for individual "j". In this case, the error terms are not distributed independently across observations. Specifically, we assume that the error term can be decomposed as

$$\varepsilon_{jt} = u_j + w_{jt} \tag{5}$$

where

   u_j represents the error associated with individuals and is assumed
      to be distributed $N(0, \sigma_u^2)$,

and

   w_{jt} represents error associated with the interaction between
      individuals and time periods and is assumed to be distributed
      $N(0, \sigma_w^2)$.

Thus, the covariance between error terms, ε_{jt}, $\varepsilon_{j't'}$, is σ^2 ($= \sigma_u^2 + \sigma_w^2$) when $j = j'$ and $t = t'$, σ_u^2 when $j = j'$ and $t \neq t'$, and 0 when $j \neq j'$.

The resultant covariance matrix for errors is

$$V = \sigma_w^2 I_{JT} + \sigma_u^2 A \tag{6}$$

where

A is the Kronecker product matrix of I_J and I_T, and

I_{JT}, I_J, and I_T are identity matrices of order JT, J, and T, respectively.

This is known as a crossed-error model (Fuller and Battese, 1974).

The best linear unbiased estimator for this model obtained by generalized least squares (GLS) is:

$$\tilde{\beta} = (X_*'V^{-1}X_*)^{-1} X_*'V^{-1}Y_* \tag{7a}$$

$$\Sigma_{\tilde{\beta}} = (X_*'V^{-1}X_*)^{-1} \tag{7b}$$

The exogenous variables generally used in trip generation models are invariant over short periods of time. Thus, the matrix

$$X_*' = [X_{11}, \; X_{12}, \ldots, X_{1T}, X_{21}, \ldots, X_{2T}, \ldots\ldots, X_{JT}] \tag{8a}$$

is equivalent to

$$X_*' = [\; X_1, \; X_1, \; \ldots, X_1, \; X_2, \; \ldots, X_2, \; \ldots\ldots, \; X_J], \tag{8b}$$

That is, the matrix of exogenous variables includes T identical observations for each respondent. In this case, it can be shown (Pas and Koppelman, 1983) that

$$X_*'V^{-1}X_* = \frac{1}{\sigma^2} \; \frac{T}{1 + a(T-1)} \; X'X \tag{9}$$

where

a is the portion of the total variance attributed to unobserved differences across individuals, $a = \sigma_u^2/(\sigma_u^2 + \sigma_w^2) = \sigma_u^2/\sigma^2$,

and

$$X_*'V^{-1}Y_* = \frac{1}{\sigma^2} \; \frac{T}{1 + a(T-1)} \; X'\bar{Y} \tag{10}$$

where

$\bar{Y}$ is a vector of the average number of trips generated by each individual over the observation period.

Thus, substituting equations 9 and 10 into 7a and b, yields

$$\tilde{\beta} = (X'X)^{-1}X'\bar{Y} \tag{11a}$$

and

$$\Sigma_{\tilde{\beta}} = (X'X)^{-1}\sigma^2 \frac{1 + a(T-1)}{T} \ . \tag{11b}$$

On the other hand, application of OLS in this case would yield (Pas and Koppelman, 1983)

$$\hat{\beta} = (X'X)^{-1}X'\bar{Y} \tag{12a}$$

and

$$\Sigma_{\hat{\beta}} = (X'X)^{-1}\sigma^2 \frac{1}{T} \ . \tag{12b}$$

That is, the parameter estimation results obtained by applying GLS and OLS to the multiday data are equivalent, but the OLS covariance estimator is biased downward by the factor $[1 + a(T-1)]$.

3.3. Comparison of single day OLS and multiday GLS estimators.

The single day OLS parameter estimator, $\hat{\beta}$ in 3a, and the multiday GLS estimator, $\tilde{\beta}$ in 11a, are equivalent in expectation. The corresponding covariance matrices (equations 3b and 11b) are related by

$$\frac{\Sigma_{\tilde{\beta}}}{\Sigma_{\hat{\beta}}} = \frac{1 + a(T-1)}{T} \ . \tag{13}$$

That is, the magnitude of the covariance matrix for the multiday sample decreases with increasing number of days of observation and is asymptotic to a. Thus, the variance reduction obtained by using a multiday sample is large when the variance attributable to differences across individuals is small. In the limit, when there is no person to person error component, $\sigma_u^2 = 0$, the variance reduction is proportional to the number of days. Alternatively, when all unexplained variation is attributable to the person to person error component, $\sigma_w^2 = 0$, there is no variance reduction; that is, no additional information is obtained by further observations on the same individual in this case. Based on the discussion in section 2, we expect both variance components to be positive. Thus, the reduction in the variance of the multiday GLS estimators will be less than proportional to the number of daily observations.

The net effect of using multiday data and estimating trip generation models with GLS is to obtain the same parameter estimates as expected with single day data. But the model using multiday data yields parameter estimates having greater precision than can be obtained by use of single day data. An additional advantage of using multiday data is that it is possible to obtain insight into the sources of unexplained variance in behavior. Specifically, the estimation process is formulated to obtain an estimator for the portion of variance attributable to unexplained differences among persons, σ_u^2, and the portion attributable to person–day interaction effects, σ_w^2.

3.4. Average multiday data.

Next, we consider an alternative approach to estimation with multiday data. We define the variables in equation 1 to represent the average of variables for each person over the daily observations. In matrix notation, this model becomes

$$\bar{Y} = X\beta + \bar{\varepsilon} \tag{14}$$

where

$\bar{Y}_i$ is an element of $\bar{Y}$ obtained by $\sum_t Y_{it}/T$, and

$\bar{\varepsilon}_i$ is an element of $\bar{\varepsilon}$ obtained by $\sum_t \varepsilon_{it}/T$.

The error term has variance given by

$$\sigma_\varepsilon^2 = \sigma_u^2 + \sigma_w^2/T$$

$$= (a + \frac{1-a}{T}) \sigma^2$$

$$= \frac{1 + a(T-1)}{T} \sigma^2 \tag{15}$$

These error terms are independent across observations. Therefore, it is appropriate to apply OLS which gives

$$\bar{\beta} = (X'X)^{-1}X'\bar{Y} \tag{16a}$$

$$\Sigma_{\bar{\beta}} = (X'X)^{-1}\sigma_\varepsilon^2$$

$$= (X'X)^{-1} \frac{1 + a(T-1)}{T} \sigma^2 \tag{16b}$$

The resultant estimator and its covariance matrix are identical to that obtained using GLS with multiday observations (equation 16a and 16b compared to equation 11a and 11b). Thus, use of average daily observations obtains identical estimation results in the case in which the explanatory variables for each person are invariant over the observation period. However, estimation by this pooled procedure loses information about the portion of total variability attributable to interpersonal and intrapersonal variability.

3.5. Day-to-day variability in trip generation.

The model of trip generation employed in the preceding analysis assumes that there is no systematic variability of behavior from day-to-day. We can test this assumption by expanding the model in one of two ways.

First, we can reformulate the basic model (equation 2) to include day effects by adding a set of day specific dummy variables. That is

$$Y = X\beta + \delta\alpha + \varepsilon \qquad (17)$$

where

δ is a matrix of row vectors of length T-1 for each individual, the elements of which are one corresponding to the day of observation and zero otherwise, and α is a column vector of parameters associated with days.

The first element of $\hat{\beta}$ in equation 3a is the intercept constant which represents an average effect on trip generation, over all observation, not explained by the exogenous variables. In the modified model, equation 17, the corresponding parameter represents the same average effect for a selected reference observation day, and the vector of α parameters represents differences in average effect between each observation day and the reference day. If individuals with given exogenous variables are observed on randomly assigned days, the expected values of the remaining parameters are unchanged. The multiday sampling model, equation 4, can be modified in the same manner. The intercept β parameter and the α parameters are interpreted in the same manner as for the single day case, and the estimators of the remaining β parameters will be identical to those obtained in the basic specification (equation 11). The existence of day to day differences in behavior can be tested by the formal hypothesis that the parameter vector, α, is equal to zero.

Second, in the multiday case, the components of error can be redefined to include a day-to-day component, v_t, as follows:

$$\varepsilon_{jt} = u_j + v_t + w_{jt} \qquad (18)$$

where

v_t represent the error associated with days and is assumed to be distributed N $(0, \sigma_v^2)$.

The resultant covariance structure is defined in Fuller and Battese (1974) and the estimators of $\tilde{\beta}$ and its covariance given in equations 7a and 7b apply with a more generally defined covariance matrix. In this case, one can test for differences in day-to-day behavior by the formal hypothesis that σ_v^2 is equal to zero.

4. EMPIRICAL ANALYSIS

The empirical analyses reported in this section are designed to (1) verify the analytic relationships developed in the preceding section and (2) estimate the relative importance of intrapersonal variability. The latter ultimately determines the improvement in parameter estimation efficiency which is obtainable by collecting multiday samples of varying duration rather than single day samples. In this section, we describe the data used, report the results of the empirical analysis, and interpret these results.

4.1. Data set.

The data used in this study were collected as part of the Reading Activity Diary Survey between January and March 1973 (Shapcott, 1978). The data were collected with a personal interview, a respondent-completed diary, and a short questionnaire. The diary required respondents to record each change of activity (both in and out of home), over a seven day period. The activity-based information was recoded into travel-activity information for the purposes of this research. The recoding procedures were validated by comparing the characteristics of the travel information in our recoded data set to the characteristics of the data collected in the Reading Travel Survey of 1971 (Downes and Wroot, 1974). The empirical results reported in this paper are based on a sample of 128 persons from the Reading Activity Diary Survey data set. For each person the data used here comprise the number of trips made on each of five consecutive weekdays, as well as the individual's gender and employment status, and the income of the individual's household.

4.2. Empirical results.

The estimation results obtained by use of multiday data with variance components (equations 11 a and b), multiday data ignoring variance components (equations 12 a and b), and averaged multiday data (equations 16 a and b) are reported in Table 1. All three procedures obtain identical parameter estimates, as expected. Further, as expected, the standard errors of estimate are identical for the variance components and average day estimations. However, the pooled multiday estimation yields standard errors which are biased downward. As indicated, by comparison of equations 11b and 12b, the downward bias in the covariance matrix when the variance components are ignored is 1 + $a(T-1)$ where $a = \sigma_u^2 / (\sigma_u^2 + \sigma_w^2) = .344$ and $T = 5$. The corresponding downward bias in standard errors is 1.54 which is exactly the bias reported in Table 1. Thus the empirical results support the analytic results reported in section 3. In particular, they verify that unbiased estimates of the parameters and their standard errors can be obtained using ordinary least squares estimation with a dependent variable which is the average over the observation period. That is, this estimation procedure provides all the information available from the variance components procedure except for information about the relative importance of intrapersonal and interpersonal variability.

The analytic results also show that unbiased parameter estimates are obtained with a sample comprising a randomly selected single day for each individual. However, in general, these parameter estimates have larger standard errors than those obtained with a multiday data set consisting of the same number of individuals. In particular, the ratio of the standard errors of the parameter estimates is related to the number of days in the multiday sample and the extent of intrapersonal variability in the observed data (see equation 13). Empirical verification of this result is undertaken by repeated selection of random one day samples from the multiday data set. That is, for each individual in the sample we randomly select one of the five available weekday observations available to be included in the sample. Using this procedure we generate 100 single day samples each consisting of one

daily observation for each individual. The means of the parameter
estimates and the root mean square of their standard errors for these
100 single day samples are reported in Table 2.

The results reported in Table 2 lead to two observations. First, the
means of the parameter estimates for the single day samples are essentially the same as those obtained with the multiday data set. This
confirms our analytic results with respect to the relationship between
the multiday GLS estimators and the expected values of the single day
OLS estimators. Second, we see that, as expected, the standard errors
of parameter estimates obtained with the single day samples are substantially larger than those obtained with multiday data. The observed
ratio of 0.68 is approximately equal to the square root of the ratio for
variance matrices shown in equation 13 when a = 0.344 and T = 5. This
value is 0.69. Thus, the empirical results obtained confirm the analytic expectations developed in section 3.

Further, the empirical results indicate the degree to which parameter
precision is improved by use of a multiday rather than single day
sample. In this case, the standard errors of the parameter estimates
are reduced by 31% through use of a five day sample rather than a single
day sample. The corresponding reductions for two, three, and four day
samples are 18%, 25%, and 29%, respectively. This indicates the
diminishing benefit of additional observations and suggests the
identification of an optimal number of sample days based on the relative
level of intrapersonal variance and the relative cost of collecting
additional daily observations. Finally, one can use this information to
determine the optimum tradeoff between sample size and number of daily
repetitions based on estimates of the relative level of intrapersonal
variability and the relative costs of additional daily observations and
of additional respondents (Pas and Koppelman, 1983).

5. SUMMARY AND CONCLUSIONS

This paper examines the estimation of least squares regression models of person trip generation using repeated observations. The paper describes the formulation and estimation of a model in which the correlations among the observations are taken into account. This model is a special case of a class of models that combine time series and cross section data, and it may be estimated using any generalized least squares program or a specialized program designed to estimate crossed error models such as TSCSREG in the SAS package (SAS, 1980). The analytic results show that if the explanatory variables are constant over the period of the observations, the same estimation results are obtained using ordinary least squares on the average number of daily trips made by each person in the sample. On the other hand, the analytic results also show that if one uses the multiday data without accounting for the crossed - error structure (i.e. if one estimates the model with ordinary least squares), one obtains the same parameter estimates as in the two cases discussed above, but the latter approach obtains downward biased estimates of the standard errors of the parameter estimates.

The analytic results also show that estimation with a randomly selected single day for each sample unit will produce unbiased parameter estimates which are less efficient than those obtained using multiday data. The reduction in the standard errors of the parameter estimates depends upon the degree to which the repeated observations provide additional information, and the number of days in the multiday observation period. This analysis shows that by employing a multiday data set rather than a single day data set we can obtain more efficient estimators and also obtain some insight into the relative importance of intrapersonal and interpersonal variability in travel behavior.

An important practical result of this work is that for a given data collection budget, one can design a multiday sampling procedure which will obtain more precise parameter estimates. Alternatively, for a specific desired level of parameter estimation precision, one can reduce the data collection budget by employing a multiday sample. In either case, the specific benefit derived from the multiday sample is dependent upon the cost structure of data collection and the relationship between intrapersonal and interpersonal variability. Empirical analyses show that in the Reading data, a five-day sample yields significant benefits relative to a single day sample (Pas and Koppelman, 1984).

Acknowledgements

The research reported here was supported in part by grants from the National Science Foundation of the United States of America (SES-8112494 and SES-8112495). James H. Dunlop, graduate research assistant in the Department of Civil and Environmental Engineering at Duke University, provided computer programming support. Merikay Smith and Marianne Thomas of the Department of Civil Engineering at Northwestern University prepared the manuscript.

REFERENCES

Chapin, F.S., Jr. (1974). Human Activity Patterns in the City: Things People Do in Time and in Space (Wiley, New York).
Chapin, F.S., Jr. (1968). Activity Systems and Urban Structure: A Working Schema. Journal of the American Institute of Planners.
Dickey, J.W. (1983). Metropolitan Transportation Planning, Second Edition (McGraw-Hill, New York).
Dobson, R. (1976). The General Linear Model Analysis of Variance: Its Relevance to Transportation Planning and Research, Socio-Economic Planning Sciences 10, pp. 231-235.
Douglas, A. (1973). Home-Based Trip End Models - A Comparison Between Category Analysis, and Regression Analysis Procedures, Transportation 2, pp. 53-70.
Downes, J.D. and Wroot, R. (1974). 1971 Repeat Survey of Travel in the Reading Area, Department of the Environment, TRRL SR 43 UC, Crowthorne, England.
Federal Highway Administration (FHWA) (1975). Trip Generation Analysis, U.S. Department of Transportation, Federal Highway Administration, Washington, D.C.
Fried, M., J. Havens, M. Thall (1977). Travel Behavior - A Synthesized Theory, Project 8-14 Final Report prepared for National Cooperative Highway Research Program (Washington, D.C.: Transportation Research Board).
Fuller, W.A., G.E. Battese (1974). Estimation of Linear Models with Crossed-Error Structure, Journal of Econometrics 2, pp. 67-78.
Hagerstrand, T. (1970). What About People in Regional Science? Papers and Proceedings of the Regional Science Association 24.
Hagerstrand, T. (1973). The Impact of Transport on the Quality of Life, Fifth International Symposium on Theory and Practice in Transport Economics, Greece.
Koppelman, F.S., A.M. Tybout, D.F. Syskowski (1978). Role Influences in Transportation Decision Making. In: Woman's Travel Issues: Research

Needs and Priorities, edited by S. Rosenbloom (Washington, D.C.: U.S. Department of Transportation, Research and Special Programs Administration), pp. 309-353.

Mitchell, R. B. and C. Rapkin (1954). Urban Traffic: A Function of Land Use (Columbia University Press, New York).

Oi, W. Y. and P. W. Shuldiner (1962). An Analysis of Urban Travel Demand. (Northwestern University Press, Evanston, Illinois).

Pas, E.I. (1980). Toward the Understanding of Urban Travel Behavior through the Classification of Daily Travel-Activity Patterns, Ph.D. dissertation, Northwestern University.

Pas, E.I., F.S. Koppelman (1984). Use of Multiday Data in Trip Generation Modeling: Estimation and Data Collection, Working Paper, Transportation Center, Northwestern University.

Reichman, S. (1976). Travel Adjustments and Life Styles - A Behavioral Approach. In: Behavioral Travel - Demand Models, edited by P.R. Stopher and A.H. Meyburg (Lexington Books: Lexington, Massachusetts), pp. 143-152.

SAS User's Guide (1979). SAS Institute Inc., Cary, North Carolina.

SAS Supplemental Library User's Guide (1980). SAS Institute Inc., Cary, North Carolina.

Shapcott, M. (1978). Comparison of the Use of Time in Reading, England with Time Use in Other Countries, Transactions of the Martin Centre for Architectural and Urban Studies 3, pp. 231-257.

Winsten, C.B. (1967). Regression Analysis Versus Category Analysis, paper presented at Planning and Transportation Research and Computation (PTRC) Seminar.

Table 1. Model estimation results with multiday data.

	Parameter Estimate (Standard Error)		
Variable[1]	Multiday Data, Variance Components Estimation[2] (Model 1)	Multiday Data, Ignoring Variance Components[3] (Model 2)	Average Day Data (Model 3)
Income[4]	.787 (.294)	.787 (.190)	.787 (.294)
Employment Status[5]	1.015 (.316)	1.015 (.205)	1.015 (.316)
Gender	.690 (.284)	.690 (.184)	.690 (.284)
σ_u^2	1.377	--	--
σ_w^2	2.630	--	--
σ^2	4.007	4.007	--
$\sigma_{\bar{\varepsilon}}^2$	--	--	1.903

(1)   All models include a constant intercept. In addition, models 1 and 2 include dummy variables to adjust for day of week differences.

(2)   Estimated with SAS procedure TSCSREG (SAS, 1980) using the Fuller-Battese Method (1974). See section 3.2.

(3)   Estimated with SAS procedure GLM (SAS, 1979) using ordinary least squares regression. See section 3.4.

(4)   Two income groups, high and low, are used to specify this variable.

(5)   Two employment status groups, employed and not employed, are used to specify this variable.

Table 2. Model estimation results: comparison of multiday and single day data.

Parameter Estimate

(Standard Error)[1]

Variable	Multiday Model (Pooled or Averaged Data)	Single Day Data Set[2]
Income	.787 (.294)	.770 (.434)
Employment Status	.690 (.284)	.674 (.420)
Gender	1.015 (.316)	.997 (.466)

(1) For the single day samples we report the mean of the parameter estimates and the root mean square of the standard errors.

(2) The single day data set comprises 100 single day samples.

Ninth International Symposium on
Transportation and Traffic Theory
© 1984 VNU Science Press, pp. 533–559

DISAGGREGATE MODELS OF MODE CHOICES: AN ASSESSMENT OF PERFORMANCE AND SUGGESTIONS FOR IMPROVEMENT

JANUSZ SUPERNAK

Department of Civil Engineering, Drexel University, Philadelphia, PA, U.S.A.

ABSTRACT

This paper investigates reasons for unsatisfactory performance of disaggregate models of mode choice in terms of there transferability, forecasting ability and policy relevance. It focuses primarily on the problem of non-choosers in MNL models of mode choice and proposers some improvements in model specification.

Similarity of modal choices of homogeneous categories of persons in Baltimore, Berlin and Hamburg, confirms the importance of a proper market segmentation built around such variables as auto availability, employment and age.

Geographic differences/temporal changes in population structure due to these variables are found to be directly related to the often postulated necessity of updating alternative specific constants in MNL models of modal choice.

1. INTRODUCTION

1.1. The problem

In spite of considerable progress which has been made recently in the
theory of discrete choice models of travel demand -- and among them
mode choice models -- there are still some disturbing facts about the
models' performance.

First, even within the same model family (e.g., MNL), the models appear
very area-specific and basically non-transferable spatially (Talvitie
and Kirschner, 1978; Louviere, 1979; Talvitie, et al., 1981; Koppelman
and Wilmot, 1982). Although it is believed that transferability should
increase with improved model specification (Ben Akiva, 1981, Louviere,
1981, Lerman, 1981) one can expect at most partial transferability
(Altherton and Ben-Akiva, 1976), or certain degree of transferability
(Lerman, 1979).

Second, the temporal stability -- and consequently forecasting ability
-- of behavioral models of travel choice can be questioned since the
prediction errors are quite high (Talvitie and Kirschner, 1978; Talvitie,
et al., 1981, 1982). The improvement of model transferability by adjusting their constants as recommended by Ben-Akiva (1981) or Dehghani
and Talvitie (1983) may be useful when transfering these models from
one area to another but is useless in forecast applications.

Third, the policy relevance of these models is also questionable since
a) policy-oriented variables such as travel time or travel cost have
often been found to have relatively little influence on travel choices
(e.g., Dehghani and Talvitie, 1983), and b) the major part of "explanatory power" of these models is associated with alternative specific
constants.

Potential improvements of the disaggregate choice models may range from
survey design and data collection issues (Stopher, 1982; Brog et al., 1983
and Wermuth et al., 1984), through selection of an appropriate model

formulation (Goudry and Degenais, 1979; McFadden, 1981; Williams and Ortuzar, 1982; Horowitz, 1982) and model specification (Koppelman, 1976; Horowitz, 1982; Talvitie et al., 1981), to calculation of choice models and estimation of model parameters (e.g., Sheffi, et al., 1982).

1.2. The objective of the paper

This paper focuses on multinomial logit (MNL) models of intracity mode choice which is one of the most common applications of the disaggregate choice models. By extending some previous works by Charles Rivers Associates (1978), Talvitie, et al. (1981), Galbraith and Hensher (1982), Deghani and Talvitie (1983) and Supernak (1983 b,c), this paper investigates the problem of non-choosers in disaggregate models of mode choice and examines the adequacy of the typical MNL modeling framework to represent mode choices. Particular attention is paid to a) appropriate market segments which reflect relatively homogeneous behavior, b) comparison of alternative specific constants across cultures, c) association between model constants and geographic differences/temporal changes in the population structures, and d) procedure of updating alternative specific constants for both geographic transfers and forecast applications of the MNL models of mode choice.

The remainder of this paper is organized as follows. Section 2 presents some representative numerical examples of the MNL mode choice models using data from Baltimore, Twin Cities and San Francisco. Section 3 presents results of travel choices of homogeneous groups (categories) of persons from Baltimore, U.S.A. and Berlin and Hamburg, Federal Republic of Germany. Section 4 offers an interpretation of findings presented in Sections 2 and 3, and a brief methodological discussion. Section 5 offers final conclusions and recommendations.

2. PERFORMANCE OF BEHAVIORAL MODELS OF MODE CHOICES

2.1. Data and major tests

For a typical example of a MNL mode choice model, a study performed at the State University of New York at Buffalo as a part of the U.S. Federal

Highway Administration Contract DOT-FH-11 in 1980-81 (Talvitie, et al., 1981), has been chosen. This study addressed: a) model specification, b) error analysis, c) market segmentation, d) spatial and temporal transferability, and e) methods of improving models' performance. Four data sets were used: data from the Minneapolis-St. Paul area (collected in 1979); the two Urban Travel Forecasting Surveys from the San Francisco Bay Area conducted before and after the introduction of BART service (collected in 1972 and 1975, respectively); and the Baltimore, Maryland Disaggregate Travel Demand data set (collected in 1977). This part of the FHWA was performed primarily by Talvitie and Dehghani, who later supplemented their research by a method of updating alternative specific model constants in order to improve the models' performance (Dehghani and Talvitie, 1983).

2.2. Model specification

Table 1 presents results of two model specifications for work trip modal split, calibrated using the Baltimore data set. The explanatory variables are defined as follows:

TTIME   total travel time in minutes
COST/INC out-of-pocket travel cost in cents/income in dollars per year
CARS   number of car owned
WACCESS   walk access to transit facility; it takes the value of 1.0
C.B.D.   dummy variable: 1.0 if trip is designated to CBD location
INC   household income in dollars per year
COST   out-of-pocket travel cost in cents

In the "complex" version of the model, variables consist of both level of service variables and socioeconomic variables specified at the household level (number of cars, income). In its "simple" version, the model employs only level of service variables (travel time and travel cost) together with alternative specific constants. The following observations can be made about the results presented in Table 1.

(a) All measures of the models' goodness-to-fit are disappointing. In particular, (1) the success index (i.e., weighted average of difference in correct prediction between the full model and the model having only alternative specific constants) and (2) the proportion

TABLE 1: Complex and Simple Model Specifications and Coefficients of the MNL Model for Baltimore, (Talvitie, et a., 1981)

Variable	Alternat. Entered*	Coeff. (t-value)	Variable	Alternat. Entered	Coeff. (t-value)
TTIME	(1-3)	-.00865 (3.30)	TTIME	(1-3)	-.00946 (3.92)
COST/INC	(1-3)	-24.881 (1.70)	COST	(1-3)	-.00139 (1.30)
CARS	(1,3)	.365 (2.80)	CONST	(1)	.534 (3.24)
WACCESS	(2)	.292 (.73)	CONST	(3)	-.842 (3.24)
C.B.D.	(1,3)	-.892 (1.87)			
INC	(1,3)	.0000199 (1.70)			
CONST	(1)	0.102 (.20)			
CONST	(3)	-1.359 (2.80			

L*(β)	-463.02	L*(β)	-477.756
Percent Right	62.5	Percent Right	61.4
Sample Size	544	Sample Size	544
Success Indix	.104	Success Index	.083
Proportion Successfully Predicted	.493	Proportion Successfully Predicted	.471
Proportion of Prediction Success Attributable to Variables other than Alternative Specific Constant	$\frac{.104}{.493} = .21$	Proportion of Prediction Success Attributable to Variables other than Alternative Specific Constant	$\frac{.083}{.471} = .18$

*Alternatives: 1) Drive Alone, 2) Bus, 3) Shared Ride ($\geq$ 2 occupants)

of prediction success attributable to variables other than alternative specific constants are both very low.

(b) Table 1 confirms studies made by Talvitie and Kirschner (1978), and Dehghani and Talvitie (1983) which show that most of the explanatory power of the models, 60-85%, is contained in the alternative specific constants, the "dummy" variables.

(c) Performance of "complex" models is nearly as disappointing as that of the "simple" ones. Since the use of complex models requires predicting many more hard-to-predict variables, there is little (if any) advantage in using such models instead of the simpler ones.

2.3. Market segmentation

Several market segmentation schemes were developed for both work and
non-work trip models of modal choice. These were based on income (low
versus high) and car ownership (0/1/2+ cars). Table 2 compares the explanatory power of modal split models for work trips in Baltimore for
different market segments (Talvitie, et al., 1981).

TABLE 2: Comparison of Explanatory Power of the Simple and Complex
          Models - Proportion of Prediction Success Attributable to
          Variables Other than Alternative Specific Constants

BALTIMORE DATA
M A R K E T   S E G M E N T S

Model Type	Everybody	Low in- come H.H.	High In- come H.H.	One-Car H.H.	Two(+) Car H.H.
Simple	0.18	0.27	0.08	0.12	0.008
Complex	0.24	0.26	0.16	0.17	0.07

Proportion Successfully Predicted:

Model Type	Everybody	Low in- come H.H.	High In- come H.H.	One-Car H.H.	Two(+) Car H.H.
Simple	.47	.49	.47	.42	.52
Complex	.51	.51	.54	.44	.55

Note that household-based market segmentation does not result in improvement in the model's performance measured by a) "proportion successfully
predicted" and b) "proportion of prediction success attributable to variables other than alternative specific constants". For example, the second
value is less than 1% for households with 2+ cars for a "simple" (recommended) model version. Furthermore, the vectors of model coefficients
-- not shown here -- were found to be significantly different for different market segments.

2.4. Prediction errors

In Talvitie, et al., (1981), studies of prediction accuracy were performed
using a) "pre-BART" model, b) "post-BART" model, and c) the Baltimore
model. These studies focused on three sources of errors: the model error
the aggregation error and transfer errors. Errors were expressed as a
fraction of the predicted share. Actual and predicted model shares were
first calculated from fifty draws (observations) selected randomly from

the appropriate data sets.

Two measures of the overall error were used. These were a) the weighted average absolute error (WAAE) and b) weighted root mean square error (WRMSE), defined as in Koppelman (1976):

$$\text{WAAE} = \sum_{j=1}^{JT} [\text{AAE}_j \cdot \frac{1}{NT} (\sum_{i=1}^{NT} P_j)] \tag{1}$$

$$\text{WRMSE} = \sum_{j=1}^{JT} [\text{RMSE}_j^2 \cdot \frac{1}{NT} (\sum_{i=1}^{NT} P_j)]^{1/2} \tag{2}$$

where the average absolute error (AAE) and the root-mean-square error (RMSE) are defined by

$$\text{AAE}_j = \frac{1}{NT} \left[\sum_{i=1}^{NT} \frac{|P_j - A_j|}{P_j} \right] \tag{3}$$

$$\text{RMSE}_j = [\text{AE}_j^2 + \text{SE}_j^2]^{1/2} \tag{4}$$

where

$$\text{AE}_j = \frac{1}{NT} \left[\sum_{i=1}^{NT} \frac{P_j - A_j}{P_j} \right] \tag{5}$$

$$\text{SE}_j = \left[\frac{\sum_{i=1}^{NT} \left(\frac{P_j - A_j}{P_j}\right)^2 - NT \cdot \text{AE}_j^2}{NT-1} \right]^{1/2} \tag{6}$$

and

JT = total number of alternatives in the choice set

NT = number of times the fifty random draws were repeated (fifty times in this research)

AE_j = average error as percent of predicted share for alternative j

SE_j = standard error for alternative j

P_j = average predicted share of alternative j calculated from fifty random draws

A_j = average observed share of alternative j calculated from fifty random draws.

The total prediction errors can be separated into three components: model specification error, aggregation error, and transfer error as follows

$$WAAE_T = WAAE_M + WAAE_A + WAAE_F \tag{7}$$

$$(WRMSE_T)^2 = (WRMSE_M)^2 + (WTMSE_A)^2 + (WRMSE_F)^2 \tag{8}$$

where the subscripts refer to T = total error, M = model specification error, A = aggregation error, and F = transfer error.

Talvitie, et al., (1981) found that "simple" models of mode choice were in most cases not worse, but better, predictors than their "complex" counterparts. Also, the transfer error component of the total error was extremely large if the modal shares between estimation and prediction data sets were substantially different (for example, between Baltimore and Twin Cities data sets).

2.5. Updating alternative specific constants

Dehghani and Talvitie (1983) proposed an iterative method of updating alternative specific constants of the multinomial logit model of choice if a) aggregate model shares S_j's and b) trip table with zonal socio-economic variables, and c) level of service variables are available for the region under study.

Table 3 shows a) total prediction errors, b) their allocation between different error components and c) the influence of the procedure of updating alternative specific constants on respective error values, Dehghani and Talvitie (1983). The results presented in Table 3 reveal the following:

TABLE 3: Total Prediction Error and Its Allocation Between Model Error, Aggregate Error, and Transfer Error Using Baltimore Models (Dehghani and Talvitie, 1983)

Error Type	Model Type	Pre-BART		Post-BART NET-LOS		Post-BART TRUE-LOS		Twin Cities	
		Alternative Specific Constants Updating							
		Before	After	Before	After	Before	After	Before	After
Total Error WAAE (WRMSE)	Complex	32.1 (35.3)	24.6 (31.0)	39.7 (42.6)	27.8 (36.4)	29.6 (37.6)	28.3 (36.6)	68.1 (64.0)	23.6 (27.5)
	Simple	15.9 (23.0)	15.2 (22.8)	26.4 (36.7)	19.8 (33.2)	16.8 (25.6)	19.1 (31.6)	49.6 (53.4)	12.1 (16.6)
Model Error % WAAE (WRMSE)	Complex	44.0 (27.0)	57.0 (35.0)	35.0 (18.0)	50.0 (25.0)	47.0 (24.0)	49.0 (25.0)	20.0 (8.0)	59.0 (44.0)
	Simple	76.0 (51.0)	80.0 (52.0)	46.0 (20.0)	61.0 (24.0)	72.0 (41.0)	63.0 (27.0)	24.0 (10.0)	100.0 (98.0)
Aggreg Error % WAAE (WRMSE)	Complex	30.0 (34.0)	39.0 (44.0)	24.0 (23.0)	34.0 (32.0)	32.0 (30.0)	14.0 (10.0)	14.0 (10.0)	41.0 (56.0)
	Simple	0.0 (0.0)	0.0 (0.0)	0.0 (0.0)	0.0 (0.0)	0.0 (0.0)	0.0 (0.0)	0.0 (0.0)	0.0 (0.0)
Transf Error % WAAE (WRMSE)	Complex	26.0 (39.0)	4.0 (21.0)	41.0 (59.0)	16.0 (43.0)	21.0 (46.0)	17.0 (44.0)	66.0 (82.0)	0.0 (0.0)
	Simple	24.0 (49.0)	20.0 (48.0)	54.0 (80.0)	39.0 (76.0)	28.0 (59.0)	37.0 (73.0)	76.0 (90.0)	0.0 (2.0)

(a) The total prediction errors are relatively large. The "simple" model specification yields consistently smaller errors than the "complex" one. This is mostly due to reduction of the "model specification" part of the total error.

(b) The transfer error component of total error is overwhelmingly large if the modal shares between estimation (Baltimore) and prediction (Twin Cities) data sets are substantially different.

(c) Updating alternative specific constants reduces the total WAAE or WRMSE errors only if the modal shares are substantially different between estimation (Baltimore) and prediction (Twin Cities) data sets. This is mainly due to elimination of the transfer errors. In many cases when the aggregate model split shares are very similar (Baltimore, pre-BART, post-BART) updating constants does not generally improve model's accuracy.

(d) A "simple" rather than "complex" model of modal choices can be recommended while transfering the model from one area to another. The alternative specific constants have to be updated if necessary. Also, level of service data have to be updated accordingly but their influence -- whether manually coded or from networks -- was found to be only marginal (Talvitie, et al., 1981; Dehghani and Talvitie, 1983).

(e) Updating of alternative specific constants for the forecasts cannot be done since the constnts have no physical meaning and the future aggregate shares are not known.

2.6. City-wide predictions by mode

The practical test of a model's performance is a check of its ability to properly predict observed modal shares in different urban environments. Table 4 shows how the different versions of the Baltimore model transfer to other cities (Dehghani and Talvitie, 1983).

The results presented in Table 4 reveal that:

(a) Overall city wide predictions are not satisfactory. For example, transfering Baltimore model in either its "complex" or "simple" version to Twin Cities without updating constants results in extremely high (about 5 times) overprediction of transit share in Twin Cities. Updating constants brings nearly perfect result for Twin Cities but "ruins" forecast for post-BART (TRUE LOS): the overall transit share and its BART part is underpredicted about 50%.

(b) The model with "simple" specification and with updated constants brings predictions which, on average, are closest to the actual shares.

(c) It is highly possible that a model based on constants only would bring comparable results to the "best" modeling results. Ironically, it should be noticed that borrowing shares directly from Baltimore to all three cases of San Francisco without any model at all would result in shares closer to the actual ones than those obtained as a result of any modeling effort undertaken. In the case of the Twin Cities the result would be practically as bad as that obtained from the "complex" model before updating constants.

TABLE 4: City-Wide Predictions of Modal Shares Using Baltimore Models Before and After Updating Constants (Dehghani and Talvitie, 1983)

Prediction Data	Complex Model		Simple Model	
	Before Updating Constants	After Updating Constants	Before Updating Constants	After Updating Constants
Pre-Bart (900)				
Predicted[a]	.53,.28,.19	.54,.24,.22	.56,.27,.17	.55,.22,.23
Actual	.55,.24,.21	.55,.24,.21	.55,.24,.21	.55,.24,.21
Post-BART (623) (NET-LOS)				
Predicted	.48,.25,.17 (.12)	.54,.19,.27 (.07)	.49,.36,.15 (.14)	.55,.17,.28 (.07)
Actual	.53,.24,.23 (.09 BART)	.53,.24,.23 (.09 BART)	.53,.24,.23 (.09 BART)	.53,.24,.23 (.09 BART)
Post-BART (565) (TRUE-LOS)				
Predicted	.54,.28,.18 (.09)	.50,.28,.23 (.09)	.56,.28,.16 (.11)	.59,.17,.24 (.07)
Actual	.55,.25,.20 (.10 BART)	.55,.25,.20 (.10 BART)	.55,.25,.20 (.10 BART)	.55,.25,.20 (.10 BART)
Twin Cities (665)				
Predicted	.56,.27,.17	.86,.05,.09	.61,.23,.16	.85,.04,.10
Actual	.86,.05,.09	.86,.05,.09	.86,.05,.09	.86,.05,.09
Baltimore				
Actual	.51,.29,.20	.51,.29,.20	.51,.29,.20	.51,.29,.20

a/ The model shares pertain to three alternatives: Drive Alone, Transit and Shared Ride, respectively.

3. ANALYSIS OF MODAL CHOICES OF HOMOGENEOUS CATEGORIES OF PERSONS: IN-TRACITY AND INTERCITY COMPARISONS

3.1. The method of analysis

This section reports results of some basic modal split analysis performed by utilizing data from Baltimore, U.S.A. and two German cities: Hamburg and Berlin. The main goal of this study was an attempt to better under-

stand the mechanism of modal choice and to find an explanation for a)
the overwhelming importance of the models' constants, and b) the success
of "simple" specifications of the modal choice models, reported in the
previous section. Two key questions were asked: (1) will similar (homogeneous) groups of persons have similar modal choice patterns under similar circumstances (options, constraints)? (2) Will an improved understanding of modal choice behavior enable one to explain the above mentioned
problems with models' specifications?

The approach used in the research reported here is an extension of previous trip generation (Supernak, et al., 1983) and car availability studies (Supernak, 1983a). These studies revealed a) high heterogeneity of
the human population due to several transportation-related choices b)
relative stability of travel behavior within homogeneous groups of persons. The following major assumptions were made while creating homogeneous
groups of persons:

(1) The groups should be based on differences among persons rather than
    households since the latter unit is always very heterogeneous by de-
    finition.
(2) The categorization process should not be made arbitrarily but result
    from a careful multivariate analysis of all characteristics of an in-
    dividual which are likely to influence his/her travel choices.
(3) The final category description should be as simple as possible leav-
    ing in the model only those variables which consistently show the
    strongest impact on trip-making behavior of individuals.
(4) Only socio-economic variables are considered at the early stage of
    the model development.

The pairwise comparison of means, analysis of variance and mode cluster
analysis were applied to create these homogeneous categories of persons
(for details see Supernak, et al., 1983). The 3-stage analysis resulted
in formulation of the following eight homogeneous person categories:

    Category 1: Persons < 18 years of age
    Category 2: Employed 18-65 with car never available
    Category 3: Employed 18-65 with car sometimes available

Category 4: Employed 18-65 with car always available

Category 5: Non-employed, 18-65, with car never available

Category 6: Non-employed, 18-65, with car sometimes available

Category 7: Non-employed, 18-65, with car always available

Category 8: Persons > 65 years of age

The "individual" approach suitable for modal split analysis required a variable which could describe the overall level of access of a given individual to the most convenient transportation mode: a car. Three levels of car availability relate respectively to the situation when (a) person cannot drive the car because he/she has no drivers' license or there is no car in the family, (b) car is shared with other drivers in the family, (c) car is exclusively used by a given person.

3.2. Intracity modal split analysis

Modal split analysis were performed for each of eight persons categories for their obligatory (work, education), and discretionary (shopping, personal business, recreational, other) trip purposes. Five main ways of traveling were introduced: drive alone, shared-ride, transit, walk and other modes. Results of this analysis generalized for the entire Baltimore Metropolitan Area presented in Table 5 reveal that auto availability level has a crucial influence on modal choices. Differences between employed and non-employed adults are significant and are in agreement with expectation.

In order to investigate the influence of area characteristics on modal choices of the eight person categories, the Baltimore Metropolitan Area was split into 3 rings: A) urban (average population density 36 persons/acre), B) urban outskirts (14 persons/acre) and C) suburbs (approximately 3 persons/acre). The overall modal splits for these three areas were highly differentiated. For example, the percentage of car driven trips was approximately 20% in Area A, over 40% in Area B and over 60% in Area C.

Table 6 presents a sample of results of analysis of variance performed to investigate the joint effect of category versus area segmentation on

TABLE 5:   Modal Shares of Eight Person Categories in Baltimore (All Shares
           in %)

| Category | MODES | | | | |
	Car driver	Car passenger	Public transit	Walk	Other Modes
			ALL TRIPS		
1	5.2	21.2	14.6	48.0	11.0
2	1.4	28.1	39.0	26.7	4.7
3	67.2	17.9	5.9	8.5	0.5
4	85.5	7.1	1.5	4.8	1.0
5	1.0	31.5	13.2	51.2	3.2
6	50.0	27.2	3.6	16.5	2.7
7	72.7	20.6	0.8	4.7	1.1
8	43.9	11.8	11.8	27.2	5.3

TABLE 6:   Analysis of Variance for Percentage of $\beta^{car\ drive}$ due to
           Category and Area Splits

Factor	Degrees of freedom	Sum of squares	Mean square	F value	$F^{0.01}$
A. Category Split	7	2.376	0.399	42.38	4.28
B. Area Split	2	0.074	0.037	4.63	6.51
Error	14	0.107	0.008		
Total	23	2.577			

modal split characteristics. The factor "category affiliation" appeared
significant for all modal shares analyzed. Public transportation shares
were the only ones significantly affected by area characteristics. This
analysis (not presented here) showed that in all three areas analyzed,
persons with car always available a) use a car as often as needed, b)
reduce their walks to a minimum, c) use public transportation only marginally (in about 1% of their trips).

3.3. Intercity modal split comparisons

Modal split characteristics of eight person categories from Baltimore
were compared with similar characteristics of identical categories from
Hamburg and Berlin. The results are presented in Tables 7 and 8.

TABLE 7: Observed Model Split of Eight Person Categories in Hamburg, Berlin and Baltimore

City	Category No.	Drive Alone	Shared Ride	Public Transport	Walk	Other Modes
HAMBURG	1	0	4.5	16.3	36.7	42.5
	2	3.1	12.1	38.0	38.0	9.0
	3	56.4	11.0	13.6	17.2	1.9
	4	78.4	0.4	5.1	13.1	3.0
	5	1.3	4.5	23.2	57.1	13.9
	6	46.9	3.1	20.4	14.3	15.3
	7	57.1	0	8.2	32.7	2.0
	8	8.6	3.6	18.4	54.6	14.8
BERLIN	1	0	3.4	42.0	39.7	14.9
	2	5.1	10.4	39.7	38.3	6.6
	3	52.3	6.9	14.3	22.8	3.7
	4	82.3	2.5	2.9	10.6	1.7
	5	1.9	12.5	30.1	48.4	7.0
	6	36.8	12.0	24.1	18.8	8.3
	7	65.9	0	4.8	15.1	14.3
	8	7.4	1.2	19.1	59.5	12.9
BALTIMORE	1	5.2	21.2	14.6	48.0	11.0
	2	1.4	28.1	39.0	26.7	4.7
	3	67.2	17.9	5.9	8.5	0.5
	4	85.5	7.1	1.5	4.8	1.0
	5	1.0	31.5	13.2	51.2	3.2
	6	50.0	27.2	3.6	16.5	2.7
	7	72.7	20.6	0.8	4.7	1.1
	8	43.9	11.8	11.8	27.2	5.3

Results from German cities presented in Tables 7 and 8 confirm regularities observed in Baltimore namely:

a) high use of car if the automobile is "always available",

b) marginal use of public transportation if the automobile is "always available",

c) decrease in walking trips as auto availability level increases.

The most visible differences in modal splits between Baltimore and German cities appear in categories 1 and 8 which are the least homogeoenous ones. Modal splits for students under 18 strongly depend on the major arrangements made for their trip to and from schools (busing or not). Modal

TABLE 8:   Observed Model Split for Employed Persons with Different Levels
of Auto Availability in Hamburg, Berlin and Baltimore

City	Category No.	Drive Alone	Shared Ride	Public Transport	Walk	Other Modes
				M O D E		
				OBLIGATORY TRIPS		
HAMBURG	2	3.1	9.4	49.6	32.6	5.4
	3	56.3	10.5	21.8	10.5	0.9
	4	82.1	0	6.6	11.0	0.3
BERLIN	2	7.2	8.3	53.2	27.0	4.3
	3	58.4	6.2	17.2	15.9	2.4
	4	86.4	2.4	3.6	7.7	0
BALTI- MORE	2	2.3	30.4	49.6	14.4	3.1
	3	65.1	18.2	9.3	7.0	0.4
	4	88.1	6.2	1.8	2.7	1.2

split for elderly depends on the actual level of auto ownership/availability for this group of persons; this level is much higher in America than in Europe. Another difference which can be observed is a higher percentage of ridesharing trips in Baltimore than in German cities.

3.4. City-wide predictions of model shares

Table 9 shows the results of city-wide predictions of modal shares made for Hamburg and Berlin. These predictions use respective modal shares from Baltimore and apply them without any change for predictions made for German cities. The resulting shares are affected only by respective differences in category percentages α_i in German cities.

The predicted shares for German cities are satisfactory for "drive alone" and "transit" mode categories. The different character of German cities (higher overall residential density) seems to be responsible for trade-off between "walking + biking" in German cities and "shared ride" modes in Baltimore. These results are still generally better than predictions made for the Twin Cities in Table 4 by transfering Baltimore model without updating constants.

TABLE 9: City-Wide Predictions of Modal Shares Using Category-Specific
Shares: Employed Persons and Their Obligatory Trips

Prediction Data	MODES			
	Drive Alone	Shared Ride	Transit	Walk and Other Modes
Hamburg				
Predicted	.45	.20	.25	.10
Actual	.41	.07	.29	.23
Berlin				
Predicted	.47	.19	.23	.11
Actual	.47	.06	.27	.20
Baltimore				
Actual	.62	.15	.15	.08

4. TRAVEL CHOICES AND THEIR REPRESENTATION IN DISAGGREGATE CHOICE
    MODELS: A PROBLEM OF NON-CHOOSERS

4.1. "Fish got to swim and birds got to fly" - or what can be learned
     from a popular song?

Previous sections provided a comparison of the performance of MNL models
of mode choice with the results of modal choice behavior by homogeneous
groups of persons in different cities located on both sides of the Atlantic.
Perhaps the shortcomings of probabilistic models and success of a very
simple approach used in Section 3, lie in the simple truth of the lyrics
of Kern's popular "Showboat" theme cited above. Let us imagine using a
disaggregate choice model on a world in which:
(a) only three groups of animals exist: fish, birds and mammals, but they
    are not identified by their "affiliation" to any family of species;
    each one is just treated as animal A;
(b) only three basic travel modes are identified: a) swimming, b) running/
    walking/jumping/other "surface" modes, and c) flying;
(c) animals make rational "modal choices" in any given situational con-
    text, i.e. they instinctively choose this "mode" which offers them

highest utility in any given circumstance;

(d) vector X of attributes defining the "behaver" (an animal) and alter-
    natives' attributes θ (e.g., speed) is known;

(e) the overall "mode choice" behavior of Earth's animals is known from
    observation of a (large enough) random samples of animals representing
    variety of species drawn in different parts of the globe.

After making these assumptions we could formally specify a disaggregate
model (MNL, for example) as in equation (9) and use the utility maximization principle for estimation of coefficients θ_i in order to explain the
observed "modal choice" behavior b_o.

$$U_i = U_i \ (\theta, X) \tag{9}$$

In order to make a prediction of the overall "modal split" we could treat
vector θ as given and use updated vector X to forecast the choice probability of alternative i under changed circumstances.

It would not be surprising to find out that such a model will:

(a) show an overwhelming explanatory power of alternative specific con-
    stants

(b) reveal small influence of other variables, either "level of service"
    ones (e.g., speed) or "behaver" characteristics (e.g., number of legs)

(c) show that a "simple" modal specification is as good (or, rather, bad)
    as the "complex" one

(d) demonstrate relatively small transferability errors for all areas
    where the existing "modal shares" are similar to those used for model
    calibration, and very large errors if the model calibrated by the data
    from the Artic Sea was transfered to, say, African savannah

(e) demonstrate satisfactory temporal stability of the model only if the
    "true" modal shares were not going to change significantly.

One can easily, of course, find an alternative solution to the problem
specified above, a much simpler and more logical, one. It is clear
that the true "modal choices" of species living on Earth will be primarily determined by the shares of different species in different areas
since it is commonly known that a) birds generally fly and seldom walk or

swim, b) mammals run/walk, seldom swim or fly, and c) fish only swim
(if we ignore flying fish). If, for some reason, proportions among
mammals, birds and fish change (different areas of life on Earth), the
overall "modal split" will also change respectively. The changes of
(modal split) behavior within each of three basic groups of animals may,
obviously, also happen as a result of changing environment or some slow
evolutionary changes in some species to better cope with a changed environment, but the influence of these changes on the overall "modal
split" will be far less important than the proportions of the species in
the population. Thus, overall "modal split" of animals will be largely
predetermined by a) relatively stable behavior within homogeneous groups
of species and b) variable proportions of representatives of these groups
in the sample.

4.2. Intracity modal split: choice or predetermination?

How much relevance could the discussion from the previous section have
for better understanding of human intracity model choices? From a conceptual point of view, our zoological analogy may not be as far-fetched
as it initially seems. International comparisons of modal splits of eight
person categories made for Baltimore, Hamburg and Berlin show clearly that
A) carless categories of persons (2, 5 and 8) are "captive non-driver"
B) groups with a choice make their travel choices according to the most
attractive mode available for them, e.g., person categories with a car
"always available" use their vehicle as their predominated mode of transport. C) The human population is as heterogeneous with respect to modal
split as is the population of animals just analyzed. The overall modal
split observed in Baltimore for employed persons $\beta_1 : \beta_2 : \beta_3$ = .17 :
.18 : .65 (Table 10) does not mean that every employee has such probabilities of 1) being a car driver, 2) being a car passenger or 3) being a
transit passenger. Table 10 shows that persons with car never available
use public transportation about 20 times more often than they drive a car
while persons with car always available drive a car nearly 50 times more
often than they travel by public transportation.

Thus, the observed modal shares β_i will always predominantly depend on
proportion of population shares $\alpha_j : \alpha_k : \alpha_l$. The actual choice of each

TABLE 10: Relationship Between Population Structure ($\alpha_2 : \alpha_3 : \alpha_4$) and
Resulting Modal Shares ($\beta_1 : \beta_2 : \beta_3$) in Baltimore

MODES	Modal Shares in % for employed, 18-65 with car available			Average Shares for EMP, 18-16
	never	sometimes	always	
Public transport	60.3	10.0	1.9	β_1 = 17.4
Car passenger	36.9	19.7	6.5	β_2 = 17.7
Car driver	2.8	70.3	91.7	β_3 = 64.9
TOTAL	100.0	100.0	100.0	100.0
Population Shares	α_2 ⬥ 21.1%	α_3 = 32.9%	α_4 = 45.0%	Total 100.0

group will fluctuate only slightly around largely predetermined choices
representing "average" behavior of a given group under normal circumstances. One should notice here that the problem of captivity in travel
choices is not limited to carless persons only. Persons with a car always available are in a sense captive auto users just as carless persons
are captive public transit riders (or pedestrians).

It is worth stressing that Brog and Erl (1983) and Wermuth (1978) found
that the fractions of "persons with options" are relatively small in real-
world situations.

Apparent predetermination of modal choices made by different segments of
the population has a lot to do with rationality of travel choice and the
utility maximization principle. In each case an individual representing
any given group G chooses the "best" mode actually available.

4.3. Homogeneity of the groups and market segmentation

The previous discussion leads one directly to the next important problem
of market segmentation. Modal split percentages are so dramatically different for persons with car a) never, b) sometimes and c) always available,
that any attempt to understand modal choices cannot ignore those fundamental differences. Looking for an analogy one could say it would be equally
difficult to describe ("travel") behavior of animals without distinguishing

between birds and fish.

For modal choices, the natural criterion of segmentation should be the
actual ability to use the most convenient transportation mode: a car.
The goal of each market segmentation is to split a heterogeneous population into a small number of homogeneous sub-populations. In order to do
that properly, in any case when the strata definition is not "obvious",
a multivariate analysis should be employed. It is clear that, for example,
an arbitrary stratification of households by income for modal split analysis does not solve the problem of population heterogeneity since differences
among members of families with the same income are often more significant
than differences among families with different incomes. Thus, this stratification does not prevent mixing "birds" with "fish" (to again use our
analogy).

One more comment refers to Table 2 presented earlier in the paper. It is
not surprising that MNL model with "simple" specification, estimated for
market segment "two (+) car households", leaves only about 1% of prediction success attributable to variables other than alternative specific
constants. This extremely "poor" index of success does not at all mean
that model is "bad"; it simply shows the following: a) In two (+) car
households, employed persons have a good access to cars, in most cases
corresponding to situation "car always available". As found before, this
situation practically "imposes" an extremely extensive use of the car.
b) In this situation the modal split is largely predetermined and model
employing only alternative specific constants can "explain" and forecast
modal shares quite well. (See, for example, consistency in category 4
modal shares in Baltimore, Hamburg and Berlin, Tables 7 and 8).

4.4. Problem with updating modal constants: prediction and policy context

The discussion in the previous chapter leads into another problem often
reported in different works devoted to disaggregate modal split modeling:
that is, a) high explanatory power of alternative specific constants, and
b) improvement of model performance after alternative specific constants
are adjusted.

There are some problems with adjusting the alternative specific constants.
First, there is no way to find out how these constants should be updated
in a long-range forecast. Second, geographic transferability using updated
constants based on known travel shares automatically guarantees "success",
but does not reflect the governing relationships.

Let us look more closely at Table 4. The most troublesome transferability
test before updating constants -- the Baltimore model to predict Twin
Cities modal shares -- became virtually perfect after updating constants.
One could say that other variables are so overwhelmed by these constants
that they are not able to spoil this prediction. In other cases (e.g.,
POST-BART, TRUE-LOS) where "other" variables are relatively more important,
the forecast is actually spoiled by the constants adjusting procedure.
Ironically, the best forecast for San Francisco would be one in which the
actual shares are directly borrowed from Baltimore, without any model at
all.

In this situation one could ask the questions: What is the basis for updating constants only? How do we know that remaining elements of vector
θ in the utility function will remain valid after this procedure? Indeed,
Ben-Akiva (1981) recommends a procedure of updating all model coefficients
using a Bayesian procedure. This, however, requires at least a few aggregate values (modal shares) for actual demand to be known.

For forecasts, where the actual shares cannot be known, the following
procedure can be considered. Suppose, that variables X contribute to the
success of prediction only marginally. If we degrade our utility functions
to constants only, utility U_1 from the vector of alternatives' utilities
$U = (U_1, U_2 \ldots U_3)$ is reduced only to

$$U_i = \theta_i \tag{10}$$

where θ_i is alternative specific constant i

and

$$P_i(\theta) = \frac{e^{\theta_i}}{\sum\limits_i e^{\theta_i}} = \frac{\sigma_i}{\sum\limits_i \sigma_i} = \beta_i \qquad (11)$$

On the other hand, using category approach we get

$$\beta_i = \frac{\sum\limits_k \alpha_k' \beta_{ik} N_k}{\sum\limits_k \alpha_k' N_k} \qquad \text{where } \sum\limits_k \alpha' = 1 \qquad (12)$$

where

β_i = overall probability of choosing alternative i

β_{ik} = probability of alternative i being chosen by persons belonging to category k

α_k' = predicted percentage of homogeneous person category k in the population

N_k = trip rate of category k

Earlier work by Supernak (1983a) suggested that β_{ik} should be a function of population density d

$$\beta_{ik} = \beta_{ik}(d) \qquad (13)$$

while α_k is a function of a) age groups in the population A, b) employment level E and c) car availability level C

$$\alpha_k = \alpha_k(A,E,C) \qquad (14)$$

Both functions β_{ik} and α_k are subject to both "natural" changes within a given metropolitan area (e.g., in population density) and changes regulated by transportation and non-transportation policies, (Supernak, 1983a). Consistent with Talvitie, et al. (1981), taking $\theta_{shared\ ride} = 0$ and knowing β_i, one can solve for all other θ_i, equation (11). Thus, alternative specific constants can be made interpretable and predictable. The updated constants can be incorporated into original model equation (15) leaving other elements of vector θ unchanged.

$$P_i(\theta,X) = e^{\bar{u}_j(\theta,X)} / \sum_{j=1}^{J} e^{\bar{u}_j(\theta,X)} \tag{15}$$

Future tests should show whether the "direct" deterministic method represented by equation (12) or the procedure described above with updating
alternative specific constants yields better results.

A third method could be an independent calibration of the MNL model for
each of the homogeneous person categories separately. In this case, the
importance of alternative specific constants should be very high and,
understandably, proportion of prediction success attributed to variables
other than alternative specific constants should be even lower than the
typical, already low, values obtained while calibrating the model for the
entire population. This should not be treated as a deficiency of the model,
since the prospect for geographic transferability of this version of the
model may appear to be better than the first two versions discussed. More
research utilizing recent ideas presented in Koppelman and Wilmot (1982),
Daly, et al., (1983), Hammerslag (1983), and Brand and Cheslow (1981) is
underway. Final recommendations can be made after several tests in different urban environments are completed.

5. CONCLUSIONS

Both theoretical discussion and empirical findings lead to the following
conclusions:

1. Unsatisfactory performance of the disaggregate models of mode choice
    is the result of a poor representation by the model of the trip-maker
    and his/her transportation-related choices, among them intracity mode
    choices. In order to improve this representation, critical patterns
    in human travel behavior have to be defined, leading to a primary
    stratification of the population into groups relevant to the analyzed
    choice.
2. Poor results for indices of a model's performance such as the "success
    index" or "prediction success attributable to variables other than

alternative specific constants" indicate the primary importance of "unobservable variables". The "success" of these variables is caused by the heterogeneity of the population with respect to intracity travel (modal) choices and their strong predetermination by the level of individual's access to the most convenient transportation mode: a car.

3. Differences in population structure (car availability level, employment status, age structure) and land use characteristics (population density) are primary reasons why disaggregate choice models are normally very poorly transferable. Dynamic changes in all elements listed above are normally not captured in the model's specification which also appears to be the primary reason for the unsatisfactory forecasting ability of these models.

4. Mode choices are strongly predetermined within homogeneous groups, and are quite similar even for international comparisons. Level of service variables appear to be of relatively little importance. Simple model specifications appear superior to complex ones, and alternative specific constants are responsible for a very high percentage of prediction success. These results should not be treated as a deficiency of the models. Rather, they demonstrate the strength of the variable "auto availability" and consistency of modal choices made by persons with car "never" and "always" available. It is worth mentioning that availability/accessibility/level of service of public transportation in a given area influence, to some extent, population structure with respect to auto availability, and thus -- indirectly -- mode choices, as well.

5. Updating constants in order to improve model's geographic transferability had been recommended recently by several authors. Since these procedures require information about existing modal shares in the new area, they cannot be used for forecasts. This paper offers a procedure for updating alternative specific constants which is applicable also for forecasts. Universal mode shares of homogeneous person categories are utilized while category shares in the forecast year have to be predicted. Thus, changes in population structure directly influence changes in alternative specific constants providing also interesting insight into the real meaning of the alternative specific constants, their instability, and their importance in forecasts and policy analyses.

6. The proposed procedure should be tested in different urban environments in order to find out whether the "improved" model of equation (15) is

better than the deterministic model (equation 12) of mode choice
specified for each homogeneous person category and given population
density limit. An alternative procedure using a series of probabilistic models calibrated for each of homogeneous population groups
separately might also be investigated.

Acknowledgements

I would like to thank Preston Luitweiler, David Schoendorfer and Marilyn
Macklin, from Drexel University for their help in preparing this paper.

REFERENCES

Atherton, T. and Ben-Akiva, M. (1976). Transferability and updating of
disaggregate travel demand models, Transportation Research Record, 610,
pp. 12–18.
Ben-Akiva, M. (1981). Issues in transfering and updating travel-behavior
models. In: New Horizons in Travel-Behavior Research (Edited by P.
Stopher, A. Meyburg and W. Brog), pp. 665–686, Lexington Books, Lexington,
MA.
Brog, W. and Erl, E. (1983). Application of a model of individual behavior
(situational approach) to explain household activity changes In: Recent
Advances in Travel Demand Analysis, (Edited by P. Jones and S. Carpenter),
pp. 350–370, Gower Publishing Co. Ltd., Aldershot, England.
Brog, W., Meyburg, A. and Wermuth, M. (1983). Development of survey in-
struments suitable for determining non-home activity patterns. Paper pre-
sented at the 62nd Annual Meeting of the Transportation Research Board,
Washington, DC.
Charles River Associates (1978). Disaggregate Travel Demand Models. Pro-
ject 8-13, Phase Report, Charles River Associates, Boston, MA.
Daly, A., Gunn, H., Barkey, P. and Pol, H. (1983). Model transfer using
data from several sources. Proceedings of the Seminar N of the 11th
Summer Annual Meeting of PTRC, pp. 177–189, University of Sussex, England.
Dehghani, Y. and Talvitie, A. (1983). Comparison of the forecasting accuracy, transferability and updating of modal constants in disaggregate
mode choice models with simple and complex specification.
Galbraith, R. and Hensher, D. A. (1982). Intra-metropolitan transferabilit:
of mode choice models. Journal of Transport Economics and Policy, XVI
(1), pp. 7–29.
Gaudry, M. J. and Degenais, M. G. (1979). The dogit model. Transporta-
tion Research, 13B, pp. 105–112.
Hammerslag, R. (1983). Population groups with homogeneous travel performance by car, public transport and bicycle. Paper presented at the World
Conference on Transport Research, Hamburg, Germany.
Horowitz, J. (1982). Specification tests for probabilistic choice models.
Transportation Research, 16A, pp. 383–394.
Koppelman, F. (1976). Methodology for analyzing errors in prediction with
disaggregate choice models. Transportation Research Record, 592, pp. 17–23.

Koppelman, F. and Wilmot, C. (1982). Transferability analysis of dis-aggregate choice models. Transportation Research Record, 895, pp. 18-23.

Lerman, S. (1981). A comment on interspatial, intraspatial, and temporal transferability. In: New Horizons in Travel-Behavior Research (edited by P. Stopher, A. Meyburg and W. Brog), pp. 628-631, Lexington Books, Lexington, MA.

Louviere, J. (1981). Some comment on premature expectations regarding spatial, temporal and cultural transferability of travel-choice models. In: New Horizons in Travel-Behavior Research (Edited by P. Stopher, A. Meyburg and W. Brog), pp. 653-664, Lexington Books, Lexington, MA.

McFadden, D. (1981). Econometric models of probabilistic choice. In: Structural Analysis of Discrete Data with Econometric Applications (Edited by C. Manski and D. McFadden), pp. 198-272, MIT Press, Cambridge, MA.

Sheffi, Y., Hall, R. and Daganzo, C. (1982). On the estimation of the multinomial probit model. Transportation Research 16A, pp. 447-456.

Stopher, P. (1982). Small-sample home-interview travel survey: applica-tion and suggested modification. Paper presented at 61st Annual Meeting of the Transportation Research Board, Washington, DC.

Supernak, J. (1983a). Car availability versus car ownership modeling: theoretical discussion and empirical findings. Paper presented at the World Conference on Transport Research, Hamburg, Germany.

Supernak, J. (1983b). Transportation modeling: lessons from the past and tasks for the future, Transportation 12, pp. 79-90.

Supernak, J. (1983c). Toward a better understanding of transportation-related choices. Paper presented at the Tenth International Transpor-tation Planning Research Colloquium. Amsterdam, The Netherlands.

Supernak, J., Talvitie, A. and DeJohn, A. (1983). Person-category trip generation model. Paper presented at the Transportation Research Board Annual Meeting, Washington, DC.

Talvitie, A., Dehghani, Y. and Anderson, M. (1982). An investigation of prediction errors in work trip mode choice models. Transportation Re-search, 16A, pp. 395-402.

Talvitie, A., Dehghani, Y., Supernak, J., Morris, M., Anderson, M., DeJohn, A., Kousheshi, B., Feuerstein, S. (1981). Refinement and application of individual choice models in travel demand forecasting, DOT Contract DOT-FH-11, State University of New York at Buffalo, Buffalo, NY.

Talvitie, A. and Kirschner, D. (1978). Specification, transferability and the effect of data outliers in modeling of the choice of mode in urban travel. Transportation 7, pp. 311-331.

Wermuth, M. (1978). Structure and calibration of a behavioral and atti-tudinal binary mode choice model between public transport and private car. Paper presented at the PTRC Summer Meeting, University of Warwick, England.

Wermuth, M., Brog, W. and Meyburg, A. (1984). Bias correction in house-hold travel survey results for non-home activity patterns. Paper pre-sented at the 63rd Annual Meeting of the Transportation Research Board, Washington, DC.

Williams, H. and Ortuzar, J. D. (1982). Travel demand and response analy-sis - some integrating themes. Transportation Research, 16A, pp. 345-362.

Ninth International Symposium on
Transportation and Traffic Theory
© 1984 VNU Science Press, pp. 561–589

MODELS OF EMPLOYEE WORK SCHEDULE

PAUL P. JOVANIS[1] and ANTHONY MOORE

Department of Civil Engineering and [1] Transportation Center, Evanston, Northwestern University, IL, U.S.A.

ABSTRACT

Despite the widespread adoption of flexible work hours, there has been limited modelling of work schedule choice, an essential element in any comparison of flex-time with other transportation management tactics. Utility theory is used to develop a conceptual structure for work schedule choice with flex-time. A beta probability density is used to model the distribution of work arrivals. The structure is empirically tested using data from the San Francisco Bay Area.

The estimated models supported much of the conceptual structure. Employees with long travel distances shifted to very early arrivals to avoid congestion and to spend evening time at home. Professionals already had informally varied their work schedules before flex-time; their adjustments after flex-time were somewhat less than non-professionals who previously had little work schedule flexibility. Variation in daily work schedule increased slightly when the worker's spouse was also employed, and decreased when young children were present in the household. The model results imply that careful consideration should be given to actual work schedules prior to flex-time implementation in order to obtain an accurate estimate of the spreading of peak travel.

1. INTRODUCTION

This paper examines the impact on the worker's arrival time choice of
adopting flextime: a program in which an employee selects his work
schedule on a day to day basis around a constraining core time, which
must be worked every day, such that a required number of hours are
worked in each reporting period. In order to properly evaluate this and
other TSM tactics, analytic tools are needed which describe the travel
impacts of the proposed tactics. At this stage, no adequate tools are
available to support a comprehensive evaluation of a flexible work hours
program so that it may be compared to other potential strategies such as
congestion pricing or carpooling (NCHRP, 1980; Jovanis, 1981). Some
research has been conducted which describes impacts of flex-time at
specific firms; these studies (e.g. Jones, et. al, 1977; Neveu and
Koeppel, 1979; and Daniels, 1980) have generally not been aimed at
development of analytic tools for work schedule prediction. The major
emphasis of this paper is to describe the development of a modelling
framework that is used to predict the work arrival time effects of
flextime.

This research was undertaken to develop and explore a structure which
describes the influences that affect the worker's choice of arrival
times with flextime. A utility-based conceptual structure is formulated
to describe the way in which a worker's choice of arrival time affects
his overall utility. This structure considers the work arrival time
choice process in terms of travel, family, workplace and individual
factors. The accuracy of this conceptual structure is explored through
the estimation of the work arrival time models, undertaken in a
probability distribution format using maximum likelihood estimation
techniques. These models investigate various specifications and examine
the use of market segmentation as a way of explaining different work
arrival time behavior by different subsets of the population.

1.1. Work Arrival Time Research Without Flextime.

There has been limited research on the development of work arrival time choice models, given fixed work hours (Cosslett, 1977; Small, 1978; and Abkowitz, 1981). These researchers all utilized multinomial logit models with alternatives chosen as 5 or 10 minute intervals around the official work arrival time. Their theoretical basis of behavior was utility maximization. Jovanis (1983) has recently developed models of work schedules without flex-time using the methodology discussed in section 4.1 of this paper.

1.2. Work Arrival Time Research With Flextime.

Under flextime, the work arrival time choice process interacts dynamically with other activity scheduling decisions of the worker himself, other household members, and friends on a daily basis. A few researchers have undertaken studies that focus more on developing an underlying behavioral structure for flextime travel choice. Ott, Slavin and Ward (1980) and Jovanis (1979) utilized regression techniques to predict mean work arrival times with socioeconomic and lifecycle characteristics identified as independent variables.

Ott et al. (1980) found that individuals who have longer travel times, higher salaries, and use transit have later mean arrivals; those who have higher numbers of children and who are older have earlier arrivals. They also found that the desire to avoid congestion and the need to coordinate with schedules of other household members are important factors in the work scheduling process. Because their data included repeated observations of individual employees, they were able to develop separate regression models of arrival time variance. They found decreasing variability in arrival times with carpooling, older workers, and the presence of children under the age of 5 and increasing variability associated with workers with higher incomes, professional and managerial employees, and the employees with children in the 5 to 13 year old age group.

Jovanis (1979) is the first researcher to outline a framework for
flextime travel decisions. His framework was based on the concept that
flextime travel behavior is driven by a complex relationship between
socioeconomic, workplace, and transportation constraints and
motivations. His results indicated the increasing levels of constraint
imposed by the household where there is a working spouse and/or young
children present.

The data set used by Jovanis is the same as the one described in Section
3 of this paper (see Table 1). Jovanis used factor analysis to reduce
the dimensionality of a data set that included over 36 variables to a
set of 17 factors (Jovanis, 1979). Based upon the 17 factors, linear
regression models were developed to predict work arrival time with flex-
time as a function of a set of predictor variables in a linear additive
structure.

$$W = a + \sum_{i=1}^{n} b_i x_i + \sum_{j=1}^{m} c_j y_j \qquad (1)$$

where: W = work arrival time with flex-time,

  x_i = socioeconomic, travel and workplace attributes of an
       individual (e.g. travel time or mode),

  y_j = dummy variables indicating responses to questions regarding
       motivations for selecting a particular work schedule,

a, b_i, c_j = parameters estimated from the data.

The models were intended to be exploratory; seeking to understand which
factors were important in the choice of work schedule. The findings
link the choice of work schedule to objective variables (such as marital
status) as well as motivational factors representing activities
occurring either before or after work. For example, a large number of
married individuals stated that spending time at home with the family
after work was a very important motivation in their choice of work
schedule.

Estimation results for three market segments are presented in Table 2
and discussed below. Choice of work arrival time (in minutes) was
regressed against the predictors shown in Table 2. The constant values

Table 1. Variable definitions for model estimation

Variable Name	Definition
PRETIME	A discrete variable giving the arrival time in minutes before flextime implementation. Its value ranges from 0 to 180 representing arrivals between 6.45 am and 9.45 am.
OCCUP	The occupational category of the worker. The values of 1 to 6 represent: professional, clerical, managerial, administrative, technical - unskilled and technical - skilled workers respectively.
AGE	A categorical variable describing the worker's age. The values of 1 to 5 represent: 18-25 years, 26-30 years, 31-35 years, 36-45 years and 46+ years respectively.
LOCAT	A categorical variable describing the travel distance from home to work. The values from 1 to 5 represent 1-10 miles, 11-20 miles, 21-30 miles, 31-40 miles and 41+ miles respectively.
SPOUSEW	A dummy variable indicating the presence of a working spouse in the household. 1 indicates a working spouse and 0 indicates no working spouse.
CHILDH	A dummy variable indicating the presence of young children in the household. 1 indicates young children and 0 indicates no young children.
MARITAL	A dummy variable that describes the marital status of the worker. 0 indicates unmarried and 1 married.
MODE	A categorical variable indicating the mode choice of the worker on his trip to work. The values from 1 to 6 represent: drive alone, shared ride, BART, bus, walk and other respectively.
DRIVE ALONE	A dummy variable indicating the use of drive alone as the mode of travel to work. 1 indicates this mode was used, 0 otherwise.
SHARED RIDE	A dummy variable indicating the use of shared ride as the mode of travel to work. 1 indicates this mode was used, 0 otherwise.

Table 1.   Variable definitions for model estimation.   (Continued)

Variable Name	Definition
FEMALE	A socioeconomic dummy variable: 1 for females, 0 for males.
LEISIMP, LEISVMP	Dummy variables, 1 if spending leisure time at home with family and friends is important or very important, respectively; 0 otherwise.
SLEEPIMP, SLEEPVMP	Similar to LEISIMP, LEISVMP but measuring the importance of a preferred waking/sleeping schedule.
SCHEDIMP, SCHEDVMP	Similar to LEISIMP, LEISVMP but measuring the importance of work, school schedules of household members.
WORKIMP, WORKVIMP	Similar to LEISIMP, LEISVMP but measuring the importance of office needs.
AMTT	A continuous variable computed from survey responses to time leaving home in the morning and the time arrived at work in the morning for a particular mode.
DDVARY	A dummy variable describing the time flexibility given an employee: 1 if the employee can vary work hours daily, 0 otherwise.
NOVARY	A dummy variable,: 1 if the employee must fix his hours in advance, 0 otherwise.

Table 2. Summary of regression findings: a, unmarried with no children;
b, married with children and non-working spouse; c, married with children
and working spouse.

Variables	a		b		c	
	Coef.	t	Coef.	t	Coef.	t
AGE	-2.2	-1.1	-1.5	-0.4	-1.6	-0.4
AMTTAUTO	0.6	-.2	-0.5	-2.2	-0.3	-1.0
AMTTBART	0.6	2.1	0.5	0.8	-0.2	-0.5
AMTTCP	0.2	1.0	-0.8	-2.9	-0.2	-1.0
AMTTOTH	0.9	1.3	-0.2	-0.2	1.0	1.8
AMTTRAN	-0.2	-0.8	-0.5	-2.4	0.6	0.3
AMTTWALK	2.0	4.3	0.3	0.3	-0.5	-0.4
DDVARY	26.0	3.0	12.9	1.0	10.4	1.0
FEMALE	1.5	0.2	-14.7	-0.6	8.0	0.9
LEISIMP	-7.7	-1.0	-12.7	-1.2	-12.3	-1.2
LEISVMP	-22.1	-2.8	-30.6	-2.4	-20.3	-2.0
NOVARY	-9.1	-1.4	-18.3	-1.8	-15.6	-1.9
SCHIMP	-11.3	-1.3	6.9	0.6	19.8	2.0
SCHVIMP	-11.7	-1.2	10.2	0.8	26.5	2.8
SLEEPIMP	16.8	2.3	-10.2	-1.0	-8.2	-1.0
SLEEPVMP	23.2	2.9	5.8	0.4	17.5	1.4
WORKIMP	9.4	1.2	-24.4	-2.3	27.0	2.7
WORKVIMP	13.9	1.9	-13.8	-1.2	10.9	1.1
CONSTANT	452		518		463	
R^2	.28		.33		.30	
N	207		90		127	

represents a time which is adjusted earlier or later for an individual depending on the value of predictors. In Model A, for example, the constant term 452 minutes represents an arrival time of 7:32 a.m. which is adjusted based upon predictor values. The results in Table 2 give insights into the causes underlying work schedule choice by linking activities with socioeconomic and travel variables.

For unmarried individuals with no children (Model A) travel time was significant for only BART and walking travelers. For BART travelers, arrival time increased with travel time so that a trip of 20 minutes arried 12 minutes later while a trip of 30 minutes arrived 18 minutes later. The desire to adjust work schedules to preferred sleep schedules was strongly significant resulting in 17 minute and 23 minute later arrivals depending on whether the factor was important or very important. Those for whom office needs were very important arrived 13.9 minutes later while those for whom spending time with family and friends was very important arrived 22.1 minutes earlier. If one considers an activity perspective these results imply that unmarried individuals adjust work schedules to arrive later to accommodate the need to sleep later. They also arrive earlier to spend more evening time with family. Schedule coordination with other household or family members caused slightly earlier arrivals, but the parameters were not significantly different from zero. The amount of flexibility available to the individual was also a determinant of work schedule choice: those with the ability to vary day to day generally chose to arrive later (by 26 minutes).

A very different set of motivations and constraints operate for married individuals with children and a non-working spouse (Model B). Travel times for auto, carpool and transit are significant and negative indicating that those with longer travel times tend to arrive earlier (opposite to the result for single individuals). Employees who find spending time with family very important tend to arrive nearly 31 minutes earlier than those who do not. Family schedule coordination has little influence, probably because the non-working spouse can attend to family needs. Not surprisingly, adjustments to work schedule to accommodate sleeping desires are unimportant for these family members. Coordination with office needs, however, results in generally earlier

arrivals. Similar results are obtained in Model C, but notice that
family schedule coordination and needs are much more important when the
spouse is employed than when spouse is at home.

2. THE CONCEPTUAL STRUCTURE

Subsequent research (Jones, 1979; Hanson and Burnett, 1979; Damm, 1979)
has focused on viewing travel as being derived from the need to
participate in activities dispersed in space and time. While this view
has been implicit in travel analyses for many years, the "activity-
based" analysis of travel recognizes the importance of activities
explicitly. Heggie and Jones (1978) have discussed at length the
domains in which current travel models are inadequate. While the data
set used in this research did not measure type of activity and duration,
the regression analyses did provide some linkage between work schedules
and activities.

The central conceptual theme of the research uses utility maximization
to represent the major constraints and motivations facing the worker in
his work arrival time choice. Continuous utility functions over time
are conceptualized to examine the tentative shape and scale of these
functions for individuals with specific characteristics. The
conceptualized utility functions help to provide a framework for the
travel choice analyses that follow. The interrelationships between the
travel, socioeconomic, workplace and individual factors are discussed in
the following four subsections and empirically explored by estimating
models of work arrival time choice.

In addition to exploring mean arrival times it is important to consider
variance in arrivals as well. The ability to vary work schedules day to
day is a formal privilege extended to flex-time workers which may be
severely constrained by workplace or household needs. Some variability
in arrivals is expected due to variations in travel time, waking-
sleeping hours, etc. These non-scheduled sources of variability exist
in addition to structural sources of variability imposed by the
individual (for leisure or recreation) or imposed by others (to meet

office needs, meet needs of children or needs for household
maintenance). The following discussion of the four major factors
influencing work schedule choice includes consideration of the influence
of each factor on the variability in work schedules.

2.1. Travel Influences

The value of time literature (e.g., Heggie, 1976; Bruzelius, 1979)
indicates that most workers attach disutility to the trip to work. It
follows that most workers will seek to minimize the travel time (or
generalized cost) of their journey to work. The general form of disutility of work travel time is hypothesized to be of the shape shown in
Fig. 1a. This curve is, effectively, a time based congestion curve,
indicating the decreasing attractiveness of commuting closer to the peak
travel period. Obviously, for particular individuals this curve will
shift later or earlier and also alter in scale depending on the actual
travel time or distance the worker is exposed to on the journey to
work. Mode choice is a further complication and suggests that a
generalized cost formulation instead of travel time may be more appropriate measure of travel disutility for public transportation modes
whose service, frequency, travel time, cost and reliability can vary
throughout the peak period.

Travel mode can have important affect on variability in work schedule.
Ride sharing clearly offers major scheduling constraints for an
individual. Those who drive alone or use transit in areas with frequent
service face much less constrained travel choices.

2.2. Family influences

A primary reason that many firms adopt flex-time is to allow employees
to better match household and work needs. The structure of the
household (the presence of a working spouse and/or young children)
represent both constraints and strong motivation for work schedule
choices (Fig. 1b).

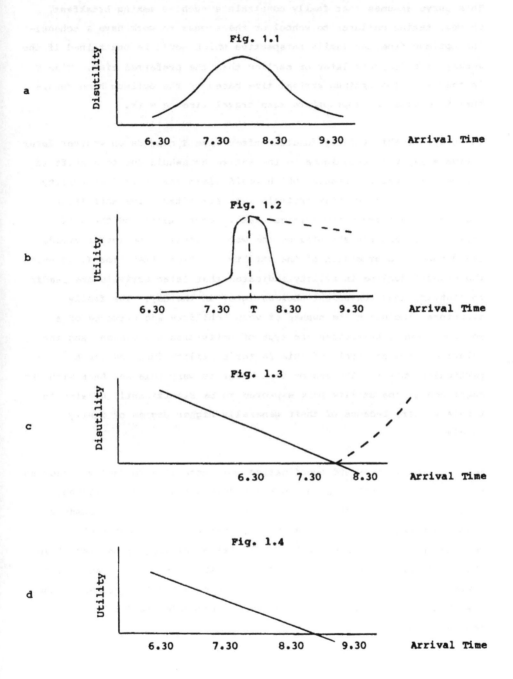

Figure 1. Hypothesized utility relationships: a. travel; b. family; c. sleeping; d. leisure.

This curve assumes that family constraints such as making breakfast, chores, taking children to school or the spouse to work have a scheduling optimum from the family perspective which would be unattained if the worker left for work later or earlier than the preferred time. Time T is therefore the optimum arrival time based on the optimum home departure time plus the appropriate mean travel time to work.

The shape of this utility function after time T depends on whether later arrivals imply inconvenience to the entire household due to a shift in the entire morning schedule (which would yield the solid line utility function) or whether later arrivals indicate alternative activities undertaken in between the critical family commitments and the work arrival (for example shopping on the way to work). The latter yields the dashed line extension of the utility function after time T, in which the gradual decline in utility indicates that later arrivals are leading to tighter, less convenient choices later in the day. The family structure, including the number of young children and presence of a working spouse, determines the type of activities undertaken, and the relative shape and scale of this factor's utility function for a particular worker. The sex of the worker is very relevant here with the magnitude of the utility peak expected to be significantly greater for female workers because of their generally higher degree of family involvement.

Meeting family needs and coordinating work schedules to family schedules represent major influences on work schedule variability. Children, particularly those with regular weekly school hours, can represent a major constraint on work schedule variability. The presence of a working spouse (presumably without flex-time privileges) can result in the use of flex-time to run errands and meet other family needs. The variability in work schedule choice may be expected to increase as the flex-time employee is expected to use his schedule freedoms to meet family needs.

2.3. Workplace influences

If the flexible work hours program is truly flexible, there should be total indifference to work schedule across the allowable range of work arrival times. Workplace constraints are important because of the probability of a "preferred" arrival time; despite flextime, the worker can feel subtle (or not so subtle) expectations from his supervisor concerning a desirable arrival time band. The worker may feel increased utility for a specific arrival time because he can impress supervisors with his punctuality or, as suggested by Owen (1979) and Jones et al. (1980), the actual choice of work arrival times may be based on coordinating activities between colleagues in the immediate work group so that the group's productivity is maintained.

Clearly, constraints imposed by superiors or co-workers hinder the employee from varying his hours day to day. An employee's position in the firm, the degree to which work is independent, and the availability of people to duplicate the employee's function all affect the disutility that an individual would perceive from varying his work hours.

2.4. Individual influences

Factors of personal preference such as the relative importance of sleeping later and the desire for greater leisure time activity also influence work schedule choice. In the case of sleeping habits generally most individuals are less happy with earlier awakening times to satisfy a work arrival time choice (see Fig. 1c). Usually, however, a time is reached beyond which three possible motivations affect the worker's choice of arrival times. These possibilities include the individual being indifferent to sleeping in later, the individual getting increasing enjoyment out of sleeping in later or (more likely in today's time precious world) the worker feeling an increasing disutility with later sleeping schedules because he feels "productive" time is being "wasted".

The desire for greater leisure time activity after work is generally associated with earlier work arrivals (Ott et al., 1980; Jovanis, 1979). This effect as illustrated in Fig. 1d indicates that the earlier one arrives at work the earlier one leaves and therefore the greater the opportunity to take advantage of leisure or personal activities after work. The above curve assumes a standard working day and assumes the worker attaches increasing utility to more leisure time.

3. THE DATA SET

The data used for the following analyses come from a survey of 689 employees working under a flextime policy at Lawrence Berkeley Laboratory, Berkeley, California. The survey data contains information about workplace trip ends before and after implementation of flextime. It also includes socioeconomic, travel and workplace characteristics for each individual. These variables include travel distance, travel time, mode choice, cars per licensed driver, cars per worker, occupation, marital status, presence of a working spouse, presence of young children, household income and age. The data also contains a set of motivational factors for each individual in which the importance of various factors in the choice of work arrival times were classified on a four point scale from "Not At All Important" to "Very Important". These factors included "Avoiding the Rush-hours", "Adjusting my Schedules for Family Needs" and "Matching Work Hours to my Preferred Waking/Sleeping Schedule". Details of exploratory analyses of these data are contained elsewhere (Jovanis, 1979; Moore, 1982).

The official start times allowed by flextime were 7:00 a.m. to 9:30 a.m. In order to allow for the inclusion of flextime workers arriving slightly outside this band, but the exclusion of part-time or shift workers inadvertantly included in the survey, the data set was screened to include only those workers with arrivals between 6:45 and 9:45. Inspection of work arrival times before and after flextime revealed that there were frequent and large changes in work schedule; 47 percent of the workforce changed their work arrival time by 15 minutes or more, 31 percent by 30 minutes or more. These shifts implied substantial readjustment of work schedules with flex-time.

A complete list of variable definitions for the analyses is contained in
Table 1. Note that three variables, Occupation, Age and Locat (travel
distance) contain 5 or 6 categories. In order to lower the cost of the
subsequent model estimations these variables were not treated as
separate dummy variables for each category. Instead, the variables are
treated as continuous, but the order is rational (Age and Locat) or, as
is the case with Occup, arranged to list the occupations with the latest
to earliest arrivals. This procedure was necessary to keep the computer
cost of estimating individual models within reasonable bounds.

4. ESTIMATION OF WORK ARRIVAL TIME MODELS

4.1. Estimation Methodology.

The discussion of the previous literature indicated the desirability of
developing a method to predict both the mean and variance of work
schedule choice. Rather than estimate separate regression equations for
the mean and variance, it seemed desirable to estimate mean and variance
parameters given a specific distribution of work schedules. Consistent
with the conceptual structure, continuous probability distributions were
considered, to explicitly deal with the continuous nature of time.
Review of histograms of observed work schedule choice (Jovanis, 1979)
revealed that the distribution of work arrival times (taken as the
indicator of work schedule) were assymetric with a slight skew toward
earlier arrivals.

Based upon the above considerations both the gamma and beta distributions were considered. Because work schedules are bounded by the limits
of the flex-time program (typically 7 - 9:30 a.m.) it was believed that
the beta distribution had stronger theoretical justification.

The beta probability density function defined over the interval (0,1) is

$$f(x; p,q) = \frac{\Gamma(p+q)}{\Gamma(p)\,\Gamma(q)} x^{p-1} (1-x)^{q-1}, \quad 0 \le x \le 1 \qquad (2)$$

$$0 \text{ elsewhere} \qquad 0 < p,\ 0 < q.$$

The mean and variance of this distribution is given by the following relationships:

$$E(x) = p/(p + q)$$
                                                                    (3)

$$Var(x) = pq/(p + q)^2 (p + q + 1).$$
                                                                    (4)

An alternative structure would have been to use the generalized beta distribution defined over an interval (a,b) for which a and b are also estimated from the data. Because the flex-time program defined work schedule limits, a and b were not estimated but merely used to transform data from a (0,1) interval to 6:45 to 9:45 a.m. interval (180 minutes).

The estimation procedure used here is a maximum likelihood estimation procedure, CHOMP (Daganzo and Shoenfield, 1978; Timmermans, 1982). The actual estimation is in terms of the mean and variance terms above where these are expressed as a function of the explanatory variables.

In considering the estimation of mean and variance parameters it is important to recall that the data set contains only one observation of work arrival time choice for each employee. Thus, the mean and variance that are calculated on this cross-sectional data set reflect the aggregate mean arrival for individuals with similar characteristics (as described by predictor variables) and the variance of the arrival times for similar groups of individuals.

4.2. Experimental design for model estimation.

For these model estimates, only linear relationships were investigated in the mean and variance formulations. Variables were selected to explore the validity of the conceptual structure described earlier as well as to reflect the findings of previous research (see Table 1 for list of variable definitions).

Pretime is chosen to provide the link between before and after flextime

arrival times. Operationally this means that work arrival time choice
with flextime can be modelled as the pre-flextime arrival time plus a
shift in arrival times associated with selected independent variables
(see models 2 and 3 in Table 3 for example). We also developed models
of work arrival time which exclude prior work arrival time to take
account of time choice where historical data are not available as for a
new employee.

Different sets of variables have been chosen for the specification of
the mean and variance based upon the conceptual structure and previous
literature. Exploratory runs were undertaken where the variables in the
specification of the mean were: travel distance, mode choice,
occupation, age, marital status and presence of young children. The
specification of the variance included travel distance, mode choice,
occupation, marital status, presence of a working spouse and presence of
young children. On the basis of these results "best" models were
obtained both with and without prior arrival time as a major explanatory
variable (Moore, 1982). The results of the exploratory specifications
were that travel distance, occupation, age, travel time and mode choice
were chosen as predictors for the mean arrival time; occupation and the
presence of a working spouse and young children were chosen for the
variance.

Histograms of actual arrivals after flextime indicated that the work
arrival time distribution may be bimodal. This is consistent with the
conceptual structure which indicates that certain of the four major time
choice factors may pre-dispose individuals to early or late arrivals.
Because the Beta Distribution is unimodal for the range of parameters
expected in this study, it is both conceptually and empirically
appealing to consider market segmentation of the data. Previous
analysis (Moore, 1982) indicated that occupation was a very important
variable in understanding the choice of work arrival time. Use of
occupation for market segmentation is conceptually appealing because it
can capture much of the influence of workplace constraints on individual
time choice.

Three base models were estimated to permit calculation of relative measures of goodness of fit. We use the ρ^2 measure of goodness of fit commonly used for disaggregate choice models. The three base models are:

1. The Null model representing uniform arrivals over the 6:45 a.m. to 9:45 a.m. period. The measure ρ_0^2, therefore gives the proportion of variability in arrival times explained by the model compared to a no knowledge case.

2. The model with constants for the mean and variance specification. This is analogous to the market shares model in choice model analyses and therefore the measure, ρ_{MS}^2, gives the additional proportion of variability explained by only knowing the overall distribution of arrivals.

3. The model as in 2 with Pretime added to the mean specification. This measure, ρ_{PRE}^2, is analagous to 2 except the model is being compared to a base case of also knowing the pre-flextime arrival times.

4.3. Findings--the pooled models.

The results discussed here are all scaled to real time so the mean is in terms of minutes where 0 represents 6:45 a.m. and 180 represents 9:45 a.m. while the variance is in units of minutes squared. The results of the three models estimated on the pooled data set are presented in Table 3. Model 1 is without Pretime, models 2 and 3 include Pretime with different specifications for the mean.

Both travel distance and occupation have the expected results: those living at greater distances and non-professional workers arrive earlier. In model 1 the parameters are highly significant with each extra category of these two variables leading to arrivals of 8 and 6 minutes earlier, respectively. Pretime is a very powerful predictor, and its parameter coupled with the constant (in models 2 and 3) indicates that arrivals with flextime generally occur earlier than they did before flextime.

Table 3. Model estimation results: pooled data set.

	Parameter Value	Parameter	Standard Error	t-Statistic
MODEL 1	a_1 (constant)	90.00	7.67	11.74
ρ_o^2 = 17.73%	a_2 (Locat)	-7.75	2.24	-3.46
	a_3 (Occup)	-6.04	1.02	-5.95
ρ_{MS}^2 = 2.06%	a_4 (Drive Alone)	2.01	2.20	0.91
	a_5 (Shared Ride)	4.26	2.12	2.01
	a_6 (constant)	1791.8	174.66	10.26
	a_7 (Occup)	-76.37	48.87	-1.56
	a_8 (Childh)	-229.09	169.92	-1.35
MODEL 2	a_1 (constant)	20.76	4.98	4.17
ρ_o^2 = 22.08%	a_2 (Pretime)	0.727	0.0362	20.11
	a_3 (Age)	-1.37	1.02	-1.34
ρ_{MS}^2 = 7.23%	a_4 (constant)	626.52	247.46	2.53
	a_5 (Spousew)	137.98	124.70	1.11
ρ_{PRE} = 0.41%	a_6 (Childh)	-365.61	120.09	-3.05
MODEL 3	a_1 (constant)	27.08	5.89	4.60
ρ_o^2 = 22.18%	a_2 (Pretime)	0.711	0.0377	18.87
	a_3 (Locat)	-4.21	1.77	-2.37
ρ_{MS}^2 = 7.36%	a_4 (constant)	550.25	49.89	11.03
	a_5 (Spousew)	74.98	61.61	1.22
ρ_{PRE} = 0.54%	a_6 (Childh)	-214.35	60.60	-3.54

Age, in model 2, has the expected sign as earlier arrivals are predicted
for older workers, however the parameter is not significant.
Exploratory analysis of the data set using the Automatic Interaction
Detector, AID, (Moore, 1982) revealed a complex interaction between age,
household location (travel distance), mode choice, and occupation. A
richer interpretation of the influence of age is contained in the
discussion of market segmentation results; the statistical
insignificance of the age variable is a reflection of these complex
interactions as well as the fact that age was modeled as a continuous
variable.

The mode dummy variables in model 1 indicate a tendency for later
arrivals with the auto-related modes although the magnitudes of the
parameters are small and only the Shared Ride variable is statistically
significant. These results are surprising as the exploratory
specifications indicated earlier arrivals with auto-based modes. Travel
distance may be explaining much of the shift to early arrivals with
carpools since ridesharing is strongly associated with long distance
travel.

The results for the specification of the variance include:

Occupation has the expected result with professional workers tending to
have greater variability in work arrival times.

The presence of young children has the consistent affect of reducing the
variability in work arrival times. An estimate of the scale of the
variability changes can be obtained by inserting predictor variable
values in the variance models and taking the square root of the result,
obtaining estimates of the standard deviation in work arrival times for
individuals characterized by different values of the predictors. The
results indicate that changes in variability due to the presence of
young children ranges from a low 3 minutes in model 1 to a value of
nearly 10 minutes in model 2.

The presence of a working spouse, however, leads to increasing variability in arrival times, perhaps indicating the greater complexity faced by households in which two workers try to coordinate their schedules. Since it is likely that the working spouse does not have flextime, the burden of adjusting work schedules to conduct family activities during the commute trip could fall more heavily on the flextime worker. The scale of the increase is 2 to 4 minutes, much smaller than the influence of young children.

The findings for Childh should not be unexpected as young children have fairly rigid time schedules during most of the year, primarily focused about school hours. Coordination of schedules with a working spouse is heavily dependent on the employment and work trip characteristics of the spouse; e.g. Is the spouse employed full or part-time? What are the spouse's work hours? Does the spouse have some work schedule flexibility? How long a commute and by what mode does the spouse travel? These details, while not obtainable in our survey, indicate that the working spouse can have a highly variable effect on work schedules choices for flextime workers.

Overall the goodness of fit measures ρ_o^2 indicates that these models explain approximately 20% of the overall variability in work arrival times, a reasonable result for disaggregate models. The ρ_{MS}^2 values of 2% to 7% indicate that only a small additional proportion of the variation is explained by the predictor variables other than the constants.

The addition of Pretime increases the goodness of fit by 5% (reflected in changes for ρ_o^2 and ρ_{MS}^2). Interestingly, the ρ_{PRE}^2 value for models 2 and 3 are very small indicating that very little additional variation is explained by the additional predictors. Investigation of the pre-flextime arrivals indicated an explanation: arrival times before flextime were already fairly spread, indicating that even before flextime workers were able to respond to family, travel, workplace and individual influences to alter their start times around the 8:00 to 8:30 a.m. regular schedule. Thus, Pretime is already capturing some of the

influence associated with occupation, travel distance, age and other
variables. It is not surprising, then, that travel distance dropped in
significance and magnitude from model 1 to model 3; Pretime was
reflecting the influence of travel distance even before flextime was
implemented. Modelling of work schedules without flextime (Jovanis,
1983) found that professionals have dispersed arrivals even before flex-
time while clerical and administrative workers are more constrained to
traditional arrival times (i.e. 8:00 a.m.).

4.4. Findings--the market segmented models

Previous research (Jovanis, 1979; Jones, et al., 1977) indicated that
workers with different job classifications faced different levels of
constraint regarding choice of work schedule. Exploratory analysis of
this data set confirmed these findings (Moore, 1982). In order to
examine the implications of occupation on work schedule choice, an
additional set of models were estimated using two market segments:
professionals (engineers and scientists) and everyone else.

In all cases Pretime is highly significant with a parameter around 0.6
to 0.8 although it is interesting that the two models for professionals
have the coefficient of Pretime at just over 0.8 with a t-statistic of
around 16 while the coefficient is 0.59 with a t-statistic around 10 for
the non-professionals. This suggests that Pretime is a slightly
stronger predictor for professionals than non-professionals. Previous
analyses (Jovanis, 1979) showed that professionals at LBL frequently
adjusted their work schedules before flex-time therefore they are more
likely to make smaller time changes with a formal flex-time policy (they
already had adjusted).

Age indicates a trend of earlier arrivals for older workers, although
the result is closer to statistical significance for the non-
professionals. The effect of age is surprisingly small, as in the
parent runs.

Travel distance is more significant for the professionals, indicating
arrivals of over 6 minutes earlier for each extra 10 miles of travel

Table 4. Model estimation results: market segmentation by occupation.

A. Specification as in Model 2

	Parameter	Parameter Value	Standard Error	t-Statistic
MODEL 2a	a_1 (constant)	12.55	8.66	1.45
(Professionals)	a_2 (Pretime)	0.832	0.0499	16.66
	a_3 (Age)	−0.609	1.68	−0.36
	a_4 (constant)	979.63	133.01	7.36
	a_5 (Spousew)	390.87	158.12	2.47
	a_6 (Childh)	−626.28	153.70	−4.07
MODEL 2b	a_1 (constant)	26.62	6.33	4.21
(Non-	a_2 (Pretime)	0.589	0.0565	10.42
Professionals)	a_3 (Age)	−1.84	1.29	−1.43
	a_4 (constant)	817.79	361.21	2.26
	a_5 (Spousew)	80.00	167.22	0.48
	a_6 (Childh)	−201.63	162.57	−1.24

ρ_o^2 = 22.63%

ρ_{MS}^2 = 7.90%

ρ_{PRE}^2 = 1.12%

B. Specification as in Model 3

	Parameter	Parameter Value	Standard Error	t-Statistic
MODEL 3a	a_1 (constant)	28.99	10.30	2.82
(Professionals)	a_2 (Pretime)	0.803	0.0515	15.58
	a_3 (Locat)	−6.82	3.30	−2.06
	a_4 (constant)	983.22	133.67	7.36
	a_5 (Spousew)	342.51	156.84	2.18
	a_6 (Childh)	−627.73	153.83	−4.08
MODEL 3b	a_1 (constant)	27.37	7.57	3.62
(Non-	a_2 (Pretime)	0.589	0.0652	10.05
Professionals)	a_3 (Locat)	−2.83	2.28	−1.25
	a_4 (constant)	1091.23	137.78	7.92
	a_5 (Spousew)	104.95	170.66	0.62
	a_6 (Childh)	−269.94	168.03	−1.61

ρ_o^2 = 22.74%

ρ_{MS}^2 = 8.02%

ρ_{PRE}^2 = 1.25%

distance from the workplace. This effect drops to under 3 minutes for
the non-professionals. A possible reason for this difference, as also
suggested by previous analyses (Moore, 1982), is that professionals have
greater flexibility in life choices such as residential location and
mode choice relative to the non-professionals and thus are more
sensitive to the effects of these variables in choosing work arrival
times.

The results of the family structure variables, Spousew and Childh, are
consistent with the previous models but there is a marked difference in
the values and significance of the parameters between the professionals
and non-professionals. Comparing the results of models 2a with 2b it is
noticeable that these two variables are highly significant for the
professionals with standard deviations of arrival times ranging from 19
minutes for the case of a non-working spouse with young children to 37
minutes for the case of a working spouse with no young children. The
parameters for the non-professionals are statistically insignificant
with the corresponding standard deviations ranging from 25 minutes to 30
minutes. The results of models 3a and 3b show similar results with the
range in standard deviations being noticeably greater for the
professionals than non-professionals. This indicates that since
professionals tend to have more flexibilty in their ability to choose
work arrival times, non-workplace factors, such as family structure,
could be more influential on their behavior than for the non-
professionals who are more constrained by workplace or other factors.

A χ^2 test indicated that market segmentation across occupation is also
an important way to separate the sample with statistically significant
better goodness of fit measures for the market segmented models.

4.5. Summary of model estimation.

Overall these models illustrate several of the most important results in
understanding the work arrival time choice process, with a consistent
shift to later arrivals among professionals, younger workers, and those

living closer to the workplace. There is also an overall decrease in
work arrival time variability for those with young children, without a
working spouse, and for the non-professional workers.

The goodness of fit measures generally indicate that the models explain
about 20% of the information and predict work arrival times with
flextime. However, the ρ_{MS}^2 values indicate that the explanatory
variables really do not add very much beyond the knowledge of the
overall distribution of arrivals. These measures indicate that previous
arrival time is by far the most significant explanatory variable,
however it must be remembered that there are circumstances when its use
is inappropriate.

5. CONCLUSIONS

Overall this research develops a conceptual structure to understand the
work arrival time choice process of a worker with flextime, then
empirically tests the structure with actual data. It is useful to
compare these results to those of Ott et al. (1980) and Jovanis (1979).

Travel influences are very important. The desire to avoid congestion is
often a reason for arrival time shifts away from the busiest half
hour. Both previous researchers also found that congestion avoidance
was an important motivation in work arrival time choice. There were
some inconsistencies in terms of the travel time and travel distance
variables, however. Ott et al. (1980) found later arrivals for those
with greater travel times, Jovanis found mixed effects with earlier
arrivals for longer trips among some of the married workers, while we
find overall there are earlier arrivals among those workers living at
greater distances from the workplace.

Family structure and the associated socioeconomic variables are also
very important in explaining work arrival time choice behavior although
degree of influence varies across occupation and travel distance. The

greater levels of family related constraint, as shown by the presence of
a working spouse and/or young children, indicate workers that have less
flexibility to alter their arrival times with flextime. The presence of
young children has a powerful influence on reducing arrival time
variability except where the children are in the teen age group where
they instead increase their parents work arrival time variability.
These results are consistent with the models of the standard deviation
of arrival times by Ott et al. (1980).

The workplace influences are represented by the occupation variable in
this study, with a result of both later arrivals and greater variability
in work arrival times for the professional workers over the non-
professionals. Ott et al. (1980) use GS salary level as a proxy for
income and occupation but nevertheless their results and those of
Jovanis are consistent with the current research.

The market segmentation model estimations here illustrate new results on
work arrival time choice modelling with flextime. The occupation market
segmentation model suggests that since professionals tend to have more
flexibility in their ability to choose work arrival times, non-workplace
factors, such as family structure and travel distance, are of greater
significance in affecting their work arrival time choice decision.

Overall these results, and the related model estimations provide a solid
basis to develop a set of comprehensive analytical tools by which
flextime can be evaluated against other TSM tactics. For companies
already located in an urban area, the models with previous arrival time
can be used, along with socioeconomic and travel data, to predict a new
arrival time pattern. For new firms, or firms relocating to an area,
the models without previous arrival time may be more applicable. Of
course, proper procedures should be used in developing aggregate
predictions from disaggregate models of this type (Koppelman, 1976).
Despite the fact that these models were developed for a single firm,
comparisons with findings from Cambridge, Massachusetts (Ott, et al.)
and from three downtown San Francisco firms (Jovanis, 1979) show
qualitatively consistent results.

Overall these empirical results indicate that many of the conceptual ideas presented here can be supported by the data with flextime introduction leading to significant changes in work arrival time behavior. This paper illustrates these changes in behavior in terms of four major categories of influences on the individual; namely: travel, family, workplace and individual influences.

During the course of the research, the authors identified several avenues of potentially fruitful future research. While the modelling results imply changes in daily activity patterns, future flextime research may want to investigate detailed changes in daily activity schedules that occur with flex-time using travel diaries or some similar measurement tool. Further, transportation system level of service (by mode), clearly an important influence in work schedule choice, could be much better modeled by time of day to more clearly explore, for example, tradeoffs between modal travel time and service frequency. Results of empirical work from other firms and locations would also help in the understanding of how flextime choice varies across firms and regions.

Acknowledgement

This research was partially supported by National Science Foundation Grant CME-8106467, "Development of a Methodology to Predict Time and Mode of Travel with Flexible Work Hours". The views expressed in the paper are those of the author, not the National Science Foundation.

REFERENCES

Abkowitz, M.D. (1981). An Analysis of the Commuter Departure Time
Decision, Transportation 10, pp. 283-297.
Abkowitz, M.D. (1981). Understanding the Effect of Transit Service
Reliability on Work Travel Behavior, Transportation Research Record 794.
Bruzelius, N. (1979). The Value of Travel Time, (Croom Helm, London).
Cosslett, S. (1977). The Trip Timing Decision for Travel to Work by
Automobile. In: Urban Travel Demand Forecasting Project, Phase 1, ITS,
University of California, Berkeley.
Daganzo, C.F., L. Schoenfeld (1978). CHOMP User's Manual, Department of
Civil Engineering, University of California, Berkeley.
Damm, D. (1979). Towards Models of Activity Schedules, Ph.D. Disserta-
tion, Department of Civil Engineering, Massachusetts Institute of
Technology.
Daniels, P. W. (1980). Flexible Work Hours and the Journey to Work to
Office Establishments, Transportation Planning and Technology, 6, pp.
55-62.
Hanson, S., K.P. Burnett (1979). Understanding Complex Travel Behavior:
Measurement Issues. In: New Horizons in Travel-Behavior Research.
(Lexington Books, Lexington, Massachusetts).
Heggie, I. (1976). Modal Choice and the Value of Travel Time.
(Clarendon Press, Oxford).
Heggie, I.G., P.M. Jones (1978). Defining Domains for Models of Travel
Demand, In: New Horizons in Travel-Behavior Research. (Lexington Books,
Lexington, Massachusetts).
Johnson, B.M. (1966). Travel Time and the Price of Leisure, Western
Economic Journal 4, pp. 133-145.
Jones, D.W., F.D. Harrison, P.P. Jovanis (1980). Work Rescheduling and
Traffic Relief: The Potential of Flex-time, Public Affairs Report,
21(1), University of California, Berkeley.
Jones, D.W., T. Nakamoto, M. P. Cilliers (1977). Flexible Work Hours:
Implications for Travel Behavior and Transport Investment Policy,
Institute of Transportation Studies, Berkeley, California.
Jones, P.M. (1979). Activity Approaches to Understanding Travel
Behavior. In: New Horizons in Travel-Behavior Research. (Lexington
Books, Lexington, Massachusetts).
Jovanis, P.P. (1979). Analysis and Prediction of Travel Responses to
Flexible Work Hours: A Socioeconomic, Workplace and System Perspective,
unpublished PhD dissertation, Department of Civil Engineering,
University of California, Berkeley.
Jovanis, P.P. (1981). Travel Impacts of Flexible Work Hours: A.
Effects on Mode Choice, B. Prospects for Highway Congestion Relief,
Transportation Center Research Report, Northwestern University.
Jovanis, P.P. (1983). Modelling the Choice of Work Schedule: Fixed
Hours and Flex-time, Applied Simulation and Modelling '83, sponsored by
International Association of Science and Technology for Development.
Koppelman, F.S. (1976). Guidelines for Aggregate Travel Prediction
Using Disaggregate Choice Models, Transportation Research Record 610.
Moore, A.J. (1982). Development of a Behavioral Structure to Predict
Work Arrival Times with Flextime, unpublished Masters Thesis,
Northwestern University, Department of Civil Engineering.
Moore, A.J., P.P. Jovanis, F.S. Koppelman (1983). Modelling the Choice
of Work Schedule with Flexible Work Hours, accepted for publication,
Transportation Science.

National Cooperative Highway Research Program [NCHRP] (1980).
Alternative Work Schedules: Impacts on Transportation, Synthesis of
Highway Practice 73.
Neveu, A. J. and K. W. P. Koeppel (1979). Who Switches to Alternative
Work Schedules and Why. New York State Dept. of Transportation,
Research Report 162.
Ott, M., H. Slavin, D. Ward (1980). Behavioral Impacts of Flexible
Working Hours, Transportation Research Record <u>767</u>.
Owen, J.D. (1979). Working Hours. (Lexington Books, Lexington,
Massachusetts).
Small, K.A. (1978). The Scheduling of Consumer Activities: Work Trips,
Princeton University.
Timmermans, G. (1982). A Continuation of the CHOMP Saga, Northwestern
University.

REFEREES

Professor Samir A. Ahmed, Stillwater, U.S.A.
Dr. Rahmi Akcelik, Nunawading, Australia
Professor Richard E. Allsop, London, United Kingdom
Professor Ir. E. Asmussen, Leidschendam, The Netherlands
Mr. Robert G.V. Baker, Sydney, Australia
Dr. K.L. Bang, Stockholm, Sweden
Professor Martin J. Beckmann, Providence, U.S.A.
Dr. Michael G.H. Bell, Karlsruhe, West-Germany
Mr. Bo Blide, Göteborg, Sweden
Professor William R. Blunden, Pymble, Australia
Mr. P. Bothner, Bremen, West-Germany
Professor David E. Boyce, Urbana, U.S.A.
Professor Dr. D. Braess, Bochum, West-Germany
Mr. David Branston, Woodbridge, United Kingdom
Professor Dr.-Ing. Werner Brilon, Bochum, West-Germany
Mr. Werner Brög, München, West-Germany
Mr. Lawrence D. Burns, Warren, U.S.A.
Mr. Ian Catling, London, United Kingdom
Professor Dr. Avishai Ceder, Haifa, Israel
Mr. B.R. Cooper, Crowthorne, United Kingdom
Mr. Ian G. Cullen, London, United Kingdom
Professor Stella Dafermos, Providence, U.S.A.
Professor Carlos F. Daganzo, Berkeley, U.S.A.
Mr. Andrew J. Daly, The Hague, The Netherlands
Dr. David Damm-Luhr, Cambridge, U.S.A.
Professor Mark S. Daskin, Evanston, U.S.A.
Professor Caroline S. Fisk, Urbana, U.S.A.
Professor Michael Florian, Montréal, Canada
Professor Dr. Terry L. Friesz, Philadelphia, U.S.A.
Professor Dr. Nathan H. Gartner, Lowell, U.S.A.
Dr. Denos C. Gazis, Yorktown Heights, U.S.A.
Dr. Peter C. Gipps, Highett, Australia
Dr. P.B. Goodwin, Oxford, United Kingdom
Dr. Yehuda Gur, Haifa, Israel
Professor Ir. Peter Hakkesteegt, Delft, The Netherlands
Dr. Susan Hanson, Worcester, U.S.A.
Professor Ezra Hauer, Toronto, Canada
Professor Dr. Heinz Hautzinger, Heilbronn, West-Germany
Mr. Donald W. Hearn, Gainsville, U.S.A.
Professor Chris Hendrickson, Pittsburgh, U.S.A.
Professor David A. Hensher, North Ryde, Australia
Professor Robert Herman, Austin, U.S.A.
Dr. Raimund K. Herz, Karlsruhe, West-Germany
Dr. Benjamin G. Heydecker, Leeds, United Kingdom
Professor Dr.-Ing. Günter Hoffmann, Berlin, West-Germany
Professor Joel Horowitz, Iowa City, U.S.A.
Dr. Howard K. Hung, Washington, U.S.A.
Mr. J. Douglas Hunt, Edmonton, Canada
Mr. T.P. Hutchinson, Coventry, United Kingdom
Dr. Peter Jones, Oxford, United Kingdom
Mr. William C. Jordan, Warren, U.S.A.
Professor K. Jörnsten, Linköping, Sweden

Professor Dr. Hartmut Keller, München, West-Germany
Mr. Howard R. Kirby, Leeds, United Kingdom
Professor Dr. Ryuichi Kitamura, Davis, U.S.A.
Professor Dr. L.H. Klaassen, Rotterdam, The Netherlands
Professor Ir. L.H. de Kroes, Delft, The Netherlands
Mr. K.S.P. Kumar, Minneapolis, U.S.A.
Professor Tenny N. Lam, Davis, U.S.A.
Mr. M.G. Langdon, Crowthorne, United Kingdom
Professor R.C. Larson, Cambridge, U.S.A.
Professor Larry LeBlanc, Nashville, U.S.A.
Professor Dr.-Ing. Wilhelm Leutzbach, Karlsruhe, West-Germany
Mr. J.J. Louviere, Iowa City, U.S.A.
Mr. Lars Lundqvist, Stockholm, Sweden
Mr. Roger L. Mackett, Leeds, United Kingdom
Mr. J.A. Martin, Crowthorne, United Kingdom
Mr. Carl D. Martland, Cambridge, U.S.A.
Professor Adolf D. May, Berkeley, U.S.A.
Professor Panos G. Michalopoulos, Minneapolis, U.S.A.
Mr. Alan J. Miller, Clayton, Australia
Mr. J.M. Morin, Arcueil-Cedex, France
Professor Gordon F. Newell, Berkeley, U.S.A.
Dr. Sang Nguyen, Montréal, Canada
Dr. Vic Outran, London, United Kingdom
Dr. Markos Papageorgiou, München, West-Germany
Mr. Harold J. Payne, San Diego, U.S.A.
Mr. M.W. Pickett, Crowthorne, United Kingdom
Dr. Laurie Pickup, Crowthorne, United Kingdom
Dr. D.E. Pitfield, Loughborough, United Kingdom
Dr. Abishai Polus, Haifa, Israel
Professor Warren B. Powell, Princeton, U.S.A.
Dr. Matthias H. Rapp, Basel, Switzerland
Professor Dr.-Ing. Hans-Georg Retzko, Darmstadt, West-Germany
Mr. Martin G. Richards, London, United Kingdom
Dr. Anthony Richardson, Clayton, Australia
Professor Dr. A.H.G. Rinnooy Kan, Rotterdam, The Netherlands
Mr. Dennis I. Robertson, Crowthorne, United Kingdom
Ir. J.P. Roos, Amersfoort, The Netherlands
Drs. Cees J. Ruijgrok, Delft, The Netherlands
Dr. R.J. Salter, Bradford, United Kingdom
Dr. S.E. Shladover, Palo Alto, U.S.A.
Professor Lars Sjöstedt, Göteborg, Sweden
Mr. M.J. Smith, Heslington, United Kingdom
Dr. Peter R. Stopher, Coral Gables, U.S.A.
Dr. J.C. Tanner, Crowthorne, United Kingdom
Professor Stan Teply, Edmonton, Canada
Professor Mark A. Turnquist, Ithaca, U.S.A.
Professor Rodney J. Vaughan, New Castle, Australia
Dr. D. van Vliet, Leeds, United Kingdom
Mr. Paul Wasielewski, Warren, U.S.A.
Dr. Marcus R. Wigan, Nunawading, Australia
Dr. Luis G. Willumsen, London, United Kingdom
Professor Dr. Sunedha C. Wiransinghe, Calgary, Canada
Dr. C.C. Wright, Enfield, United Kingdom
Professor Sam Yagar, Waterloo, Canada
Dr. Henk J. van Zuylen, Tilburg, The Netherlands

AUTHORS

Dr. Michael G.H. Bell
Universität (TH) Karlsruhe
Lehrstuhl und Institut für Verkehrswesen
Kaiserstrasse 12
D-7500 Karlsruhe, West-Germany

Professor Moshe Ben-Akiva
Massachusetts Institute of Technology
Department of Civil Engineering
Room 1-181
Cambridge, Massachusetts 02139, U.S.A.

Mr. M.J. Bergman
Ministry of Transport
Rijkswaterstaat
P.O. Box 20906
2500 EX The Hague, The Netherlands

Mr. Dimitrios E. Beskos
University of Patras
Department of Civil Engineering
Patras, Greece

Professor Dr. Avishai Ceder
Technion-Israel Institute of Technology
Department of Civil Engineering
Technion, Technion City
Haifa 32 000, Israel

Mr. Shih-Miao Chin
Greenman-Pedersen, Inc.
245 Lark Street
Albany, New York 12210, U.S.A.

Professor Dr. Takeshi Chishaki
Kyushu University
Faculty of Engineering
Hakozaki, 6-10-1 Higashi-ku
Fukuoka 812, Japan

Dr. Janusz Chodur
Cracow Technical University
Roads, Railways and Bridges Institute
ul.Warszawska 24
31-155 Krakow, Poland

Professor Dr. Michael Cremer
Technische Universität Hamburg-Harburg
Lohbrügger Kirchstrasse 65
2050 Hamburg 80, West-Germany

Professor Stella Dafermos
Brown University
Division of Applied Mathematics
Providence, Rhode Island 02912, U.S.A.

Mr. Andrew J. Daly
Cambridge Systematics Europe BV
Laan van Meerdervoort 32
2517 AL The Hague, The Netherlands

Professor Amir Eiger
Rensselaer Polytechnic Institute
Department of Civil Engineering
Troy, New York 12181, U.S.A.

Professor Caroline S. Fisk
University of Illinois at Urban-Champaign
Newmark Civil Engineering Laboratory
208 North Romine Street
Urbana, Illinois 61801, U.S.A.

Professor Dr. Terry L. Friesz
University of Pennsylvania
Department of Civil and Urban Engineering
113 Towne Building D3
Philadelphia, PA 19104, U.S.A.

Dr. Masaharu Fukuyama
Mitsubishi Research Institute, Inc.
Social Systems Department
Time & Life Building
3-6, Otemachi 2-Chome, Chiyoda-ku
Tokyo 100, Japan

Professor Patrick T. Harker
University of California
Department of Geography
Santa Barbara, CA 93106, U.S.A.

Professor Ezra Hauer
University of Toronto
Department of Civil Engineering
35 St. George Street
Toronto, Ontario M5S 1A4, Canada

Professor Dr. Heinz Hautzinger
Fachhochschule Heilbronn
Institüt für angewandte Verkehrs- und Tourismusforschung e.V.
Max-Planck-Strasse 39
D-7100 Heilbronn, West-Germany

Professor Chris Hendrickson
Carnegie-Mellon University
Department of Civil Engineering
Schenley Park
Pittsburgh, Pennsylvania 15213, U.S.A.

Dr. Benjamin G. Heydecker
Transport Studies Group
University College London
Gower Street
London WC1E 6BT, United Kingdom

Professor V.F. Hurdle
University of Toronto
Department of Civil Engineering
35 St. George Street
Toronto, Ontario M5S 1A4, Canada

Professor Paul P. Jovanis
Northwestern University
The Technological Institute
Evanston, Illinois 60201, U.S.A.

Professor Dr. Hartmut Keller
Technische Universität München
Fachgebiet Verkehrsplanung und Verkehrswesen
Arcisstrasse 21
D-8000 München, West-Germany

Professor Dr. Ryuichi Kitamura
University of California
College of Engineering
Davis, California 95616, U.S.A.

Professor Frank S. Koppelman
Northwestern University
The Technological Institute
Evanston, Illinois 60201, U.S.A.

Dr. Reinhart D. Kühne
AEG-Telefunken
Forschungsinstitut
Physikalische Technologie
Postfach 1730
D-7900 Ulm (Donau), West-Germany

Professor Tenny N. Lam
University of California
College of Engineering
Davis, California 95616, U.S.A.

Mrs. Sue McNeil
Garmen Associates
57, Whippany Road
Whippany, NJ 07981, U.S.A.

Professor Panos G. Michalopoulos
University of Minnesota
Department of Civil and Mineral Engineering
500 Pillsbury Drive, S.E.
Minneapolis, Minnesota 55455, U.S.A.

Mr. Anthony Moore
Northwestern University
The Technological Institute,
Evanston, Illinois 60201, U.S.A.

Professor Anna Nagurney
University of Massachusetts
School of Management
Amherst, Massachusetts 01003, U.S.A.

Mr. Yoshiharu Namikawa
Kyoto University
Department of Transportation Engineering
Sakyo-ku
Kyoto 606, Japan

Professor Iwao Okutani
Shinshu University
Department of Civil Engineering
500 Wakasato
Nagano City 380, Japan

Professor Dr. Eric I. Pas
Duke University
Department of Civil Engineering
Durham, North Carolina 27706, U.S.A.

Professor Warren B. Powell
Princeton University
Civil Engineering Department
Room 408
Princeton, New Jersey 08544

Mr. Rohit Ramaswamy
Cambridge Systematics Europe BV
Laan van Meerdervoort 32
2517 AL The Hague, The Netherlands

Mr. Geoffrey Rose
Northwestern University
The Transportation Center
2001 Sheridan Road
Evanston, Illinois 60201, U.S.A.

Professor Tsuna Sasaki
Kyoto University
Department of Transportation Engineering
Sakyo-ku
Kyoto 606, Japan

Mr. Thomas Schwerdtfeger
Universität (TH) Karlsruhe
Lehrstuhl und Institut für Verkehrswesen
Postfach 6380
D-7500 Karlsruhe 1, West-Germany

Professor Yosef Sheffi
Massachusetts Institute of Technology
Department of Civil Engineering
Room 1-180
77 Massachusetts Avenue
Cambridge, Massachusetts 02139, U.S.A.

Mr. M.J. Smith
University of York
Department of Mathematics
Heslington, York YO1 5DD, United Kingdom

Professor Helman I. Stern
Ben-Gurion University of the Negev
Department of Industrial Engineering and Management
P.O. Box 653
Beer Sheva 84105, Israel

Professor Dr. Janusz Supernak
Drexel University
College of Engineering
Philadelphia, Pennsylvania 19104, U.S.A.

Mr.Youichi Tamura
Yamaguchi University
Faculty of Engineering
Tokiwadai
Ube, Yamaguchi 755, Japan

Mr. Sebastien Thiriez
Massachusetts Institute of Technology
Department of Civil Engineering
77 Massachusetts Avenue
Cambridge, Massachusetts 02139, U.S.A.

Dr. Marian Tracz
Cracow Technical University
Roads, Railways and Bridges Institute
ul.Warszawska 24
31-155 Krakow, Poland

Professor Rodney J. Vaughan
University of Newcastle
Department of Mathematics, Statistics and Computer Science
Newcastle, New South Wales 2308, Australia

Dr. Luis G. Willumsen
University College London
Transport Studies Group
Gower Street
London WC1E 6BT, United Kingdom

Milton Keynes UK
Ingram Content Group UK Ltd.
UKHW021931071024
449327UK00022B/1767